Developments in Industrial Microbiology Series

Microbial Metabolites
Volume 32

Editors

Dr. Claude Nash
Series Editor

Dr. Jennie Hunter-Cevera
Senior Editor

Dr. Ray Cooper
Senior Editor

Dr. Douglas E. Eveleigh
Senior Editor

Dr. Robert Hamill
Senior Editor

WCB Wm. C. Brown Publishers
Dubuque, Iowa • Melbourne, Australia • Oxford, England

Copyright © by Wm. C. Brown Communications, Inc. All rights reserved

A Times Mirror Company

Library of Congress Catalog Card Number: 92–75577

ISBN: 0–697–16688-0

No part of this publication may be reproduced, stored in a retrieval system, or transmitted, in any form or by any means, electronic, mechanical, photocopying, recording, or otherwise, without the prior written permission of the publisher.

Printed in the United States of America by Wm. C. Brown Communications, Inc., 2460 Kerper Boulevard, Dubuque, IA 52001

10 9 8 7 6 5 4 3 2 1

Contents

1
Immunosuppressants of Microbial Origin 1
J. J. Sanglier, G. Baumann, M. Dreyfuss, T. Fehr, R. Traber, and M. H. Schreier

2
The Biosynthesis and Enzymology of an Immunosuppressant, Immunomycin. Produced by *Streptomyces hygroscopicus var. ascomyceticus* 29
K. M. Byrne, A. Shafiee, J. B. Nielsen, B. Arison, R. L. Monaghan and L. Kaplan

3
WS 9482, An Immunomodulator of Microbial Origin 47
Shizue Izumi, Tsutomu Kaizu, Michio Yamashita, Masanori Okamoto, and Masakuni Okuhara

4
Trichosporin, Ion Channel Forming Peptide, and ISP—I, An Immunosuppressive Amino Acid 59
Tetsuro Fujita

5
The Role and Value of Microbial Screening for Agrochemicals 73
Keith A. Powell and Sarah B. Rees

6
Herbicidal Amino Acids of Microbial Origin 79
Paul B. Lavrik, Barbara G. Isaac, Stephen W. Ayer and Richard J. Stonard

7
Isolation and Insecticidal Activity of Verruculogens and Tryptoquivalines 93
John B. Pillmoor, Brian D. Bush, and Duncan A. Gates

8
Screening of Herbicides based on Antimicrobial Activity—Cellulose Synthesis Inhibitors 101
Yoshitake Tanaka, Haruo Tanaka, and Satoshi Omura

9
Discovery and Identification of a Novel Fermentation-Derived Insecticide 109
Herbert A. Kirst, Karl H. Michel, Eddie H. Chio, Raymond C. Yao, Walter M. Nakatsukasa, LaVerne D. Boeck, John L. Occolowitz, Jonathan W. Paschal, and Jack B. Deeter

10
Novel Ayermectins by Mutational Biosynthesis 117
K. S. Holdom, D. Beck, C. J. Dutton, S. P. Gibson, A. C. Goudie, E. W. Hafner, B. W. Holley, S. E. Lee, H. A. I. McArthur, M. S. Pacey, J. C. Ruddock, R. G. Wax, W. C. Wernau, J. D. Bu'Lock, and M. K. Richards

11
Biological Modification of Cyclosporins by the Multifunctional Enzyme Cyclosporin Synthetase 125
Horst Kleinkauf and Alfons Lawen

12
Semduramicin: Design of a New Anticoccidial Inonophore 133
E. A. Glazer, W. P. Cullen, G. M. Frame, A. C. Goudie, D. A. Koss, J. A. Olson, A. P. Ricketts, E. J. Tynan, N. D. Walshe, W. C. Wernau, and T. K. Schaaf

13
Stereospecific Microbiological Chiral Reduction of Ketone Groups by A Novel *Aspergillus* Species 141
Ellen S. Baron and Michael Homann

14
Searching for Anthelmintics in the Post-Avermectin Era 155
Christopher L. Haber, David P. Thompson, Howard A. Whaley, B. Lamar Lee, Thomas F. Brodasky, Charles K. Marschke, Jerry W. Bowman, Timothy G. Geary, Eileen M. Thomas, Richard D. Conklin, Veronica H. Wiley, Charlotte L. Heckaman, Derek L. Rosa, and Raymond J. Zielinski

15
Paraherquamide—A Novel Antiparasitic Agent; Production and Activity 169
Prakash S. Masurekar, Michel M. Chartrain, Kodzo Gbewonyo, Robert T. Goegelman, John G. Ondeyka, Maragaret S. Sosa, Louis Kaplan, Dan A. Ostlind, Wesley L. Shoop

16
Enzyme and Receptor Assays as Natural Product Screens 185
R. Murray Tait and Martyn N. Banks

17
TAN-950A, A Novel Glutamate Receptor Agonist produced by *Streptomyces platensis* A-136 199
Seiji Hakoda, Sigetoshi Tsubotani, Tenichi Tamura, Toshi Iwama, Setsuo Harada, and Takashi Iwasa

18
Inhibitors of Proline Specific Peptidases and their Possible Applications in Medicine 207
Takaaki Aoyagi, Yasduhiko Muraoka, and Tomio Takeuchi

19
Screening, Isolation and Subsequent Chemical Modifications of Low Molecular Weight Cysteine Proteinase Inhibitors from Micro-Organisms 215
M. Saito, N. Higuchi, and T. Tanaka

20
Microbial Metabolites as Gastric H^+–K^+ ATPase Inhibitors 227
Jon S. Mynderse, Dennis R. Berry, Rosanne Bonjouklian, Otis W. Godfrey, Frederick P. Mertz, Walter N. Nakatuskasa, Raymond C. Yao, Ann H. Hunt, Jack B. Deeter, Jaswant S. Gidda, and Anne H. Dantzig

21
Microbial Secondary Metabolites which Regulate Mammalian Cell Growth 237
Hiroyuki Osada and Kiyoshi Isono

22
The Starfish Embryo Assay useful for Screening of New Inhibitors of RNA Synthesis 247
Susumu Ikegami

23
New Anthracyclines effective on Adriamycin-Resistant Cell Lines 253
Tskeshi Uchida, Noboru Otake, and Tomio Takeuchi

24
Progress in Esperamicin Research 261
Kin Sing Lam, Salvatore Forenza, Judith A. Veitch, Donald R. Gustavson, Jerzy Golik, and Terrence W. Doyle

25
Oxetanocins, Antiviral Nucleosides 275
Tomoshi Takita

26
High-Volume Screening of Natural Products for IL-1 Receptor Level Antagonists 285
A. L. Laborde, J. A. Shelly, S. E. Truesdell, V. P. Marshall, J. I. Cialdella, W. F. Liggett, D. A. Yurek, D. G. Chirby, J. W. Paslay, C. K. Marschke, M. S. Kuo

1

Immunosuppressants of Microbial Origin

J. J. Sanglier, G. Baumann, M. Dreyfuss, T. Fehr, R. Traber and M. H. Schreier
(Preclinical Research, Sandoz Pharma Ltd, CH-4002 Basel, Switzerland)

INTRODUCTION

This article reviews the different immunosuppressants of microbial origin, both the microbes and the modes of action of the immunosuppressants. The breakthroughs made during the two last years are highlighted.

THE IMMUNE SYSTEM— FUNDAMENTALS

The mammalian immune system has evolved to detect and destroy foreign molecules and cells. Recognition is mediated by T and B lymphocytes (Klein, 1990; Roitt et al., 1989).

In the immunology of transplantation, the helper lymphocytes (T_h) play the central role (Fig. 1.1). These cells respond to the signal of a processed antigen in association with the major histocompatibility complex (MHC) class 2 presented at the surface of other cells such as macrophages and dendritic cells, called antigen presenting cells (APCs). This binding event initiates an ordered cascade of biochemical changes and generates signals that are transmitted from the cell surface to the nucleus leading to transcription of lymphokine genes (growth and proliferation factors) (Altman et al., 1990). This is a complex process beginning with modifications in the membrane (formation of inositol-3-phosphate and diacylglycerol by hydrolysis of phosphatidyl inositol biphosphate), changes in the cytoplasm (increase in the Ca^{++} content, transient de-/phosphorylation of proteins, activation of protein kinases, phosphatases) and finally expression of the genes coding for lymphokines. Interleukin 2 (IL-2) released from T_h, binds to Il-2 receptors and thereby leads to clonal proliferation. Only activated T cells express complete high affinity Il-2-receptors (autocrine system). This process is stimulated by IL-1 produced by the macrophages. Th also release other lymphokines (IL-2, IL-3, IL-4, IL-5, GM-CSF,...), which affect T cells, B cells and other cells involved in the immune system. Furthermore, precursor cytotoxic lymphocytes (PCTL) are activated by processed antigens presented by MHC class I and express Il-2R. IL-2

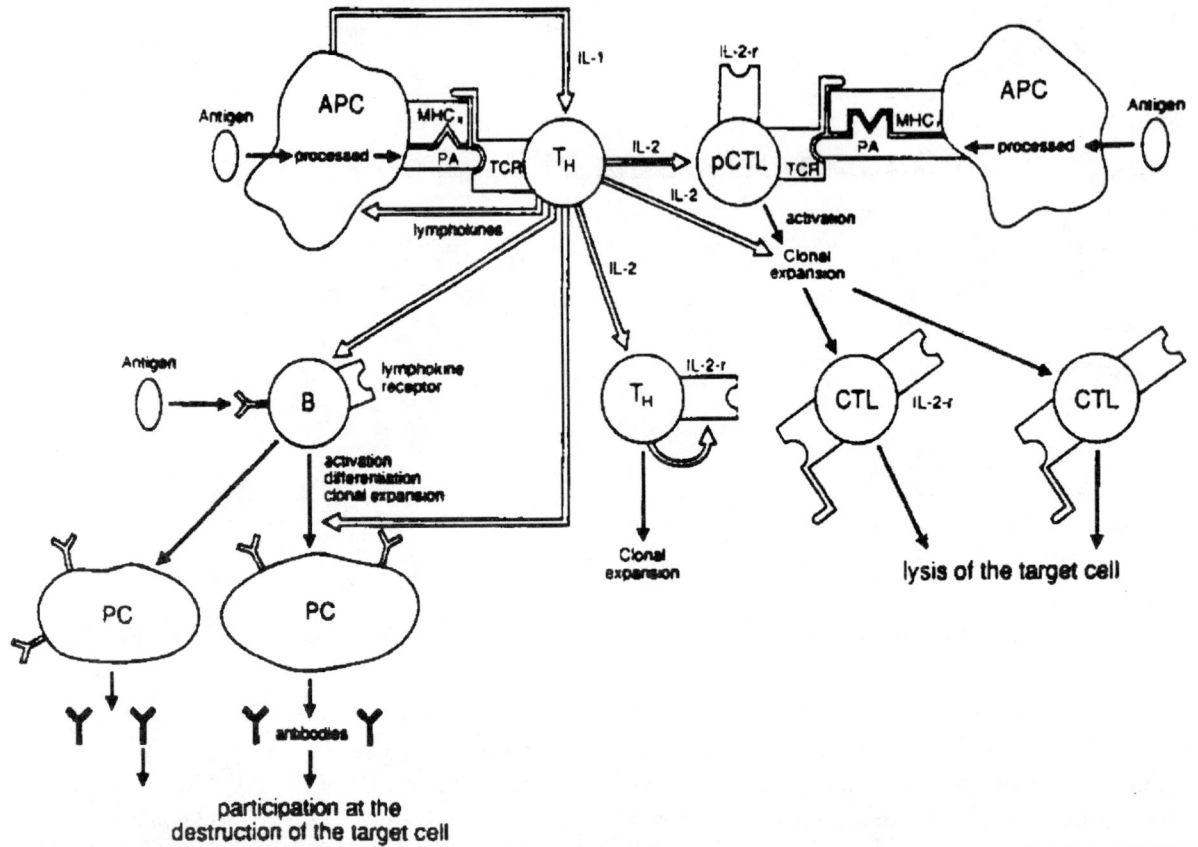

Figure 1.1 / The immune response.

induces the clonal amplification of the activated CTL. Surface receptors of the CTL recognize the H-2 determinants present on the surface of the target cells. These target cells will be lysed. This specific cellular immune response is the main process occuring in organ allograft rejection.

The humoral immune response against the graft is initiated by antibodies. B cells in the lymphoid organs encounter soluble antigens from the graft and Th cells sensitized to these antigens. B cells recognize antigens in their native conformation either free or on the surface of cells, using surface immunoglobulins as specific antigen receptor. As a result the B cells are activated and clonal proliferation and differentiation often occur to yield plasma cells which produce antibodies (immunoglobulins). The processes of activation, proliferation and differentiation of B cells are modulated by the macrophage mediators Il-1, TNF and Il-6 and T-cell derived lymphokines (IL-2, Il-4, Il-5, IL-6, and possibly others). The antibodies generated in the humoral immune response contribute to the destruction of the graft in different ways (opsonic adherence, immune adherence, antibody-dependent cell mediated cytotoxicity, adherence through C3 receptors and complement-mediated lysis) (Klein, 1990).

In autoimmune diseases antibodies directed against self-structures (autoantibodies), and T lymphocytes, named autoreactive T lymphocytes, play the dominant role in different cases (juvenile-onset diabetes mellitus, uveitis, psoriasis, rheumatoid arthritis,...).

IMMUNOSUPPRESSANTS

An immunosuppressant is any compound that inhibits the proliferation, the differentiation or the general function of lymphocytes. A selective immunosuppressant has an activity that is specific to lymphocytes and has little influence on hematopoiesis, shows no or little toxicity and should not compromise the host to bacterial or fungal infection or severely inhibit tumor defense mechanisms.

In the early days of transplantation, nonspecific immunosuppressants with cytotoxic agents, such as cyclophosphamide, were used. Later, more selective immunosuppression was achieved using azathioprine and, in particular, steroids. The first selective immunosuppressant drug was cyclosporin A (CsA) which specifically suppresses T-lymphocyte proliferation without inhibiting other immunological functions (Drews, 1990).

Since then, the introduction of CsA (Sandimmun) has revolutionized organ transplantation (Borel, 1986; Borel, 1989; Borel et al., 1989; Kahan, 1989) and has greatly increased the success of transplantation not only of kidneys but also of heart, liver, pancreas and bone-marrow successfully transplanted on a routine basis (Mandel, 1989). In addition to the more than 120,000 transplantations (Table 1.1) that have been performed since the introduction of CsA (Sandimmun), it is being increasingly used to treat a variety of autoimmune diseases such as uveitis, psoriasis, nephrotic syndrome and rheumatoid arthritis (Bach, 1989; Cruse and Lewis, 1989). Renal dysfunction remains a problematic side effect (Mason, 1989) but can be significantly reduced by careful drug-monitoring (Quesniaux, 1989).

FK-506 is a recently discovered immunosuppressant with a selective action on T lymphocytes

Table 1.1 / Increased usage of cyclosporin for organ transplants

Year	Kidney	Heart	Liver
1983	12,161	296	256
1988	20,028	3,283	2,474

(Worldwide figures, Sandoz Pharma information.)

(Kino et al., 1987b; Thomson, 1989). Despite some strong reservations concerning the safety of FK-506 (Collier et al, 1989) and especially of a new molecule in clinical transplantation, progress in the development of FK-506 has been very rapid, including the first trial in transplant patients at the University of Pittsburg in 1989 (Starzl et al., 1989; Starzl et al., 1990) and an evaluation in certain experimental autoimmune diseases such as uveoretinitis (Kawashima et al., 1988). FK-506 shows a toxicological profile that varies with animal species. In man, severe arterial hypertension as well as some neurological side-effects have been observed in early studies (Starzl et al., 1990) and definitive assessment of the drug will only be possible after systematic toxicological studies.

Both CsA (Sandimmun) and FK-506 are of microbial origin. This has led to reassessment of microbial products as immunosuppressive compounds. Immunosuppressants discovered in the last years are mainly previously described microbial products.

CYCLOSPORIN

ORIGIN AND STRUCTURE

Cyclosporin A (CsA) was discovered in the beginning of the seventies as metabolic products of two Fungi Imperfecti, *Tolypocladium inflatum* (formerly described as *Trichoderma polysporum*) and the second identified as *Cylindrocarpon lucidum* (Dreyfuss et al., 1976; Regger et al., 1976). CsA

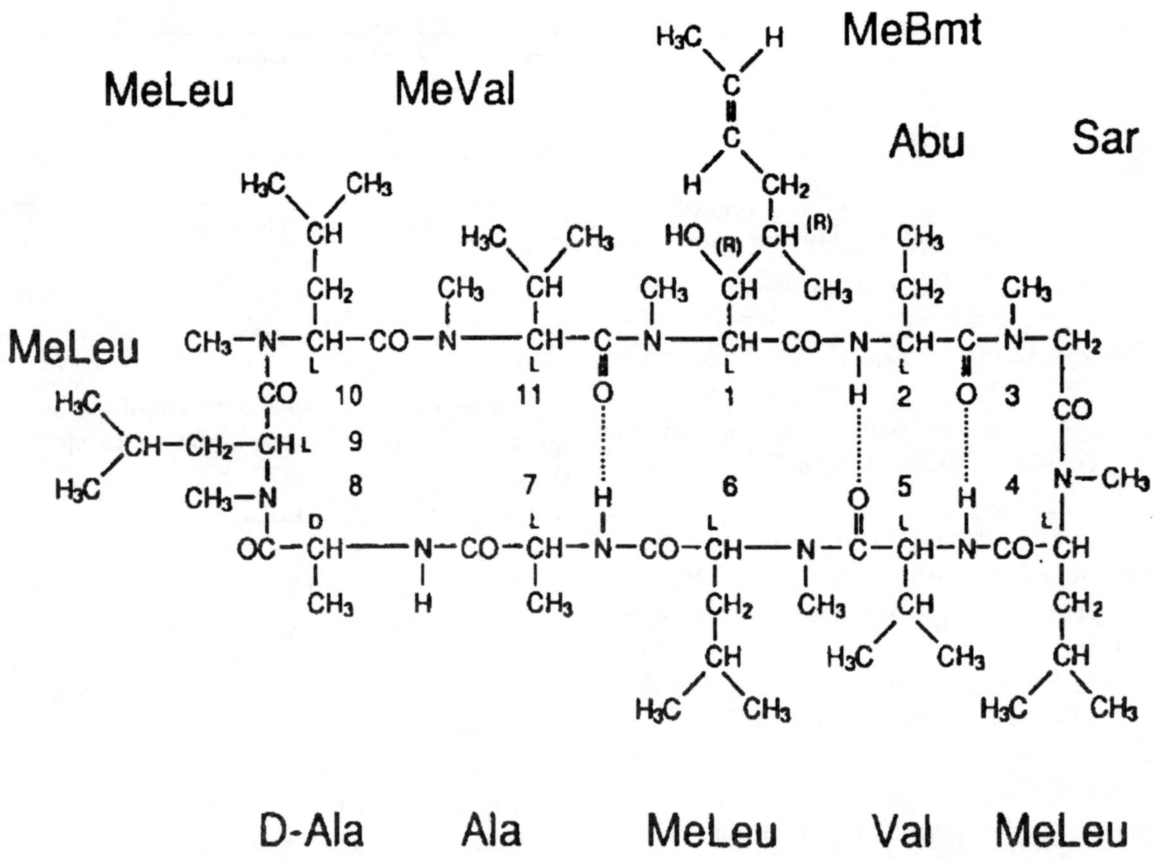

Figure 1.2 / Cyclosporin A (= Sandimmun).

(Fig. 1.2) is a neutral, highly lipophilic, cyclic peptide composed of 11 amino acids, all having the S-configuration, except for the D-alanine in position 8, which has the R-configuration, and sarcosine in position 3. Seven amino acids are N-methylated. One amino acid was hitherto unknown; this is the amino acid in position 1, which is designated (4R)-4-[(E)-2-butenyl]-4,N-dimethyl-L-threonine (abbreviation MeBmt) according to IUPAC/IUB rules and is also reported as the "C-9 amino acid." CsA has a molecular weight of 1202 (see ref. in Fliri and Wenger, 1990).

Besides cyclosporin A, thirty natural related congeners have been isolated so far (Fig. 1.3) (von Wartburg and Traber, 1988). Variations in the amino acid sequence occur in all positions except 3 (sarcosine) and 8 (D-alanine). In positions 6, 9 and 10, the lack of the N-methyl (Me) groups is the only modification compared to cyclosporin A. The greatest variability in the cyclosporin molecule is observed in position 2 which can be occupied with L-2-aminobutyric acid (cyclosporin A), L-alanine (cyclosporin B), L-threonine (cyclosporin C), L-valine (cyclosporin D) or L-norvaline (Nva) (cyclosporin G). The total synthesis of cyclosporin has been reported (e.g., Wenger, 1983).

In the biosynthesis of cyclosporins, all catalytic reactions leading to cyclosporin formation from the amino acid precursors, except the epimerization of L-alanine, occur on a multienzyme polypeptide (Billich and Zocher, 1987). This polypeptide is non-glycosylated and has a molecular weight

Metabolite	Amino acid composition 1	2	3	4	5	6	7	8	9	10	11
Cy A	MeBmt	Abu	Sar	MeLeu	Val	MeLeu	Ala	D-Ala	MeLeu	MeLeu	MeVal
Cy B	MeBmt	Ala	Sar	MeLeu	Val	MeLeu	Ala	D-Ala	MeLeu	MeLeu	MeVal
Cy C	MeBmt	Thr	Sar	MeLeu	Val	MeLeu	Ala	D-Ala	MeLeu	MeLeu	MeVal
Cy D	MeBmt	Val	Sar	MeLeu	Val	MeLeu	Ala	D-Ala	MeLeu	MeLeu	MeVal
Cy E	MeBmt	Abu	Sar	MeLeu	Val	MeLeu	Ala	D-Ala	MeLeu	MeLeu	MeVal
Cy F	Desoxy-MeBmt	Abu	Sar	MeLeu	Val	MeLeu	Ala	D-Ala	MeLeu	MeLeu	Val
Cy G	MeBmt	Nva	Sar	MeLeu	Val	MeLeu	Ala	D-Ala	MeLeu	MeLeu	MeVal
Cy H	MeBmt	Abu	Sar	MeLeu	Val	MeLeu	Ala	D-Ala	MeLeu	MeLeu	D-MeVal
Cy I	MeBmt	Val	Sar	MeLeu	Val	MeLeu	Ala	D-Ala	MeLeu	Leu	MeVal
Cy K	Desoxy-MeBmt	Val	Sar	MeLeu	Val	MeLeu	Ala	D-Ala	MeLeu	MeLeu	MeVal
Cy L	Bmt	Abu	Sar	MeLeu	Val	MeLeu	Ala	D-Ala	MeLeu	MeLeu	MeVal
Cy M	MeBmt	Nva	Sar	MeLeu	Nva	MeLeu	Ala	D-Ala	MeLeu	MeLeu	MeVal
Cy N	MeBmt	Nva	Sar	MeLeu	Val	MeLeu	Ala	D-Ala	MeLeu	Leu	MeVal
Cy O	MeLeu	Nva	Sar	MeLeu	Val	MeLeu	Ala	D-Ala	MeLeu	MeLeu	MeVal
Cy P	Bmt	Thr	Sar	MeLeu	Val	MeLeu	Ala	D-Ala	MeLeu	MeLeu	MeVal
Cy Q	MeBmt	Abu	Sar	Val	Val	MeLeu	Ala	D-Ala	MeLeu	MeLeu	MeVal
Cy R	MeBmt	Abu	Sar	MeLeu	Val	Leu(?)	Ala	D-Ala	MeLeu	Leu(?)	MeVal
Cy S	MeBmt	Thr	Sar	MeLeu	Val	MeLeu	Ala	D-Ala	MeLeu	MeLeu	MeVal
Cy T	MeBmt	Abu	Sar	MeLeu	Val	MeLeu	Ala	D-Ala	MeLeu	Leu	MeVal
Cy U	MeBmt	Abu	Sar	MeLeu	Val	Leu	Ala	D-Ala	MeLeu	MeLeu	MeVal
Cy V	MeBmt	Abu	Sar	MeLeu	Val	MeLeu	Abu	D-Ala	MeLeu	MeLeu	MeVal
Cy W	MeBmt	Thr	Sar	MeLeu	Val	MeLeu	Ala	D-Ala	MeLeu	MeLeu	Val
Cy X	MeBmt	Nva	Sar	MeLeu	Val	MeLeu	Ala	D-Ala	Leu	MeLeu	MeVal
Cy Y	MeBmt	Nva	Sar	MeLeu	Val	Leu	Ala	D-Ala	MeLeu	MeLeu	MeVal
Cy Z	Me-2-Amino-octanoic acid	Abu	Sar	MeLeu	Val	MeLeu	Ala	D-Ala	MeLeu	MeLeu	MeVal

Bmt = (2S,3R,4R,6E)-2-amino-3-hydroxy-4-methyl-6-octenoic acid

Figure 1.3 / Chemical structure of cyclosporins A-Z.

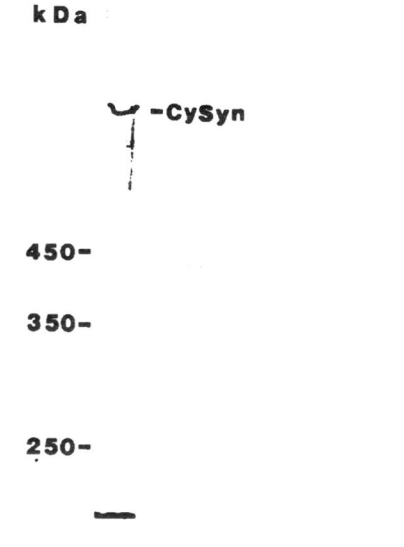

Figure 1.4 / Molecular mass estimation of cyclosporin synthetase (with kind permission of Lawen and Zocher).

between 650 and 800 kD (Fig. 1.4) (Lawen and Zocher, 1990).

More than 200 *Tolypocladium inflatum* strains and 30 *T. geodes* strains have been isolated in our group or obtained from culture collections; all these isolates are cyclosporin producers. Producers have also been found or reported in the following taxa: *Acremonium* sp., *Aphanocladium* sp., *Beauveria bassiana*, *B. brongniarti*, *Chaunopycnis alba*, *Fusarium solani*, *Isaria felina*, *Neocomospora vasinfecta*, *Paecilomyces* sp., *Tolypocladium cylindrosporum*, *T. nubicola*, *T. terricola*, *T. tundrense*, *Verticillium* sp. (Dreyfuss, 1986; Dreyfuss, unpublished results; Jegorov et al., 1990), but in these cases cyclosporin production is restricted to singular and rare strains. All wild strains produce cyclosporins in low amounts under standard conditions; CsA is always the main component (65 to 95%), followed by CsC and CsB. Recent

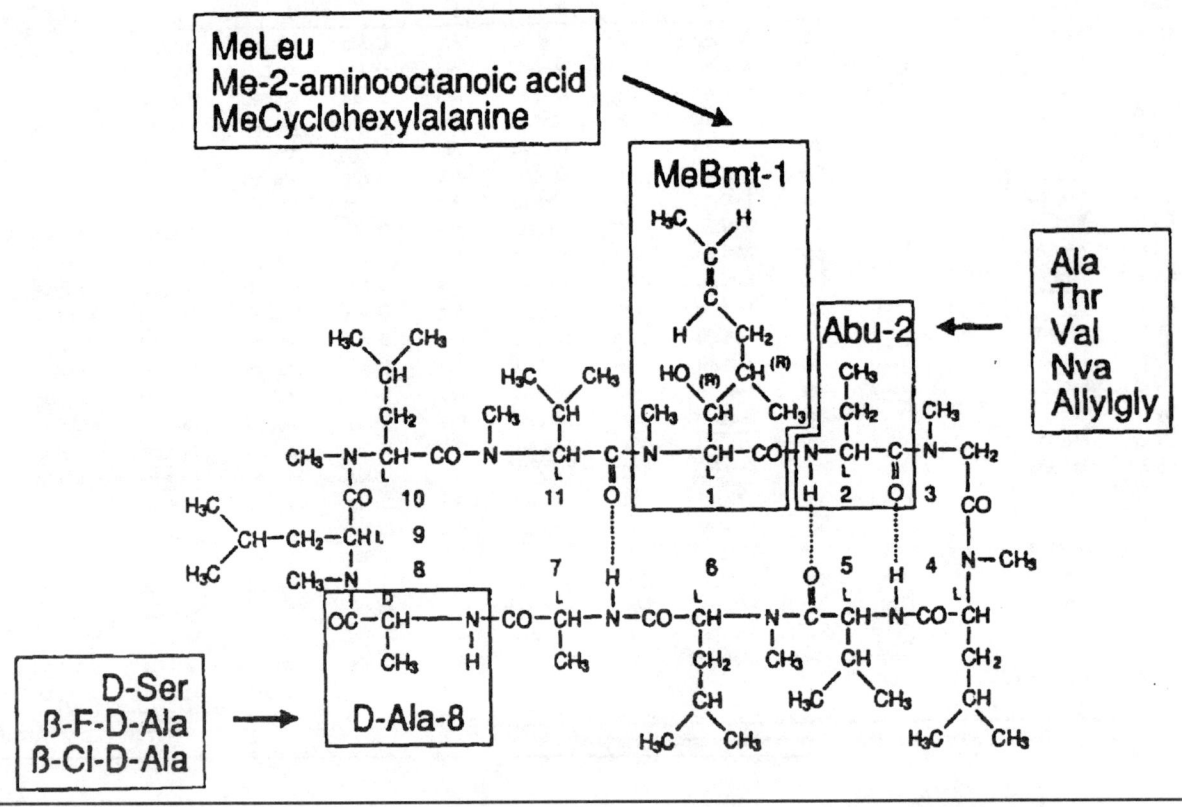

Figure 1.5 / Modifications of cyclosporin by amino acid precursor directed biosynthesis.

ly a strain of *Neocomospora vasinfecta* was reported to produce only CsC (Nakajima et al., 1989). In *Fusarium* and *Neocomospora* strains, Cs-production is limited to static culture.

Various pigmented variants of *T. inflatum*, with different capacities to produce cyclosporins, have been isolated (Aarnio and Agathos, 1990). Sequential addition of two carbon substrates such as sorbose followed by maltose allows an improvement of the production (Agathos et al., 1986). Exogenous L-valine exerts a strong enhancing effect on both rate of production and final titer of CsA (Lee and Agathos, 1989). Addition of zinc also enhances CsA productivity (Agathos et al., 1988). Immobilisation of *T. inflatum* spores in porous Celite beads causes a three-fold higher yield than the free cell culture (Chun and Agathos, 1989).

NEW CYCLOSPORINS BY DIRECTED BIOSYNTHESIS

By supplemention of the fermentation medium with excess natural or synthetic amino acids several new cyclosporins were obtained such as [N-methyl-L-ß-cyclohexylalanine^1]cyclosporin, [L-allyl-glycine^2]cyclosporin and [D-serine^8]cyclosporin (Kobel and Traber, 1982; Traber et al., 1989) (Fig. 1.5). The incorporation of ß-fluoro-D-alanine and of the corresponding α-deuterated precursor in position 8 was recently reported (Patchett et al., 1988). However, *in vivo* incorporation of foreign amino acids occurs only with a relatively limited number of specific amino acids. The isolation and purification of the multienzyme polypeptide cyclosporin synthetase from *T. inflatum* (Billich and Zocher, 1987) and the production of the key amino acid Bmt by a blocked mutant (Sanglier et al., 1990) enabled the total *in vitro* synthesis of CsA (Billich and Zocher, 1987) and, by appropriate adjustment of the amino acid composition, of several naturally occuring congeners (cyclosporins B, C, D, G, M, O, Q, U, V) as well as a series of analogues known to be produced by the fungus via precursor directed biosynthesis. Furthermore, by using the enzyme system, *in vitro* synthesis of new cyclosporins was achieved which were not obtainable by fermentation; examples are [N-methyl-(+)-2-amino-3-hydroxy-4,4-dimethyloctanoic acid^1] cyclosporin, [L-norvaline^{2,5}, N-methyl-L-norvaline^{11}]cyclosporin, [L-norvaline^5, N-methyl-L-norvaline^{11}]cyclosporin, [L-allo-isoleucine^5, N-methyl-L-allo-isoleucine^{11}]cyclosporin, [L-allo-isoleucine^{5,11}]cyclosporin, [D-2-aminobutyric acid^8] cyclosporin (Lawen et al., 1989). All new cyclosporins were tested in various *in vitro* assays for their immunosupressive effects, but none of the compounds modified in positions 2, 5, 8 and/or 11 reached the activity level of CsA (Lawen et al., 1989).

MODE OF ACTION OF CYCLOSPORIN

CsA suppresses cell-mediated and humoral immune responses, although its activity against cell-mediated reactions predominates (Drews, 1990). It affects other systems such as the hematopoietic system only partially or indirectly (Tasato et al., 1982). The main cellular target of CsA is the helper lymphocyte (Th). CsA inhibits T cell proliferation when added at the initial stage of activation (within two hours) (Borel et al., 1977). Once a cell has progressed beyond the early G1 phase, addition of CsA is without effect. It inhibits the synthesis of Il-2 as well as of other lymphokines normally produced by the antigen activated Th and thus blocks the further steps of the immunological cascade, proliferation of Th, differentiation of CTL and of plasma cells (Borel, 1989; Di Padova, 1989; Ryffel, 1989). Numerous studies on the production of Il-2 in both mitogenic and antigen-stimulated primary T-lymphocyte cultures, T-cell lines and antigen-specific T-cell clones have all shown a clear inhibition of synthesis of Il-2 (Fig. 1.6, lanes 1–5) and mRNA for IL-2 (Nordmann et al., 1989). Similar CsA sensitivity has also been demonstrated for several other lymphokines, such as IFN-γ, IL-3, IL-4 and IL-6, whereas IL-1α, IL-1β and GM-CSF are CsA resistant (for review see Borel, 1989; Di Padova, 1989; Ryffel, 1989). The inhibitory activity on the expression of Il-2 receptors is controversial, perhaps due to the presence of two types of receptors (Foxwell and Ryffel, 1990). CsA inhibits B cell responses that depend on lymphocyte secretion by T cells. In addition CsA has a direct inhibitory effect on some B cell subsets (Di Padova, 1989). CsA inhibits the activation of resting B cells by agents causing rapid breakdown of phosphatidylinositol biphosphate and mobilisation of intracellular Ca^{2+} such as anti-Ig antibody (Klaus, 1988) but has low effectiveness in its responses to LPS (Stoeck et al., 1985). The action of CsA in this context is reversible.

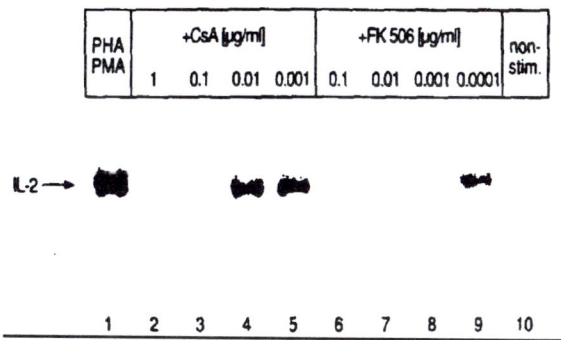

Figure 1.6 / Inhibition of IL-2 production in Jurkat cells. Lane 1: stimulation by PHA (phytohemagglutinin) and PMA (phorbol 12-myristate 13-acetate.

Recent work has shown that CsA interferes with the function of intracellular molecules that transmit calcium-associated signals between the T cell receptor and the activation of genes in the nucleus (Ullmann et al., 1990). CsA does not however bind to DNA or RNA. Specifically, expression of the IL-2 is inhibited at the level of transcription and this appears to be an important mechanism by which CsA impairs T cell function (Krönke et al., 1984). Transcriptional regulation of gene expression is generally modulated by the combination of specific interactions of nuclear proteins (transcription factors) with unique regulatory sequences (cis-elements). These DNA/protein interactions in the promoter and/or enhancer together with RNA polymerase II are crucial in determining whether the expression of a specific gene is activated or repressed (Mitchell and Tjian, 1989).

In order to understand the mechanism by which transcription of the IL-2 gene is modulated by CsA, it is necessary to elucidate the molecular events at the level of DNA/protein interaction. In human T cells the activity of the IL-2 gene is mainly controlled by a 275 base pair long promoter extending from position -326 to -52 (with respect to the start site of transcription at position +1) (Fujita et al., 1986; Durand et al., 1988). Transient transfection experiments with a linked indicator gene in the Jurkat T cell line have shown that this promoter region spans the majority of control elements necessary for T lymphocyte specific induction and CsA-directed suppression of the IL-2 gene (Ullman et al., 1990). Further dissection of this promoter revealed a modular organisation with three functional domains, which are all affected to some extent by CsA:

1. The functional element at position (-285/-255) forms a lymphocyte specific DNA/protein complex designated NF-AT (Nuclear factor of activated T cells) (Durand et al., 1988; Shaw et al., 1988). CsA was found to specifically inhibit the appearance of DNA-binding activity of NF-AT. In addition, CsA abolished the ability of a multimerized NF-AT-binding site to activate a linked promoter in transfected T lymphocytes (Emmel et al., 1990). Binding of nuclear proteins from mouse EL4 lymphoma cells interacting with the so called purine boxes of the mouse IL-2 promoter is also affected in CsA treated cells (Randak et al., 1990). These purine boxes have striking homology to the NF-AT site in the human IL-2 promoter.

2. The second regulatory site in the IL-2 promoter between positions -72 and -80 shares sequence homology with the so called "octamer" sequence of immunoglobulin genes and binds a ubiquitous nuclear factor present in nuclear extracts of both activated and non-activated T cells (Nabel et al., 1988). By analysing the chromatin structure from isolated nuclei by DNaseI treatment, a new DNaseI hypersensitive site appears upon activation of T cells at this site of the IL-2 promoter. CsA blocks the induction of this new site in suppressed Jurkat cells, establishing also that this drug acts before the chromatin rearrangement of the IL-2 gene (Siebenlist et al., 1986).

3. The third region extending from position -206 to -195 exhibits homology with regulatory sequences in the k light chain enhancer of immunoglobulins, the IL-2 receptor α-chain promoter and the human immunodeficiency virus (HIV) type 1 promoter (Hoyos, 1989). A related sequence element is also present in the promoter of several lymphokine genes including GM-CSF, IL-4 and IFN-γ (Shannon et al., 1990). The inducible transcription of these lymphokine genes can be suppressed by CsA (Tocci et al., 1989). Nuclear proteins interacting with this sequence element in the IL-2 promoter have been shown to be affected by CsA at concentrations paralleling its immunological activity *in vivo* (Baumann et al., 1991). The results of this reverse approach, that is to start with the transcriptional control region of the IL-2 gene and work toward the T cell receptor, clearly indicate that at least the appearance of inducible nuclear proteins interacting with the IL-2 promoter at position (-206/-195) and (-285/-255) are affected by CsA. Isolation and further characterization of the described DNA-binding proteins will be required to finally identify the primary target(s) of CsA in

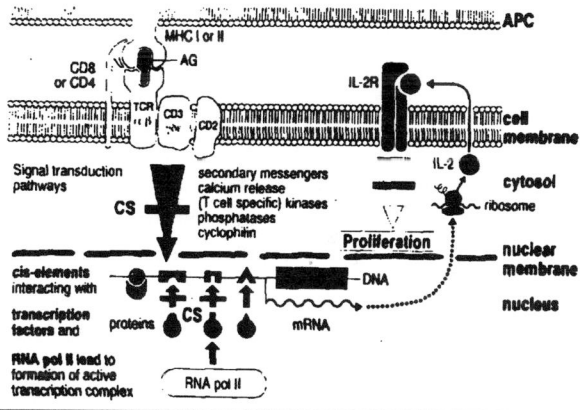

Figure 1.7 / Model of the molecular pathways involved in the process of T cell activation and site of action of cyclosporin A.

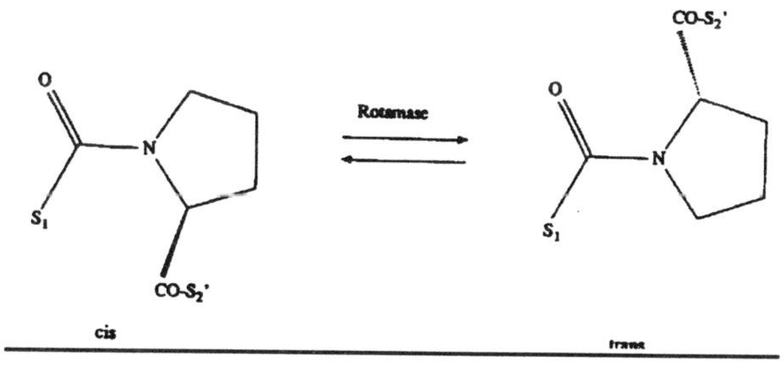

the signal transduction pathway emanating from the T cell receptor and resulting in the DNA/protein complex formation required for IL-2 transcription.

In the search for an explanation of the link between the so-called second messengers and the transduction of information to the nucleus (Fig. 1.7), various studies reported the binding of the compound to several cytosolic proteins, such as calmodulin (Colombani et al., 1985) and to a series of other proteins of molecular weight up to 200 kD (review in Ryffel, 1989). The major interest has however focused on a cyclosporin binding-protein, named cyclophilin, that has been initially identified in thymus cells (Handschumacher et al., 1984) and therafter in a variety of eucaryotic cells, including fungi (Tropschung, 1988; Haendler et al., 1989). This soluble, cytosolic protein has a molecular weight of 17 kD (163 amino acid residues) (Harding et al., 1986). Major and minor isoforms have been identified. Cyclophilin is present in quite high concentrations (0.1 to 0.4% of total cytosolic proteins) in mammalian tissues, especially in T cells (Koletsky et al., 1986) and shows a remarkable evolutionary conservation of amino acid sequence. Cyclophilin binds CsA in a 1:1 ratio with a binding constant of approximatively 10 nM. The

Figure 1.8 / Rotamase activity of cyclophilin.

binding of cyclophilin showed highly specificity for the peptide-ring conformation of CsA, which is not the case for calmodulin (Quesniaux et al., 1988). Studies with more than 100 CsA analogues indicate that immunosupressive activity of a compound parallels with affinity for cyclophilin, although there are a few exceptions to this rule (Quesniaux et al., 1987). Cyclophilin reacts preferentially with CsA-residues 1, 2, 10 and 11, which, together with residues 3 and 9, are the residues known to contribute to the immunosuppressive activity of CsA. The inhibitory effects of CsA in fungi are mediated by cyclophilin; yeast and *Neurospora crassa* that have been selected for CsA resistance possess defects in their cyclophilin genes (Tropschug et al., 1989).

In an independent approach, cyclophilin has been characterized as a peptidyl-prolyl-cis-trans-isomerase (= PPIase) (Fischer et al., 1989; Takahashi et al., 1989). PPIases catalyse the cis-trans-isomerisation of peptide bonds containing the nitrogen of proline (= rotamase) (Fig. 1.8) and thereby accelerate the slow, rate-limiting steps in the folding of several proteins required for proper cellular function (Lang et al., 1987). This enzymatic activity can be easily measured using a peptide linked with a chromogenic substrate and a second enzyme, chymotrypsin, which can only degrade the transform of the peptide. In the presence of CsA both catalysis of cis-trans isomerisation of proline-containing oligopeptides and the acceleration of protein folding are inhibited dose-dependently (Fig. 1.9a) (Harrison and Stein, 1990) at concentra-

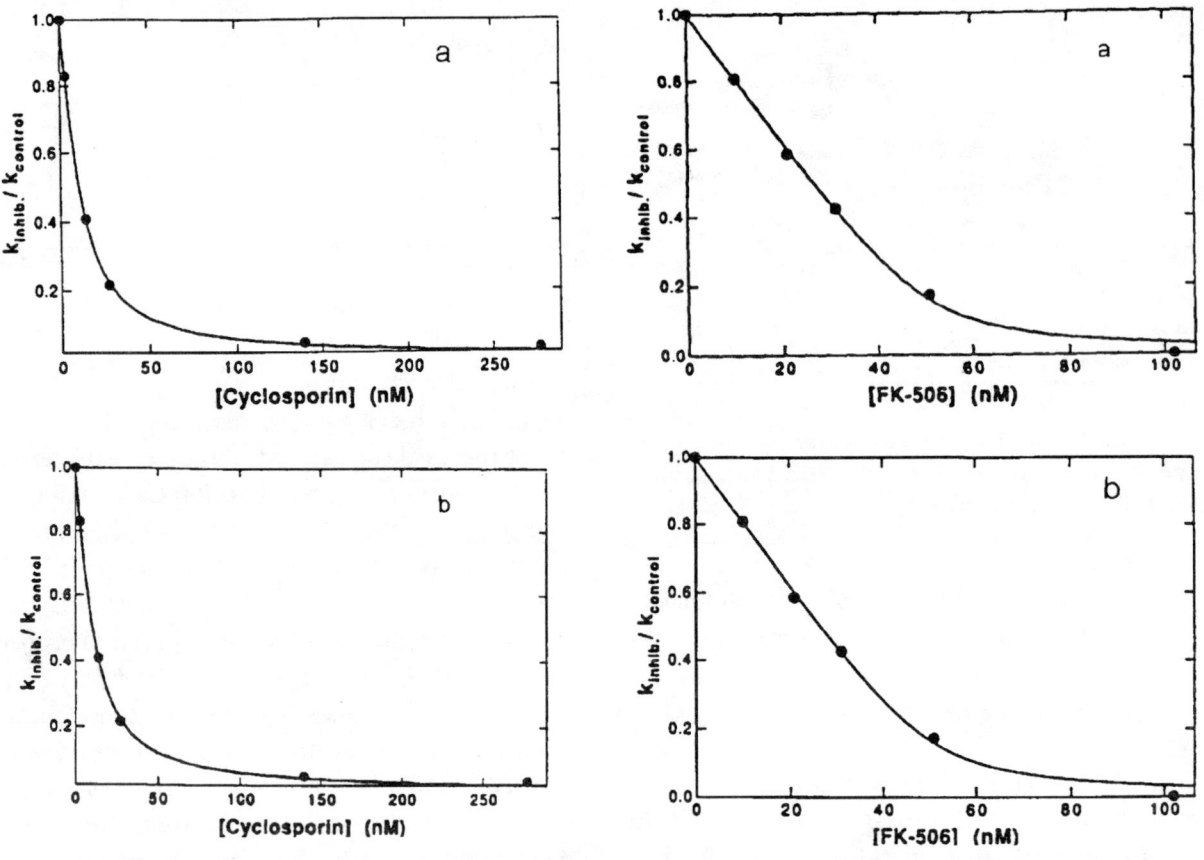

Figure 1.9 / Inhibition of the rotamase activity (a) by Cyclosporin A, (b) by FK-506 (with kind permission of Harrison and Stein).

tions similar to those required for immunosuppression. A close correlation was established between inhibition of rotamase activity and cyclophilin binding of different Cs analogues (Sigal et al., 1991). The relevance of this rotamase activity will be discussed later. Cyclophilin has been cloned and overexpressed in *Escherichia coli* (Liu et al., 1990).

CYCLIC PEPTOLIDE SDZ 214-103

ORIGIN AND STRUCTURE

SDZ 214-103 is a new immunosuppressant discovered in our microbiology group (Dreyfuss er al., 1988). SDZ 214-103 is the major component of aseries of cyclic peptolides which are closely related to cyclosporins (Fig. 1.10). The main structural difference consists of the presence of D-2-hydroxy-isovaleric acid in position 8 in place of D-alanine in cyclosporins, resulting in an ester linkage instead of an amide bond. Like CsA, SDZ 214-103 contains the characteristic C9-amino acid in position 1. Using the conventional nomenclature

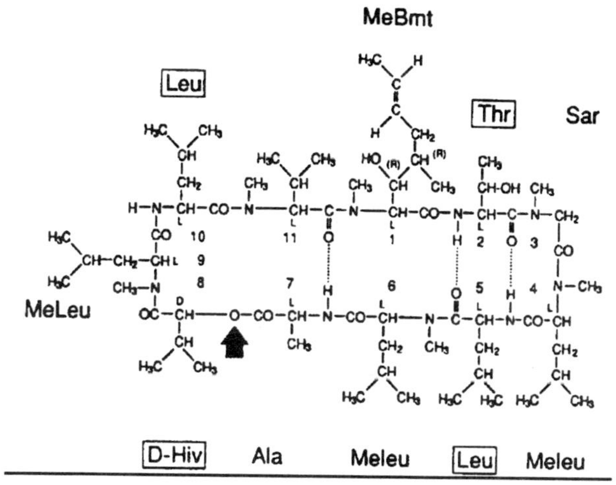

Figure 1.10 / Cyclic peptolide SDZ 214-103.

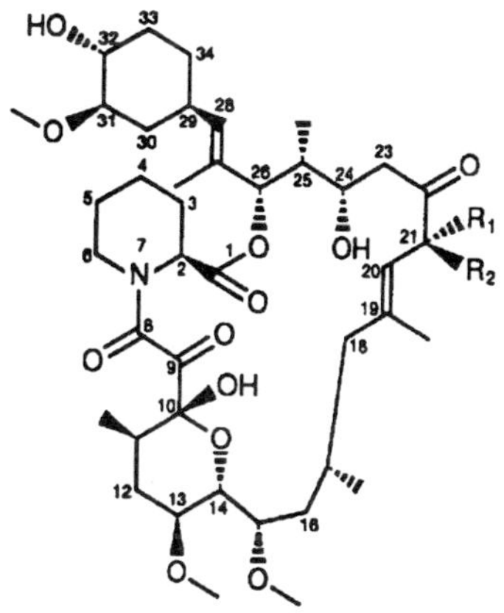

of cyclosporins based upon the structure of CsA, SDZ 214-103 may be represented as [Thr2, Leu5, D-Hiv 8, Leu10]cyclosporin. The compound SDZ 214-103 possesses similar chemical features to CsA.

In addition, five minor metabolites were isolated from large scale fermentations. The spectroscopic characterization revealed the presence of aD-Hiv unit in position 8 and structural modifications in positions 3 and 4 (N-demethylation), 5 (valine or isoleucine in place of leucine) and 1 (8'-hydroxy-MeBmt).

These cyclic peptolides are produced by *Cylindrotrichum oligospermum* (Corda) Bonorden (Syn. *Chaetopsis oligospermum*), strain NRRL 18320. This strain was isolated from maple tree litter collected near Waldenburg in the Swiss Jura. In contrast to all other known cyclosporin-producing fungi, which are hyaline, *C. oligospermum* is dematiacous and on this does not seem to be closely related to the previous group.

A partially purified mycelial enzyme synthetizes the peptolide SDZ 214-103 under addition of the constitutive aminoacids, D-2-hydroxyisovaleric acid, ATP and S-adenosyl-L-methionine (Lawen et al., 1990). The peptolide synthetase does not synthetise CsA, while cyclosporin synthetase is unable to produce the peptolide SDZ 214-103. The molecular weight of both synthetases is in the same range

FK-506	$R_1 = CH_2-CH=CH_2$	$R_2 = H$
21-EPI-FK-506	$R_1 = H$	$R_2 = CH_2-CH=CH_2$
Ascomycin	$R_1 = CH_2-CH_3$	$R_2 = H$
FR-900523	$R_1 = CH_3$	$R_2 = H$
21-EPI-FR-900523	$R_1 = H$	$R_2 = CH_3$

Figure 1.11 / FK-506 and related macrolides.

and both appear to be non-glycosylated (Lawen et al., unpublished data).

MODE OF ACTION

This new fungal metabolite is an approximatively equipotent immunosuppressant to cyclosporin, binds with equal affinity to cyclophilin and the mode of action is thought to be analogous to CsA (Schreier et al., unpublished results). SDZ 214-103 is still under evaluation.

FK-506 AND RELATED MACROLIDES

ORIGIN AND STRUCTURE

FK-506 was obtained from a new species *Streptomyces tsukubaensis*, strain FERM BP-927, obtained from soil collected in the north of Japan (Kino et al., 1987a). Structurally, FK-506 is a macrolide. Its structure is however unique in that it includes a pipecolic acid (positions 1-7) and three carbonyl groups in a series and bridges a lateral cycle derived from shikimic acid (positions 29-34) (Fig. 1.11).

The solubility of FK-506 is similar to CsA. It is fairly lipid soluble, very soluble in alcohol, chloroform, ethyl acetate and acetone but poorly soluble in water and hexane. Its molecular weight is 804 (Kino et al., 1987). Total synthesis of FK-506 has been achieved (Jones et al., 1989; Jones et al., 1990a,b; Nakatsuka et al., 1990).

Related macrolides have also been isolated (Fig. 11.1). FR-900520, now renamed immunomycin, has similar immunosuppressive activities and the same basic structure as FK-506. The difference is that FK-506 has an allylic side chain at position 21 (incorporation of propionate plus one acetate) and FR-900520 an ethyl group (incorporation of a butyrate). FR-900520 is produced by some strains of *Streptomyces hygroscopicus* (e.g., Ferm BP-928, ATCC14891). In fact this compound had already been discovered in the early sixties as an antifungal agent and was named ascomycin (Arai et al., 1962). A third compound in this series, FR-900523, is characterized by a methyl group at position 21 (Hatanaka et al., 1988).

In the course of our screening from soils, we have discovered a new producer of FK-506 isolated from Kyoto, Japan, a new producer of ascomycin from Ischia, Italy, and a strain producing mainly FR-900523 from Iguacu, Brasil. This illustrates the large geographical diversity of the producers of this family of compounds.

Recent biosynthetic studies (Byrne et al., these proceedings) have shown that the cyclic C7 unit is derived from the shikimic acid pathway while the pipecolate arises from lysine catabolism. An ATP dependent enzyme, with a MW of 170 kD, activating L-pipecolate and and to a lesser degree L-proline, has been isolated from cell free extracts. The methyl moieties of the 3 methoxy groups originate from methionine. An O-methyl transferase was purified from cell extracts of *S. hygroscopicus*, strain ATCC 14891.

As is the case for many secondary metabolites, a series of related natural derivatives could be isolated as minor metabolites:

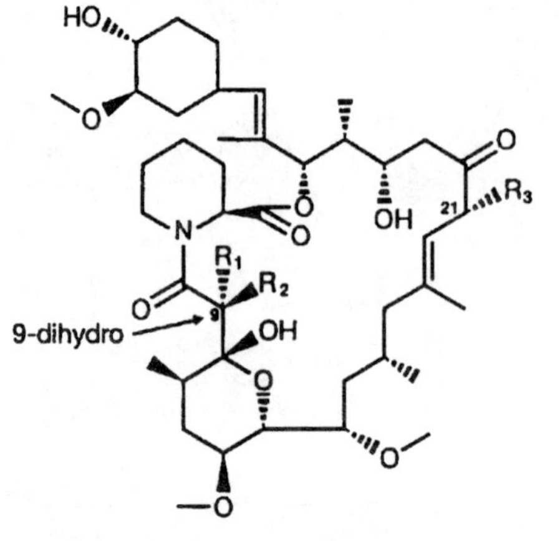

S9993/A-10	R_1 = OH	R_2 = H	R_3 = CH_2–CH=CH_2
S88-30112/A-10	R_1 = OH	R_2 = H	R_3 = CH_2–CH_3
S88-30112/A-18	R_1 = H	R_2 = OH	R_3 = CH_2–CH_3
S88-30112/A-20	R_1 = H	R_2 = OH	R_3 = CH_3

Figure 1.12 / 9-dihydro derivatives of FK-506 and related macrolides.

- compounds with the side chain in the epi position at C-21,
- 9-dihydro compounds, bearing a hydroxyl group instead of a carbonyl at position 9 (Fig. 1.12),
- dehydro-derivatives, with a double bond between C-23 and C-24 (Fig. 1.13).

In addition, these last compounds can also be obtained by biotransformation of the corresponding macrolide by *Actinomucor elegans* (ATCC 6476) or *Rhizopus oryzae* (ATCC 11145).

- compounds with a ring contraction by formation of a new lactone bound at position 24 (Fig. 1.14) (Fehr and Sanglier, unpublished results).

L-prolin added in excess to the fermentation medium is incorporated in place of the L-pipecolic acid (Fig. 1.15) (Fehr and Sanglier, unpublished results). Such compounds have also been isolated

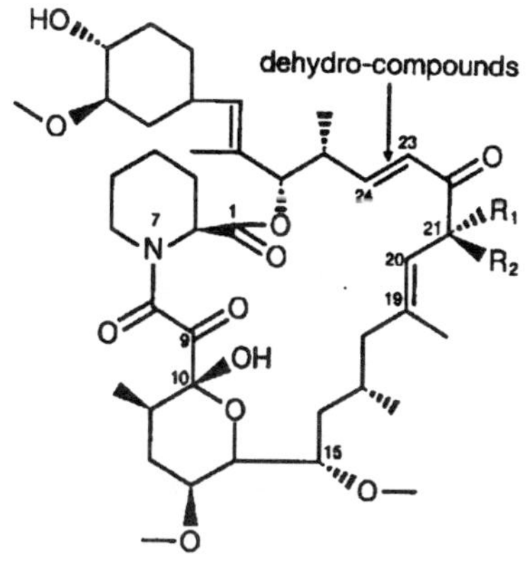

S9993/A-4	R₁ = CH₂–CH=CH₂	R₂ = H
S9993/A-2	R₁ = H	R₂ = CH₂–CH=CH₂
S88-30112/A-14	R₁ = CH₂–CH₃	R₂ = H
S88-30112/A-1,2	R₁ = H	R₂ = CH₂–CH₃

Figure 1.13 / Dehydro-derivatives of FK-506 and related macrolides.

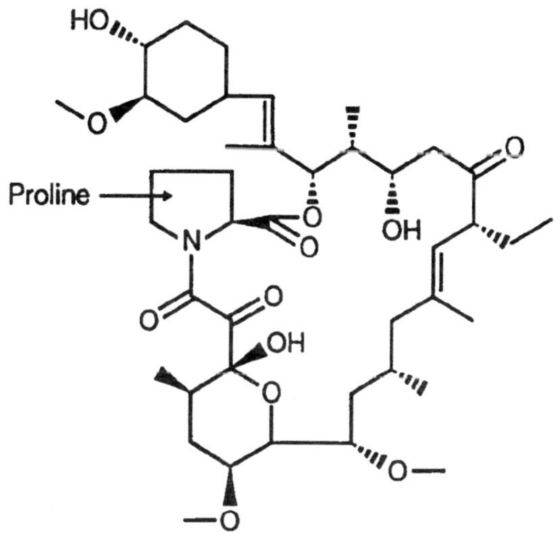

S88-30112/A-13

Figure 1.15 / Ascomycin derivative with proline instead of pipecolic acid.

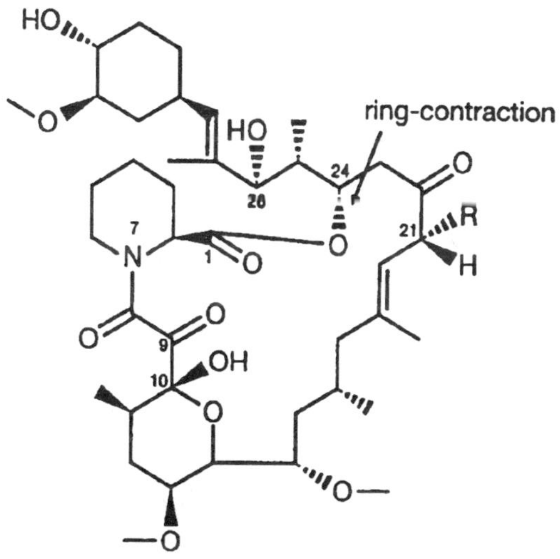

S9993/A-13	R = CH₂–CH=CH₂
S88-30112/A-16	R = CH₂–CH₃

Figure 1.14 / FK-506 and Ascomycin derivatives presenting a ring contraction.

as minor metabolites from broths of *Streptomyces tsukubaensi* (Hatanaka et al., 1989).

By biotransformation with the *Actinoplanacete* sp ATCC 53771, the demethylated analogs of ascomycin and FK-506 were produced (Chen et al., 1990). Also with the same organism or with another actinomycetes strain (ATCC 53828) mono- and bisdemethylated, ring rearranged derivatives of FK-506 and ascomycin could also be isolated from the reaction broth (Arison et al., 1990a,b,c) (Fig 1.16).

MODE OF ACTION

Although totally different in structure from CsA, FK-506 and related macrolides exhibit similar selective immunosuppressive activity (Johanson and Möller, 1990; Kino et al., 1987b; Kay, 1989; Yoshimura, 1989a,b). Analogously to CsA, FK-506 inhibits the proliferation of B-lymphocytes induced by goat antimouse IgM but does not affect the induction of proliferation by bacterial lipo-

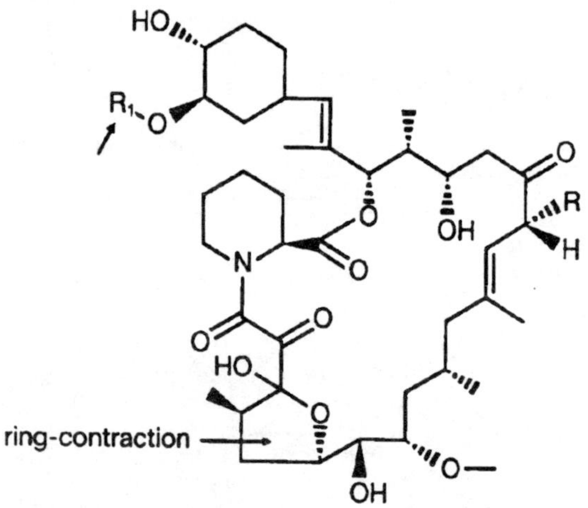

R = CH₂–CH₃ or CH₂–CH=CH₂
R₁ = H or CH₃

Figure 1.16 / Examples of FK-506 and Ascomycin derivatives obtained by biotransformation.

polysaccharide (LPS) (Walliser et al., 1989). The potency of FK-506 *in vitro* is about 100 fold higher than CsA (Kino et al., 1987c; Johansson and Moeller, 1990) but *in vivo* this difference is less pronounced (Hiestand and Schreier, unpublished results; Kino et al., 1987b,c).

FK-506, like CsA interferes with the process of T cell activation by blocking the transcription of IL-2 and other lymphokines (Tocci et al., 1989). FK-506 inhibits IL-2 mRNA accumulation at concentrations of 0.01–0.001 µg/ml (Fig. 1.6, lanes 6–9). As our studies and those of others clearly demonstrate, both immunosuppressive compounds affect the binding of the same transcription factors to the IL-2 promoter (see New Cyclosporins by Directed Biosynthesis, mentioned earlier) at concentrations paralleling their immunosuppressive activity. The decrease in DNA/protein interaction induced by FK-506 and CsA correlates with a decrease in IL-2 production (Baumann et al., 1991). FK-506 is 10 to 100 times more potent than CsA in its ability to inhibit sequence specific DNA-binding and IL-2 mRNA accumulation. This difference in activity may reflect a higher affinity of FK-506 for its intracellular receptor. However, it remains unknown whether FK-506 and CsA utilise identical primary targets in the same or in separate pathways to influence DNA-binding and to bring about the inhibition of IL-2 expression.

A cytoplasmic binding protein has also been discovered for FK-506 : FKBP. This binding protein is distinct from cyclophilin but possesses similar enzymatic activity (rotamase) (Harding et al., 1989; Siekierka et al., 1989). It has a molecular weight of 12 KD. Like cyclophilin, FKBP is ubiquitous and abundant. It has also been isolated from fungal cells (Tropschug, 1990). FK-506 totally inhibits the activity of the FKBP (Fig. 1.9b) (Harrison and Stein, 1990). In this case there is also a good correlation between the affinity of FK-506 analogues to FKBP and the immunosuppressive activity. FK-506 does not bind cyclophilin and CsA does not inhibit FKBP. However, the structural related macrolide rapamycin also binds FKBP and inhibits the rotamase activity without affecting the transcription of IL-2. FKBP and cyclophilin have no apparent sequence similarity (Standaert et al., 1990; Tropschug et al., 1990). FKBP has been cloned and overexpressed in *Escherichia coli* (Standaert et al., 1990). This will facilitate an analysis of the binding interactions and may aid the further selection and design of more specific drugs.

The findings that cyclophilin and FKBP are both PPIases, that there is a good correlation between affinity to immunophilins and immunological activity and that CsA and FK-506 act on their corresponding immunophilins at concentrations similar to those required for immunosuppression, suggest that the PPIase activity of these proteins might play a critical role in T cell activation. The specificity of the reaction in T cells may be partially explained by the existence of cell specific forms of PPIases as it is suggested by the potential existence of a family of cyclophilin genes (Danielson et al., 1988; Haendler et al., 1987). Is the blocking of the rotamase activity the key aspect? It is now generally accepted that the rate-limiting step in major refolding reactions of proteins can be associated with proline isomerisation. At the present time, only two enzymes that catalyse slow steps in protein folding are known, the second one being protein disulfide isomerase (Lang and Schmid, 1988). Folding/unfolding events are necessary for regulated intracellular protein traffic. Unfolded proteins can traverse intracellular membranes (Rassow et al., 1988), which is not the case for fully folded proteins. On the other hand such proteins must be refolded in the organelles; for example a heat shock protein group, hps60, in the case of mitochondrial import of F1-ATPase subunits and other proteins of the inner mitochondria (Eilers et al., 1988). In addition, it has been shown that cyclophilin-like molecules participate in a phototransduction pathway (Shieh et al., 1989). There is however a series of arguments against the importance of inhibition of the rotamase activity. Doubts about the efficiency of such catalysis have been expressed (Lin et al., 1988). Recent investigations (Harrison and Stein, 1990) demonstrated that cyclophilin and FKBP have dramatically different substrate specificities and suggested the existence of a family of different enzymes. Even if we hypothesize a concentration of 0.4% of the protein contents for cyclophilin and a concentration of CsA of 3 to 10 nM, only about 10% of the binding sites on cyclophilin would be occupied. Complete inhibition of the T cell proliferation is obtained with either CsA or FK-506 alone. The inhibition of the rotamase activity of FKBP is insufficient for mediating the immunological effect of FK-506 (Bierer et al., 1990). The fact that rapamycin also binds to FKBP, blocks PPIase activity (Albers et al., 1990) and shows a different type of immunological activity is an additional argument against a key role of the inhibition of the rotamase activity of the binding proteins. It has been suggested that the common part of FK-506 and rapamycin is the binding domain while the other one is the effector domain (Schreiber, 1991). It is more probabe that the complex immunosuppressant—immunophilin plays the key role. A plausible model is that immunophilins bind a component of the transcriptional apparatus and thereby control the activity of transcription factors (Schreiber, 1991) in a way similar to the cytoplasmic anchoring protein systems (Hunt, 1989).

RAPAMYCIN

ORIGIN AND STRUCTURE

Rapamycin (RAP) is a lipophilic macrolide antibiotic (M.W.914) produced by *Streptomyces hygroscopicus*, strain NRLL 5491. It was discovered in 1975 on the basis of growth inhibition of yeasts and some dermatophytes (Seghal et al., 1975). Rapamycin contains a 1,2,3-tricarbonyl moiety linked to a pipecolic acid lactone, a characteristic shared by FK-506. It differs from FK-506 by its larger size (29 versus 21-membered ring) and by a triene segment (Fig. 1.17). Demethoxy-rapamycin is co-produced with RAP (Seghal et al., 1983).

The majority of the macrolide ring of RAP is formed from six acetate and seven propionate units. Methionine and glycine label the three methoxy groups of rapamycin to a high degree (Paiva et al., 1991).

MODE OF ACTION

Although immunosuppressive activity of RAP was described in 1977 (Martel et al., 1977), its use in organ allografting with animals was not reported until 1989 (Calne et al., 1989; Meiser et al., 1989). It has been found that RAP given i.p. at low doses resulted in a profound prolongation of graft survival in rats and that it has a high therapeutic index (Meiser et al., 1990). Obviously, the discovery of FK-506 stimulated a renewed interest in the immunological properties of RAP. Like CsA and FK-506, RAP suppresses the proliferation of T cells by lectins or antibodies. However RAP exhibits a different bioactivity pattern in that it does not

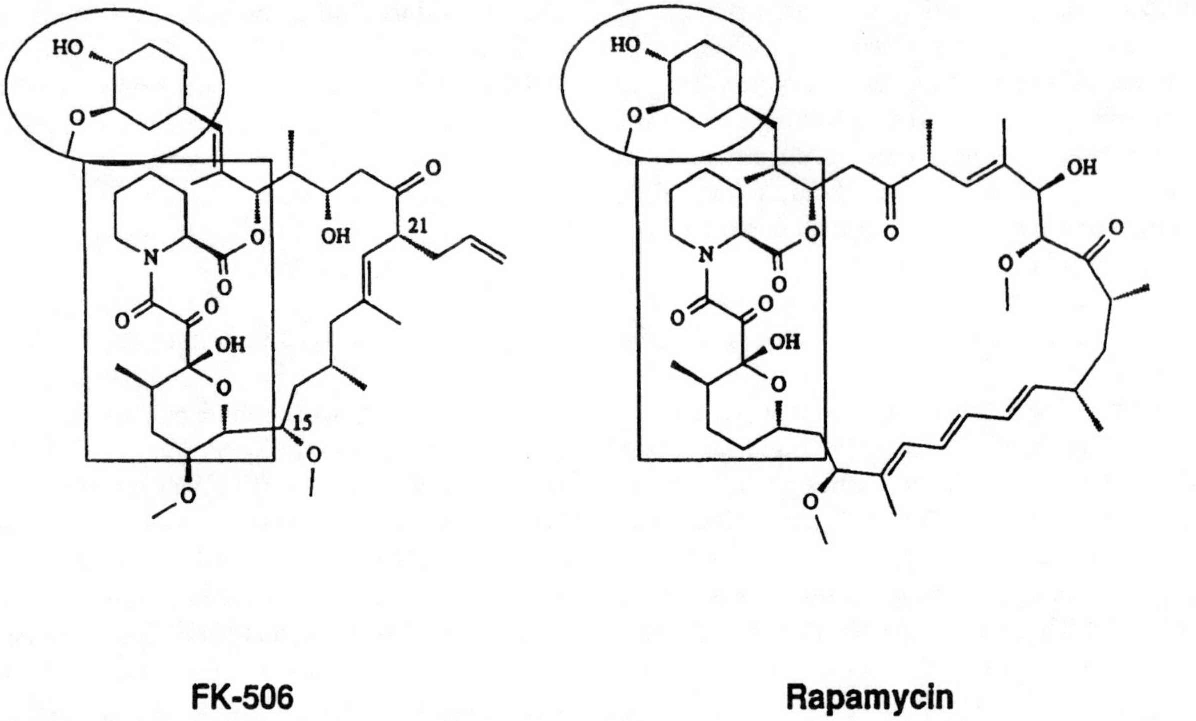

FK-506 **Rapamycin**

Figure 1.17 / Comparison of the structure of FK-506 and Rapamycin.

affect lymphokine production or Il-2 receptor expression but can markedly suppress Il-2 and Il-4-driven T cell proliferation. The effect of RAP on T cell proliferation cannot be reversed by addition of exogenous IL-2. This suggests that RAP immunosuppression is mediated primarily by an impairment of T cell response to growth-promoting lymphokines. However, unlike CsA and FK-506, RAP does not show T lymphocyte specificity but inhibits the proliferation of other cell types as well. For inhibition, CsA and FK-506 need to be added within 3 h after stimulation. In contrast, RAP can excert inhibitory activity even when added 12 h after initiation of the cultures. This suggests that RAP alters events later in the activation process than CsA or FK-506 (Dumont et al., 1990b; Metcalfe and Richards, 1990). RAP potentiates the effect of CsA on lymphocyte proliferation but acts as an antagonist of FK-506 suppression (Dumont et al., 1990a). Inhibition of IL-2 mRNA accumulation by FK-506 can be reversed by the addition of an excess of RAP (100 fold) but this is not the case for the inhibition by CsA (Baumann et al., unpublished data). This suggests that CsA and FK-506 act by binding to different cellular receptors. As mentioned previously, RAP binds to FKBP with the same affinity as FK-506.

DEOXYSPERGUALIN

ORIGIN AND STRUCTURE

Spergualin is a polyamine with terminal primary amine and guanidine groups, from *Bacillus laterosporus* (Takeuchi et al., 1981). 15-deoxyspergualin (Fig. 1.18) (DSG) is the most active derivative and can be chemically synthesized (Iwasawa et al., 1982)

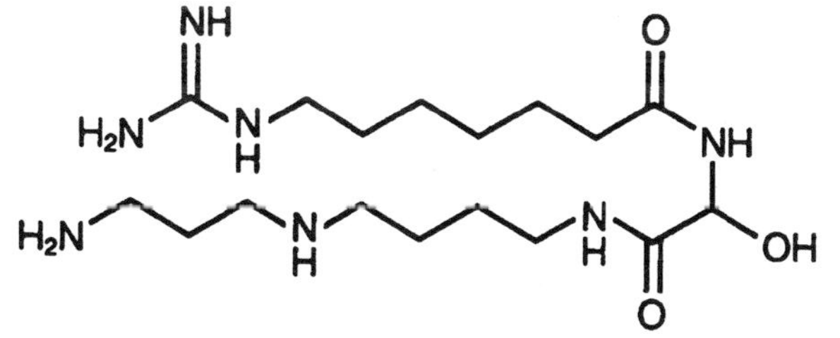

Figure 1.18 / 15-Deoxyspergualin.

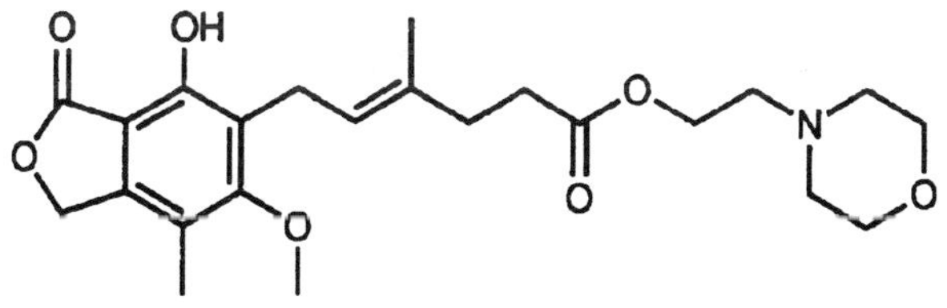

Figure 1.19 / Mycophenolic acid morpholinoethylester (RS-61443).

MODE OF ACTION

Previously described as an antitumor antibiotic (Muller et al., 1987), immunosuppressive activity has now been demonstrated in different transplantation systems (Morris and Yuh, 1990) and autoimmune diseases (Nemoto et al., 1988). In animal models, 15-deoxyspergualin significantly prolongs allograft survival in a dose-dependent manner. In these models, DSG is about six times more potent than CsA. The immunosuppressive effect continues well beyond the administration period (Morris and Yuh, 1990). However in dogs, DSG appears to have severe side effects and no clear benefit (Collier et al., 1988).

Spergualins exhibit suppressive effects on LPS induced blastogenesis and antibody production to TD and TI antigens *in vitro* and *in vivo* indicating that they affect the proliferation and differentiation of B cell lineage directly (Fuji et al., 1990). They also have a suppressive effect on cytotoxic T lymphocytes but even at high concentrations the inhibition is not total. Spergualins have little effect in the early stage of a MLR response. It has been suggested that they act on proliferation and differentiation of T cells which respond to growth factors such as IL-2 (Fuji et al., 1989).

MYCOPHENOLIC ACID

ORIGIN AND STRUCTURE

Mycophenolic (MPA) (M.W.320) was first isolated in 1896 from a strain of *Penicillium brevicompactum*, its structure elucidated in 1933, its antibiotic activity recognized in 1945 (Glasby, 1979). It is also produced by other Penicillia (*P. roqueforti* and *P. stolonifcrum*). Its synthesis has been described (Danheiser et al., 1986).

MODE OF ACTION

MPA has a potent antimitotic effect in mammalian cells. It selectively inhibits, non-competitively and reversibly, inosine monophosphate dehydrogenase and guanine monophosphate synthetase biosynthetic enzymes for guanosine monophosphate (Sweeney et al., 1972). MPA-induced depletion of guanosine and deoxyguanosine nucleotides has greater antiproliferative effects in T and B lymphocytes than in other cell types because of the greater dependence of lymphocytes on *de novo* purine biosynthesis (Morris et al., 1990). MPA blocks the transition from the G1 to the S phase and also inhibits the proliferation of promonocytes (Eugi et al., 1990). A practical problem with MPA is that it is cleared rapidly from the plasma. Mycophenolic acid morpholinoethylester, RS-61443, (Fig. 1.19) is a semisynthetic derivative (Nelson et al., 1988) that has a better bioavailability than MPA. After oral administration, RS-61443 is progressively hydrolysed to liberate MPA (Lee et al., 1990).

RS-61443 prolongs the survival of histo-incompatible organ allografts in rat recipients (Morris et al., 1989) and allows islet allografting in mice (Hao et al., 1990). Maximum efficacy is achieved when the compound is administered during the period of antigenic stimulation. RS-61443 seems to lack severe toxicity *in vivo* (Almquist et al., 1990).

BREDININ (MIZORIBINE)

ORIGIN AND STRUCTURE

Bredinin is a nucleoside-related compound (Fig. 1.20) from broths of *Eupenicillium brefeldianium* (Mizumo et al., 1974). It is soluble in water but insoluble in most organic solvents (Hillier and Castaner, 1978). Its synthesis has been described (Fukukawa et al., 1986).

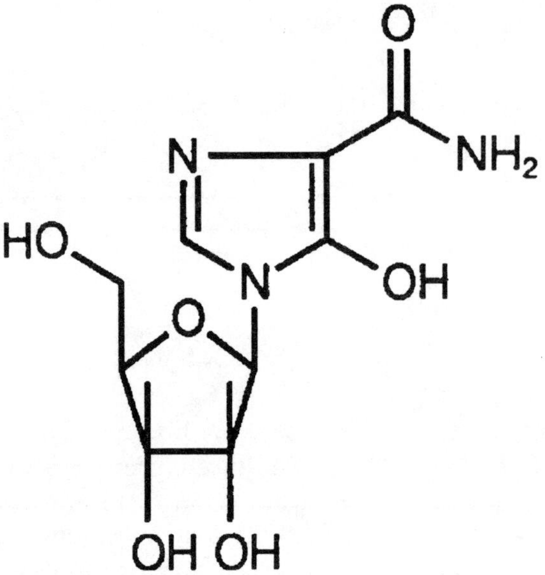

MODE OF ACTION

The immunosuppressive effect of bredinin is based on the blockage of the conversion of inosinic acid to guanylic acid. Within cells, bredinin is converted to the 5'-phosphate, which in contrast to bredinin itself, competitively inhibits both IMP dehydrogenase and GMP synthetase (Kusumi et al., 1989). T cell as well as B cell responses, are inhibited by the administration of bredinin (Kamata et al., 1983). Synergistic activity with CsA has been reported (Amemiya et al., 1989). Teratogenic effects of bredinin have been detected (Okamoto et al., 1978)

VARIA

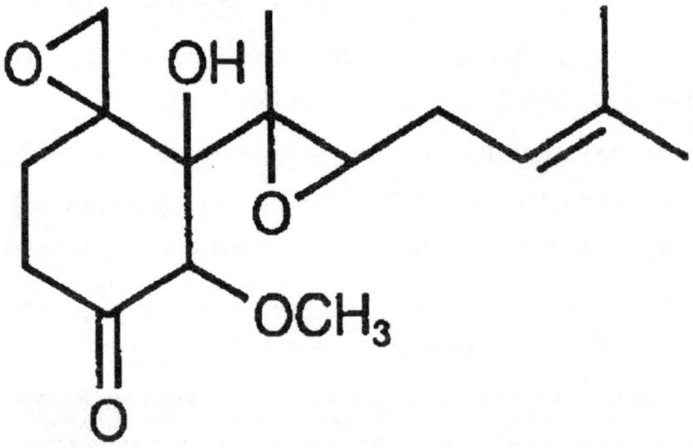

Ovalicin (Fig. 1.21) is the first immunosuppressive compound of microbial origin which has been described (Lazary and Staehelin, 1968; Bollinger et

Figure 1.20 / Bredinin.

Figure 1.21 / Ovalicin.

al., 1973). This sequiterpene is produced by a strain of the fungus *Pseudeurotium ovalis*. It acts as a very potent inhibitor of DNA synthesis in proliferating lymphocytes and in lymphoma cells. Due to toxicological problems, the development of this drug had to be stopped but nevertheless, ovalicin opened new opportunities in immunology and paved the way for the discovery of cyclosporin.

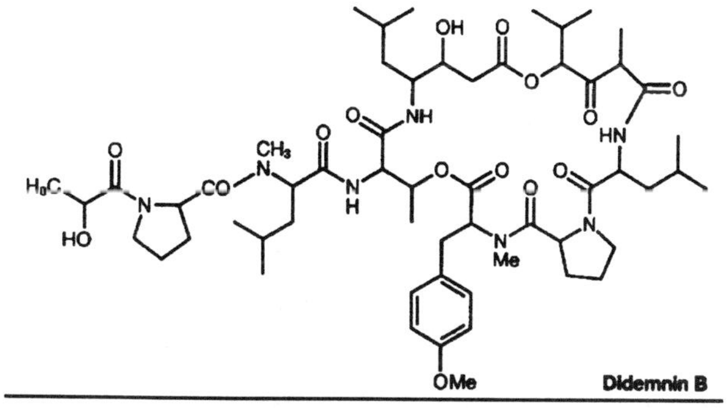

Figure 1.22 / Didemnin B.

Didemnins are lipophilic depsipeptides originally isolated from a Carrabbean tunicate of the family Didemnidae, *Tridemnum solidum* (Rinehart et al., 1981). They consist of an unusual depsipeptide ring structure that contains one unit of 2-(2-hydroxyisovaleryl)propionate and one unit of an isomer of the amino acid statine (Rinehart, 1988). It has been proposed that these compounds are in fact produced by a cyanobacterium (*Synephocystis tridemni*) living in the tunicate. The most active compound is didemnin B (Fig. 1.22) which is a potent inhibitor of protein and RNA synthesis (Li et al., 1984; Legrue et al., 1988). Didemnin B is primarly an anticancer drug (Rinehart et al., 1988; Rinehart these proceedings). In several *in vitro* lymphoproliferative assays, DB is 100 to 1000 times more effective than CsA (Montgomery and Zulowski, 1985) but in fact DB functions as an antiproliferative and not as a specific immunosuppressive compound (Legrue et al., 1988). A cumulative toxicity is observed at minimal therapeutic dosages (Yuh et al., 1989).

Tetranectin (Fig. 1.23) is a macrotetrolide, hydrophobic antibiotic produced by *Streptomyces aureus* (Ando et al., 1971). Tetranectin suppresses *in vitro* lymphoproliferation and the generation of cytotoxic cells (Callewaert et al., 1988).

Brefeldin A (Fig. 1.24) a 13-membered macrocyclic lactone with antifungal activity produced by a strain of *Eupenicillium brefeldianum* (Haerri et al., 1963) should be mentioned as an important tool in immunopharmacology. It inhibits presentation

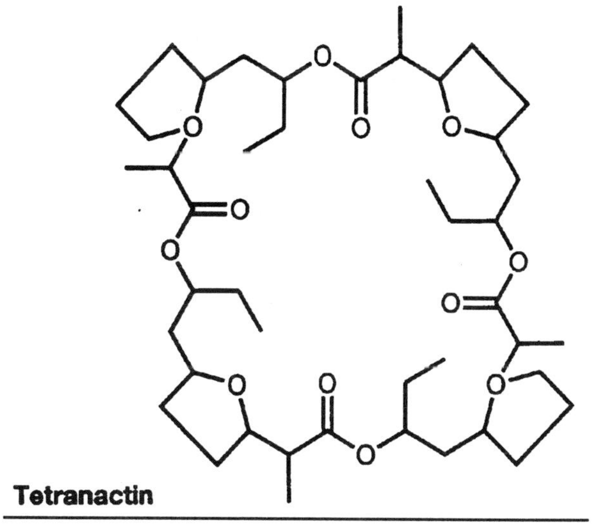

Figure 1.23 / Tetranactin.

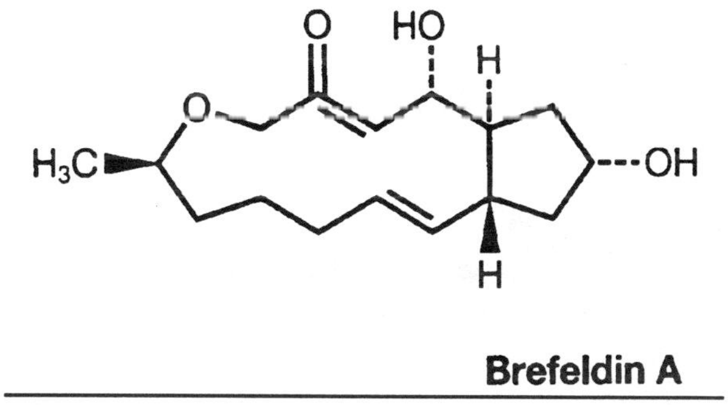

Figure 1.24 / Brefeldin.

of protein antigens (Nuchtern et al., 1989; Yewdell and Bennink, 1989) but could not be developed as immunomodulator because it acts by blocking the transport of proteins from the endoplasmatic reticulum to the Golgi apparatus, a key function in all eucaryotic cells.

Some new compounds have been recently published without detailed information: NK 86-1086 from a streptomycetes (Seiichi et al., 1989), myriocin from a strain of *Isaria sinclarii* (Fujita et al., 1989), FR651814, a demethyl derivative of epi-fumagillol, isolated from *Penicillium jensenii* (Hatanaka et al., 1988), cammunomicin from a strain of streptomycetes (Franco et al., 1989), gliooxin from strains of *Aspergillus fumigatus* (Waring and Müllbacher, 1990), prodigiosin from *Serratia marcescens* (Tsijii et al., 1990), the compound WS 9482 from a strain of *Kitasatosporia kifunense* (Okuhara, these proceedings), depsidomycin from a strain of *Streptomyces lavandulae* (Isshiki et al., 1990), discodermolide from the marine sponge *Discodermia dissoluta* (Gunasekera et al., 1990).

CONCLUSION

Cyclosporin has allowed major progress in organ transplantation and it remains the only selective immunosuppresive drug widely used in the clinic. Its value in various autoimmune diseases has also been recognized. In the case of FK-506, the therapeutic index still needs to be exactly determined. CsA and FK-506 have proven extremely useful tools in basic immunology and cell biology. The detailed elucidation of the mode of action of CsA and FK-506 will increase our understanding of lymphocyte activation and the signalling pathway leading to gene expression.

The isolation of the CsA-synthetase has recently allowed the controlled biosynthesis of new compounds. Enzymes for the FK-506 pathway have also been isolated. These enzymes of CsA and FK-506 will be of wide application in the production of a series of novel immunosuppressants.

The selectivity of CsA and FK-506 is high. This is not the case for the other immunosuppressants and their therapeutic value will have to be evaluated in detail in comparison and in combination with CsA or FK-506. Of importance is also their potential use in autoimmune diseases. These compounds, especially rapamycin, will also contribute to a better understanding of the immune response. The discovery of new immunosuppressants has been rare. However new assays now possible due to the progress in molecular biology will focus on targets such as the MHC, the processing or the presentation of antigens, the synthesis of lymphokines other than IL-2 and the blockade of lymphokine receptors. Compounds active at such sites would provide immunosuppressants with new pharmacological profiles and screening of compounds of microbial origin will continue to play a central role in these endeavours.

ACKNOWLEDGMENTS

We thank S. Cottens and T. Payne for critically reviewing the manuscript and P. Bollinger, M. Egi, B. Huegi and H. Neubauer for bibliographical data.

REFERENCES

1. Aarnio, T. H. and Agathos, S. N. 1990. Pigmented variants of *Tolypocladium inflatum* in relation to Cyclosporin A production. *Appl. Microbiol. Biotechnol.* 33: 435–437.
2. Agathos, S. N., Chun, G. T. and Lee, J. 1988. The physiology of Cyclosporin production in submerged cultures of *Tolypocladium inflatum* in: Abstract book of the "Second international symposium on overproduction of Microbial Products," p. 58.
3. Agathos, S. N., Marshall, J. W., Moraiti, C., Parekli, R. and Mathosingh, C. 1986. Physiological and genetic factors for process development of Cyclosporin fermentations. *J. Ind. Microbiol.* 1: 39–48.
4. Albers, M. W., Walsh, C. T. and Schreiber, S. L. 1990. Substrate specificity for the human rotamase FKBP: a view of FK-506 and Rapamycin as Leucine-(twisted amide)-proline mimics. *J. Org. Chem.* 55: 4984–4986.
5. Almquist, S., Chellman, G., Dunne, J. and Eugui, E. 1990. Analysis of immune function in cynomolgus monkey treated with a novel immunosuppressive agent. *FASEB J.* 4: 423.
6. Altmann, A., Coggeshall, K. M., and Mustelin, T. 1990. Molecular events mediating T-cell activation. *Adv. Immunol.* 48: 227–314.

7. Amemiya, H., Suzuki, S., Watanabe, H., Hayashi, R., Niija, S. 1989. Synergistically enhanced immunosuppressive effect by combined use of cyclosporine and mizoribine. *Transplant. Proc.* 21: 956–958.
8. Ando, K., Murakami, Y. and Nawata, Y. 1971. Tetranactin, a new miticidal antibiotic. *J. Antibiot.* 24: 418–423.
9. Arai, T., Kayama, Y., Suenaga, T. and Honda, H. 1962. Ascomycin, an antifungal antibiotic. *J. Antibiot.* 15: 231–232.
10. Arison, B. H., Inamine, E. S., Chen, S-S. T., Wicker, L. S. 1990a. Microbial transformation of L-679–934. Europ. Patent Appl. 0 378 317.
11. Arison, B. H., Inamine, E. S., Chen, S-S. T., Wicker, L. S. 1990b. Microbial transformation product. Europ. Patent Appl. 0 378 320.
12. Arison, B. H., Inamine, E. S., Chen, S-S. T., Wicker, L. S. 1990c. Microbial transformation product of L-683–590. Europ. Patent Appl. 0 378 321.
13. Bach, J. 1989. Cyclosporine in autoimmune diseases. *Transplant. Proc.* 21: 97–113.
14. Baumann, G., Geisse, S., Sullivan, M., 1991. Cyclosporin A and FK-506 both affect DNA binding of regulatory nuclear proteins to the human Interleukin-2 promoter. *The New Biologist* 3: 270–278.
15. Beveridge, T., 1986. Clinical transplantation: Overview. *Prog. Allergy* 38: 269–2.
16. Bierer, B. E., Somer, P. K., Wandless, T. J., Burakoff, S. J., Schreiber, S. L. 1990. Probing immunosuppressant action with nonnatural immunophilin ligands. *Science* 250: 556–559.
17. Billich, A. and Zocher, R., 1987. Enzymatic synthesis of cyclosporin A. *J. Biol. Chem.* 262: 17258–17259.
18. Bollinger, P., Sigg, H. P., Weber, H. P. (1973). Die Struktur von Ovalicin. *Helv. Chim. Acta* 56, 818–830.
19. Borel, J. F. (Ed.) 1986. Ciclosporin: Progr. *Allergy 38.*
20. Borel, J. F. 1989. The cyclosporins. *Transplant. Proc.* 21: 810–815.
21. Borel, J. F. 1989. Pharmacology of Cyclosporine (Sandimmune). IV. Pharmacological properties *in vivo*. *Pharmacol. Rev.* 41: 259–372.
22. Borel, J. F., Di Padova, F., Mason, J., Quesniaux, V., Ryffel, B. and Wenger, R. 1989. Pharmacology of Cyclosporin (Sandimmune) I. Introduction. *Pharmacol. Rev.* 41: 239–242.
23. Borel, J. F., Feurer, C., Magne, C., Stahelin, H. 1977. Effects of the new anti-lymphocyte peptide cyclosporine A in animals. *Immunology* 32: 1017–1025.
24. Byrne, K. M., Shafiee, A., Nielsen, J. B., Arison, B. H., Monaghan, R. L. and Kaplan, L. 1991. The biosynthesis and enzymology of an immunosuppressant, Immunimycin, produced by *Streptomyces hygroscopicus* var. *ascomycetus*. Abstract S-32, Biotechnology of Microbial Products, SIM, Oct. 90.
25. Callewaert, P. M., Radcliff, G., Tanouchi, Y., and Shichi, H. 1988. Tetranactin a macrotetrolide antibiotic suppresses *in vitro* proliferation of human lymphocytes and generation of cytotoxicity. *Immunopharmacology* 16: 25–32.
26. Calne, R. Y., Collier, D. J., Lim, S., Pollard, S. G., Samaan, A., White, D. J., Thiru, S. 1989. Rapamycin for immunosuppression in organ allografting. *Lancet 1989 (2)*: 227.
27. Chen, S-S. T., Inomine, E. S., Arison, B. H., Garvity, G. M., Mochales, S., Wicker, L. S. 1990. New immunosuppressant-producing culture. Europ. Patent App. 0 353 827.
28. Chun, G. T. and Agathos, S. N. 1989. Immobilisation of *Tolypocladium inflatum* spores into porous celite beads for Cyclosporin production. *J. Biotechnol.* 9: 237–254.
29. Collier, D. S., Calne, R., Thiru, S., Friend, P. J., Lims., White, D. J., Kohno, H. and Levickis, J. 1989. FK-506 in experimental renal allografts in dogs and primates. *Transpl. Proc. 20*, Suppl. 1: 226–228.
30. Collier, D. S., Calne, R., Thiru, S., Kohno, H., Levickis, J. 1988. 15-Deoxyspergualin in experimental dog renal allografts. *Transplantation Proc.* 20: 240–241.
31. Colombani, P., Robb, A. and Hess, A. 1985. Cyclosporin A binding to Calmodulin: a possible site of action on T lymphocytes. *Science* 228: 337–339.
32. Cruse, M., Lewis, E. 1989. Ciclosporin therapy of autoimmune diseases. *Concepts Immunopathol.* 7: 1–19.
33. Danheiser, R. L., Gee, S. K. and Perez, J. L. 1986. Total synthesis of mycophenolic acid. *J. Am. Chem. Soc. 108*: 806–819.
34. Danielson, P. E., Forss-Petter, S., Brow, M. A., Calavetta, L., Douglas, J., Milner, R. J., Sutcliffe, J. G. 1988. p1B15: a cDNA clone of the rat mRNA encoding cyclophilin. *DNA* 7: 261–267.

35. Di Padova, F. E. 1989. Pharmacology of Cyclosporine (Sandimmune). V. Pharmacological effects on immune function: *in vitro* studies. *Pharmacol. Rev.* 41: 373–406.
36. Drews, J. 1990. Immunopharmacology. Springer-Verlag, Berlin.
37. Dreyfuss, M. 1986. Neue Erkenntnisse aus einem pharmakologischen Pilz-Screening. *Sydowia* 29: 22–36.
38. Dreyfuss, M., Härri, E., Hofmann, H., Kobel, H., Pache, W. and Tscherter, H., 1976. Cyclosporin A and C, new metabolites from *Trichoderma polysporum* (Link et Pers.) Rifai. *Eur. J. Appl. Microbiol* 3: 125–133.
39. Dreyfuss, M., Schreier, M. H., Tscherter, H. and Wenger, R. 1988. "Cyclic peptolides" Europ. Pat. Appl. 0 296 123 A2.
40. Dumont, F. J., Melino, M. R., Staruch, M. J., Koprak, S. L., Fischer, P. A. and Sigal, N. H. 1990a. The immunosuppressive macrolides FK-506 and Rapamycin act as reciprocal antagonists in murine T-cells. *J. Immunol.* 144: 1418–1424.
41. Dumont, F. J., Staruch, M. J., Koprak, S. L., Melino, M. R. and Sigal, N. H. 1990b. Distinct mechanisms of suppression of murine T-cell activation by the related macrolides FK-506 and Rapamycin. *J. Immunol.* 144: 251–258.
42. Durand, D. B., Shaw, J. P., Bush, M. R., Replogle, R. E., Belagaje, R. and Crabtree, G. R. 1988. Characterization of antigen receptor response elements within the Interleukin-2 enhancer. *Mol. Cell. Biol.* 8: 1715–1724.
43. Eilers, M., Verner, K., Hwang, S., Schatz, G. 1988. Import of proteins into mitochondria. *Phil. Trans. R. Soc. Lond. (Biol. Sci.)* 319: 121–126.
44. Emmel, E. A., Verweij, C. L., Durand, D. B., Higgins, K. M., Lacy, E. and Crabtree, G. R. 1989. Cyclosporin A specifically inhibits function of nuclear proteins involved in T-cell activation. *Science* 246: 1617–1620.
45. Eugui, E. M., Nelson, P. and Allison, A. C. 1990. Mycophenolic acid and analogs as immunosuppressive drugs. Paper presented at the 200th ACS National Meeting, Washington, Aug. 1990.
46. Fischer, G., Wittmann-Liebold, B., Lang, K., Kiefhaber, Th., Schmid, F. X. 1989. Cyclophilin and peptidyl-prolyl cis-trans isomerase are probably identical proteins. *Nature* 337: 476–479.
47. Fliri, H. and Wenger, R. 1990. Cyclosporine: Synthetic studies, structure-activity relationships, biosynthesis and mode of action. In "Biochemistry of Peptide Antibiotics." (Kleinkauf, H. and von Döhren, H, eds.) pp. 245–287, W. De Gruyter: Berlin.
48. Foxwell, B. M. and Ryffel, B. 1990. The mechanisms of action of Cyclosporine. *Cardiol. Clin.* 8: 107–117.
49. Franco, C. M., Chatterjec, S., Vijayakumar, E. K., Ganguli, B. N. and Rupp, R. H. 1989. Ein neues Antibiotikum, Cammunocin, ein Verfahren zu seiner Herstellung und seiner Verwendung als Arzneimittel. Europ. Pat. Appl. 0 327 045.
50. Fujii, H., Takada, T., Nemoto, K., Abe, F., Fujii, A. and Takeuchi, T. 1989. *In vitro* immunosuppressive properties of Spergualins to murine T-cell response. *J. Antibiot.* 42: 788–794.
51. Fujii, H., Takada, T., Nemoto, K., Yamashita, T., Abe, F., Fujii, A. and Takeuchi, T. 1990. Deoxyspergualin directly suppresses antibody formation *in vivo* and *in vitro*. *J. Antibiot.* 43: 213–219.
52. Fujita, T., Shibuya, H., Ohashi, T., Yamanishi, K. and Taniguchi, T. 1986. Regulation of human interleukin-2 gene: functional DNA sequences in the 5′ flanking region for the gene expression in activated T-lymphocytes. *Cell* 46: 401–407.
53. Fujita, T., Toyama, R., Sasaki, S., Okumoto, T. Chiba, K. 1989. Myriocin and its manufacture with *Isaria*. Patent: Jpn Kokai Tokkyo Koho 89104087.
54. Fukukawa, K., Shuto, S., Hirano, T. Ueda, T. 1986. Synthesis of Bredinin from 1-beta-d ribofuranosyl-5-aminomidazole-4-carboxamide by a photoreaction. *Chem. Pharm. Bull.* 34: 3653–3657.
55. Glasby, J. S. 1979. *Encyclopaedia of Antibiotics*. John Wiley: New York.
56. Gunasekera, P., Gunasekera, M., Longley, R. E. and Schulte, G. K. 1990. Discodermolide: a new bioactive polyhydroxylated lactone from the marine sponge *Discodermia dissoluta. J. Org. Chemie* 55: 4912–4915.
57. Haendler, B., Hofer-Warbinek, R., and Hofer, E. 1987. Complementary DNA for human T-cell cyclophilin. *EMBO J.* 6: 947–950.
58. Haendler, B., Keller, R., Hiestand, P. C., Kocher, H. P., Wegmann, G., Movva, R. 1989. Yeast Cyclophilin: isolation and characterization of the protein cDNA and gene. *Gene* 83: 39–46.

59. Haerri, E., Loeffler, W., Sigg, H. P., Staehelin, H. und Tamm, C. 1963. Ueber die Isolierung nueuer Stoffwechselprodukte aus *Penicillium brefeldianum*. *Helv. Chim. Acta* 46:1235–1239.
60. Hamada, Y., Kondo, Y., Shibata, M., Shioiriti, T. 1989. Efficient total synthesis of Didemnins A and B. *J. Am. Chem. Soc.* 111:669–673.
61. Handschumacher, R. E., Harding, M. W., Rice, J., Drugge, R. J. 1984. Cyclophilin: A specific cytosolic binding protein for Cyclosporin A. *Science* 286:544–547.
62. Hao, L., Lafferty, K. S., Allison, A. C. and Eugui, E. M. 1990. RS-61443 allows islet allografting and specific tolerance induction in adult mice. *Transplant. Proc.* 22:876–879.
63. Harding, M. W., Handschumacher, R. E. and Speicher, D. W. 1986. Isolation and amino acid sequence of Cyclophilin. *J. Biol. Chem.* 261:8547–8555
64. Harding, M. W., Galat, A., Uehling, D. E. and Schreiber, S. L. 1989. A receptor for the immunosuppressant FK-rot is a cis-trans peptidyl-prolyl isomerase. *Nature* 341:758–760.
65. Harrison, R. K. and Stein, R. L. 1990. Substrate specificities of the peptidyl prolyl Cl3-trans isomerase activities of Cyclophilin and FK-506 binding protein: evidence for a family of distinct enzymes. *Biochem.* 29:3813–3816.
66. Hatanaka, H., Kino, T., Asano, M., Goto, T., Tanaka, H. and Okuhara, M. 1989. FK-506 related compounds produced by *Streptomyces tsukubaensis* No. 9993. *J. Antibiot.* 42:620–622.
67. Hatanaka, H., Kino, T., Hashimoto, M., Tsurumi, Y., Kuroda, A., Tanaka, H. 1988. FR65814, a novel immunosuppressant isolated from a *Penicillium* strain. *J. Antibiot.* 41:999–1008.
68. Hatanaka, H., Kino, T., Miyata, S., Inamura, N., Kuroda, A., Goto, T., Tanaka, H., Ohuhara, M. 1988. FR-900 520 and FR-900 523 novel immunosuppressants isolated from a streptomyces. II. Fermentation isolation and physico-chemical and biological characteristics. *J. Antibiot.* 41:1592–1601.
69. Hillier, K. and Castaner, J. 1978. Bredinin. *Drugs Future* 3:567–568.
70. Hitoshi, S., Yoshihito, T., Yoshio, K. 1989. Immunosuppressive agent. U.S. Patent 4.843.092.
71. Hoyos, B., Ballard, D. W., Böhnlein, E., Siekevitz, M. and Warner, C. G. 1989. Kappa B-specific DNA binding proteins: role in the regulation of human interleukin-2 gene expression. *Science* 244:457–460.
72. Hunt, T. 1989. Cytoplasmic anchoring proteins and the control of nuclear localization. *Cell* 59:949–951.
73. Isshiki, K., Sawa, T., Nagahana, H., Koizumi, Y., Matsuda, N., Hamada, M. 1990. Depsidomycin, a new immunomodulating antibiotic-immunosuppressive production by *Streptomyces lavendofoliae*. *J. Antibiot.* 43:1195–1198.
74. Iwasawa, H., Kondo, S., Ikeda, D., Takenchi, T., Umezawa, H. 1982. Synthesis of (-)-15-Deoxyspergualin and (-)-Spergualin-15-phosphate. *J. Antibiot.* 35:1665–1669.
75. Jegorov, A., Matha, V. and Weiser, J., 1990. Production of Cyclosporins by entomopathogenic fungi. *Microbiol. Lett.* 45:65–70.
76. Johanson, A. and Möller, E. 1990. Evidence that the immunosuppressive effects of FK-506 and cyclosporine are identical. *Transplantation* 50:1001–1007.
77. Jones, K., Mills, G., Reamer, A., Askin, D., Desmond, R., Volante, P. and Shinaki, I. 1989. Total synthesis of the immunosuppressant (-)-FK-506. *J. Am. Chem. Soc.* 111:1157–1159.
78. Jones, A., Villalobos, Linde II, G. and Danishefsky J. 1990a. A formal synthesis of FK-506. Exploration of some alternatives to macrolactamization. *J. Org. Chem.* 55:2786–2979.
79. Jones, T. K., Reamer, R. A., Desmond, R. and Mills, S. G. 1990b. Chemistry of tricarbonyl hemiketals and application of Evan's technology to the total synthesis of the immunosuppressant (-)-FK-506. *J. Am. Chem. Soc.* 112:2998–3017.
80. Kahan, B. D. 1989. Drug therapy-Cyclosporine. *N. Engl. J. Med.* 321:1725–1738.
81. Kamata, K., Okubo, M., Ishigamori, E., Masaki, Y., Uchida, H., Watanabe K., Kashiwagi, N. 1983. Immunosuppressive effects of Bredinin on cell mediated and humoral immune reactions in experimental animals. *Transplantation* 35:144–149.
82. Kawashima, H., Fujino, Y., Mochizuki, M. 1988. Effects of a new immunosuppressive agent, FK-506, on experimental autoimmune uveoretinitis in rats. *Investig. Ophtalm. and Visual Sciences* 29:1265–1271.

83. Kay, J. E., Benzie, C. R., Goodier, M. R., Wick, C. J. and Doe, S. E. A. 1989. Inhibition of T-lymphocyte activation by the immunosuppressive drug FK-506. *Immunol. 67:* 473–477.
84. Kino, T., Hatanaka, H., Hashimoto, M., Nishiyama, M., Goto, T., Okuhara, M., Kosaka, M., Aoki, H., Imanaka, H. 1987a. FK-506, a novel immunosuppressant isolated from a Streptomyces. I. Fermentation, isolation and physiochochemical and biological characteristics. *J. Antibiot. 40:* 1249–1255.
85. Kino, T., Hatanaka, H., Miyata, S., Inamura, N., Nishiyama, M., Yajima, T., Goto, Tl, Okuhara, M., Kohsaka, M., Aoki, H. and Ochii, T. 1987b. FK-506 a novel immunosuppressant isolated from a streptomyces II. Immunosuppressant effect of FK-506 *in vitro. J. Antibiot. 40:* 1256–1265.
86. Kino, T., Inamura, N., Sakai, F., Nkahar, T., Goto, T., Okuhara, M., Kohsaka, M., Aoki, H. and Ochiai, T. 1987c. Effect of FK-506 on human mixed lymphocyte reaction *in vitro. Transplant. Proc. 19* (Suppl. 6):36–39.
87. Klaus, G. G. 1988. Cyclosporine-sensitive and cyclosporine-insensitive modes of B cell stimulation. *Transplantation 46* (Suppl.): 11–14.
88. Klein, J. 1990. *Immunology.* Blackwell: Oxford.
89. Kobel, H. and Traber, R. 1982. Directed biosynthesis of Cyclosporins. *Eur. J. Appl. Microbiol. Biotechnol. 14:* 237–240.
90. Koletsky, A. J., Harding, M. W. and Handschumacher, R. E. 1986. Cyclophilin: Distribution and variant properties in normal and neoplastic tissues. *J. Immunol. 137:* 1054–1059.
91. Krönke, M., Leonard, W. J., Depper, J. M., Arya, S. K., Wong-Staal, F., Gallo, R. C., Waldmann, T. A. and Greene, W. C. 1984. Cyclosporin A inhibits T-cell growth factor gene expression at the level of mRNA transcription. *Proc. Natl. Acad. Sci. (USA) 81:* 5214–5218.
92. Kusumi, T., Tsuda, M., Katsunuma, T., Yamamura, M. 1989. Dual inhibitory effect of Bredinin. *Cell. Biochem. Funct. 7:* 201–204.
93. Lang, K. and Schmid, F. X. 1988. Protein-disulphide isomerase and prolyl isomerase act differently as catalysts of protein folding. *Nature 331:* 453–455.
94. Lang, K., Schmid, F. X. and Fischer, G. 1987. Catalysis of protein folding by prolyl isomerase. *Nature 329:* 268–270.
95. Lawen, A. and Zocher, R., 1990. Cyclosporin Synthetase. *J. Biol. Chem. 265:* 11355–11360.
96. Lawen, A., Traber, R., Geyl, D., Zocher, R. and Kleinkauf, H. 1989. Cell-free biosynthesis of new cyclosporins. *J. Antibiot. 42:* 1283–1289.
97. Lawen, A., Traber, R. and Geyl, D. 1990. *In vitro* biosynthesis of [Thr^2, Leu^5, D-HIV^8, Leu^{10}]-Cyclosporin, a Cyclosporin-related peptolide with immunosuppressive activity. Abstract of the conference ``Cyclosporin, immunosuppression and proteinfaltung," Nalmitz, Germany, 8.-12.10.90.
98. Lazary, S. and Stähelin, H. 1968. Immunosuppressive and specific antimitotic effects of ovalicin. *Experientia 24,* 1171–1173.
99. Lee, J. and Agathos, S. N. 1988. Effect of amino acids on the production of Cyclosporin A by *Tolypocladium inflatum. Biotechn. Lett. 11,* 77–82.
100. Lee, W. A., Gu, L., Mikszal, A. R., Chu, N., Leung, K., Nelson, P. H. 1990. Bioavailability improvement of mycophenolic acid through amino ester derivatization. *Pharm. Res. 7:* 161–166.
101. Legrue, S. J., Shen, T. L, Carson, D. D., Loidlaw, J. L., Sanduja, S. K. 1988. Inhibition of T-lymphocyte proliferation by the cyclic polypepdide Didemnin B: no inhibition of lymphokine stimulation. *Lymphok. Res. 7:* 21–30.
102. Li, L. H., Timmins, L. G., Wallace, T. L., Krueger, W. C., Prainic, M. D. and Im, W. B. 1984. Mechanism of action of Didemnin B, a depsipeptide from the sea. *Cancer Lett. 23:* 249–288.
103. Lin, L., Hasumi, H., Brandts, J. F. 1988. Catalysis of proline isomerization during protein-folding reaction. *Biochim. Biophys. Acta 956:* 256–261.
104. Liu, J., Albers, M. W., Chen, C.-M., Schreiber, S. L., Walsh, C. T. 1990. Cloning expression and purification of human cyclophilin in *Escherichia coli* and assessment of the catalytic role of cysteins by site-directed biosynthesis. *Proc. Natl. Acad. Sci. (USA) 87:* 2304–2308.
105. Mandel, T. E. 1989. Transplantation. *Immunol. Today 10:* 1–3.
106. Martel, R. R., Klicius, J., Galet, S. 1977. Inhibition of the immune response by Rapamycin, a new antifungal antibiotic. *Can. J. Physiol. Pharmacol. 55:* 84–51.

107. Mason, J. 1989. Pathophysiology and toxicology of Cyclosporin in humans and animals. *Pharmacol. Rev. 41:* 423–434.
108. Meiser, B. M., Billingham, M. E., Morris. R. E. 1990. The effect of Cyclosporine and two immunosuppressive macrolides FK-506 and Rapamycin on heart graft rejection and graft coronary atherosclerosis. *J. Heart Transplant. 9:* 55.
109. Meiser, B. M., Wang, J. and Morris, R. E. 1989. Rapamycin: a new and hightly active immunosuppressive macrolide with an efficiency superior to Cyclosporin. *Progr. Immunol. 7:* 1195–1198.
110. Metcalfe, S. M., Richards, F. M. 1990. Cyclosporine FK-506 and Rapamycin: some effects on early activation events in serum-free mitogen-stimulated mouse spleen cells. *J. Heart Transplant. 9:* 55.
111. Mitchell, P. J. and Tjian, R. 1989. Transcriptional regulation in mammalian cells by sequence-specific DNA binding proteins. *Science 245:* 371–378.
112. Mizuno, K., Tsujino, M., Takada, M., Hayashu, M., Atsumi, K., Asano, K, and Matsuda, T. 1974. Studies on Bredinin I. Isolation, characterization and biological properties. *J. Antibiot. 27.* 775–779.
113. Montgomery, D. W., Celniker, A. and Zukoksi C. F. 1987. Didemnin B—an immunosuppressive cyclic peptide that stimulates murine hemagglutinating antibody responses and induces leukocytosis in vivo. *Transplantation 43:* 133–139.
114. Morris, R. E. and Yuh, D. D. 1990. 15-Deoxyspergualin: an immunopharmacologic and mechanistic analysis of an effective immunosuppressant. *J. Heart Transplant. 9:* 73.
115. Morris, R. E., Hoyt, E. G., Eugui, E. M. and Allison, A. C. 1989. Prolongation of rat heart allograft survival by RS-614443. *Surg. Forum 40:* 337–338.
116. Morris, R. E., Hoyt, E. G., Murphy, M. P., Eugui, E. M., Allison, A. C. 1990. Mycophenolic acid morpholinoetylester (RS-61443) is a new immunosuppressant that prevents and halts heart allograft rejection by selective inhibition of T- and B-cell purine synthesis. *Transplant. 22:* 1659–1662.
117. Muller, W., Weissmann, N., Maidhof. 1987. Deoxyspergualin, a potent antitumor agent: further studies on the dytobiological mode of action. *J. Antibiot. 40:* 1028–1031.
118. Nabel, G. J., Gorka, C. and Baltimore, D. 1988. T-cell-specific expression of Interleukin 2: Evidence for a negative regulatory site. *Proc. Natl. Acad. Sci.* (USA) *85:* 2934–2938.
119. Nakajima, H., Hamasaki, K., Nishimura, T., Kondo, Y., Kimura, Y., Udagawa, S. and Sato, S., 1988. Isolation of 2-acetylamino-3-hydroxy-4-methyloct-6-enoic acid, a derivate of the "C9-amino acid" residue of cyclosporins, produced by the fungus *Neocosmospora vasinfecta.* E. F. Smith. *Agric. Biol. Chem. 52:* 1621–1623.
120. Nakatsuka, M., Ragan, J. A., Sammakia, T., Smith, D. B., Uehling, D. E. and Schreiber, S. L. Total synthesis of FK-506 and an FKBP probe reagent, (C_8, C_{9-13}, C_2)-FK506. *J. Am. Chem. Soc. 112:* 5583–5601.
121. Nelson, P. H., Gu, C. L., Allison, A. C., Eugui, E. M., Lee, W. A. 1988. Preparation of morpholinoethyl esters of mycophenolic acid and pharmaceutical compositions containing them as immunosuppressive and anti-inflammatory agents. US Patent 4753935.
122. Nemoto, K., Hayashi, M., Sugawara, Y., Ito, L., Abe, F, Takito, T., Nakamura, T. and Takenchi, T. 1988. Biological activities of Deoxyspergualin in autoimmune disease mice. *J. Antib. 41:* 1253–1259.
123. Nordmann, R., Andersen, E., Trussardi, R. and Mazer, N. A. 1988. Kinetics of Interleukin 2 mRNA and protein produced in the human T-cell line jurkat and effect of Cyclosporin A. *Biochem. 28:* 1791–1797.
124. Nuchtern, J. G., Bonifacino, J. S., Biddison, W. E. and Klausner, R. D. 1989. Brefeldin A implicates egress from endoplasmic reticulum in class I restricted antigen presentation. *Nature 339:* 223–226.
125. Okuhara, M., 1990. WS 9482, an immunomodulator of microbial origin. Abstract S-30, Biotechnology of Microbial Products, SIM, Oct. 1990
126. Okamato, K., Kobayashi, Y., Yoshida, K., Nozaki, Y., Kawai, Y., Kawano, H., Mayumi, T., Hamat, T. 1978. Teratogenic effects of Bredinin, a new immunosuppressive agents in rats. *Jpn. Congentital Anom. 18.* 227–234.
127. Paiva, N. L., Demain, A. L., Roberts, M. F. 1991. Incorporation of acetate, propionate and methionine in rapamycin by *Streptomyces hygroscopicus. J. Nat. Prod. 54:* 167–177.

128. Patchett, A. A., White, R. F., Goegelman, R. T. 1990. Cyclosporin derivatives with modified "8-amino acid." Eur. Pat. Appl. 0 373 260.
129. Quesniaux, V. F. 1989. Pharmacology of cyclosporine (Sandimmune) III. Immunochemistry and monitoring. *Pharmacol. Rev. 41:* 249–258.
130. Quesniaux, V. F., Schreier, M. H., Wenger, R. M., Hiestand, P. C., Harding, M. W. and Van Regenmortel H. V. M. 1988. Molecular characteristics of Cyclophilin-Cyclosporine interaction. *Transplantation 46:* 23–28.
131. Quesniaux, V. F., Schreier, M. H., Wenger, R. M., Hiestand, P. C., Harding, and Van Regenmortel M. H. 1987. Cyclophilin binds the region of Cyclosporin involved in its immunosuppressive activity. *Eur. J. Immunol. 17:* 1359–1365.
132. Randak, C., Brabletz, T., Hergenröther, M., Sobotta, I. and Serfling, E. 1990. Cyclosporin A suppresses the expression of the Interleukin 2 gene by inhibiting the binding of lymphocyte-specific factors to the IL-2 enhancer. *EMBO 9:* 2529–2536.
133. Rassow, J., Guiard, B., Wienhues, U., Herzog, V., Hartl, F. U., Neupert, W. 1989. Translocation arrest by reversible folding of a precursor protein imported into mitochondria: A means to quantitate translocation contact sites. *J. Cell. Biol. 109:* 1421–1428.
134. Rinehart, K. L. 1988. Didemnins and its biological properties. In "Peptids: Chemistry and Biology" (Marshall, G. R., ed) pp. 626–631: Escom Science Publ. Leiden.
135. Rinehart, K. L. 1990. Antitumor and antiviral agents from Tunicates. Abstract S-8. Biotechnology of Microbial Products, S.I.M., Oct. 90.
136. Rinehart, K. L., Glover, J. B., Hughes, R. G., Renis, H. T., McGovern, J. P., Swynenberg, E. B., Stringfellow, D. A., Kuentzel, S. L., Li, L. H. 1981. Didemnins: antiviral and antitumor depsipetides from a Caribean tunicate. *Science 212:* 933–935.
137. Rinehart, K. L., Kishore, V., Bible, K. C., Sakai, R., Sullins, D. W., Lih-K. M. 1988. Didemnins and tunichlorin novel natural products from the marine tunicate *Trididemnumn solidum. J. Nat. Prod. 51:* 1–21.
138. Roitt, I., Brostoff, J. and Male, D. 1989. Immunology—2nd edition, Gower Medical Publ: London-New York.
139. Regger, A., Kuhn, M., Lichti, H., Loosli, H. R., Huguenin, R., Quiquerez, C., und von Wartburg, A. 1976. Cyclosporin A, ein immunosuppressiv wirksamer Peptidmetabolit aus *Trichoderma polysporum* (Link ex Pers.) Rifai. *Helv. Chim. Acta 59:* 1075–10.
140. Ryffel, B. 1989. Pharmacology of Cyclosporine. VI. Cellular activation: Regulation of intracellular events by Cyclosporin. *Pharmacol. Rev. 41:* 407–422.
141. Sanglier, J. J., Traber, R., Buck, R. H., Hofmann, H. and Kobel, H. 1990. Isolation of (4R)-4-[(E)-2-Butenyl]-4-Methyl-L-Threonine, the characteristic structural element of cyclosporins, from a blocked mutant of *Tolypocladium inflatum. J. Antibiot. 43:* 707–714.
142. Schreiber, S. L. 1991. Chemistry and biology of immunophilins and their immunosuppressive ligands. *Science 251:* 283–287.
143. Sehgal, S. N., Baker, H., Vezina, C. 1975. Rapamycin (Ay-22, 989), a new antifungal antibiotic: II. Fermentation, isolation and characterization. *J. Antibiot. 28:* 727–732.
144. Sehgal, S. N., Baker, H., Eng, C. P., Signh, K. and Vzina, C. 1983. Demethoxyrapamycin (AY-24,668), a new antifungal antibiotic. *J. Antibiot. 36:* 351–354.
145. Seichii, S., Kyoichi, S., Masakuni, Y., Takash, H., Nobuyoshi, S. 1989. Immunization-inhibiting substance NK86–0186 and its manufacture with *Streptomyces.* Patent: Jpn. Kokai Tokkyo Koho 01108987.
146. Shannon, M. F., Pell, L. M., Lenardo, M. J., Kuczek, E. S., Occhiodoro, F. S., Dunn, S. M. and Vadas, M. A. 1990. A novel tumor necrosis factor responsive transcription factor which recognizes a regulatory element in hemopoietic growth factor genes. *Mol. Cell. Biol. 8:* 2950–2959.
147. Shaw, J. P., Utz, P. J., Durand, D. B., Toole, J. J., Emmel, E. A. and Crabtree, G. R. 1988. Identification of a putative regulator of early T-cell activation genes. *Science 241:* 202–205.
148. Shieh, B. H., Stamnes, M. A., Seavello, S., Harris, G. L. and Zuker, C. S. 1989. The ninaA gene required for visual transduction in *Drosophila* encodes a homologue of Cyclosporin A-binding protein. *Nature 338:* 67–70.

149. Siebenlist, U., Durand, D. B., Bressler, P., Holbrook, N. J., Norris, C. A., Kamoun, M., Kant, J. A. and Crabtree, G. R. 1986. Promoter region of Interleukin-2 gene undergoes chromatin structure changes and confers inducibility on chloramphenicol acetyltransferase gene during activation of T-cells. *Mol. Cell. Biol. 6:* 3042–3049.

150. Siekierka, J. J., Hung, H. Y., Poe, M., Lin, C. S., Sigal, N. H. 1989. A cytoxolic binding protein for the immunosuppressant FK-506 has peptidylprolyl isomerase activity but is distinct from cyclophilin. *Nature 431:* 755–757.

151. Sigal, H., Dumont, F., Durette, F., Siekierka, J. J., Peterson, L., Rich, D.H., Dunlapp, B. E., Staruch, M. J., Melino, M. R., Koprak, S. L., Williams, D., Wirzel, B. and Pisano, J. M. 1991. Is cyclophilin involved in the immunosuppressive and nephrotoxic mechanism of action of Cyclosporin A? *J. Exp. Med. 173:* 619–626.

152. Standaert, R. F., Galat, A., Verdine, G. L. and Schreiber, S. L. 1990. Molecular cloning and overexpression of the human FK-506-binding protein FKBP. *Nature 346:* 671–674.

153. Starzl, T. T., Fung, J., Venkatarammen, R., Todos., Demetris, A. J., Jain, A. 1989. FK-506 for liver, kidney and pancreas transplantation. *Lancet 1989 (2):* 1000–1004.

154. Starzl, T. T., Fung, J., Jordan, M., Shapiro, R., Tsakis, A., McCanley, J., Jahrston, J., Iwaki, Y., Jain, A., Alessiani, M., Toto, S. 1990. Kidney transplantation under FK-506. *Jama 264:* 63–67.

155. Stoeck, M., Wildhagen, K., Szamel, M., Lowett, D. and Resch, K. 1985. Studies on the mechanism whereby cyclosporin A inhibits T-lymphocyte activation. *Immunology 169:* 83–96.

156. Sweeney, M. J., Hoffman, D. H., Esterman, M. A. 1972. Metabolism and biochemistry of mycophenolic acid. *Cancer Res. 32:* 1803–1809.

157. Takahara, S., Fukunishi, T., Kokado, Y., Ichikawa, Y., Ishibashis, M., Nagano, S., Sonoda, T. 1988. Combined immunosuppression with low-dose Cyclosporine Mizoribine prednisolone. *Transplant. Proc. 20,* Suppl. 3:147–151.

158. Takahashi, N., Hayano, T. and Suzuki, M. 1989. Peptidyl-prolyl cis-trans isomerase is the Cyclosporin A-binding protein Cyclophilin. *Nature 337:* 473–475.

159. Takeuchi, T., Tihuma, H., Kunimoto, S., Masude, T., Ishizuka, M., Takeuchi, M., Hamado, N., Nagamania, H., Kondo, S. and Umezawa, H. 1981. A new antibiotic "Spergualin": Isolation and antitumor activity. *J. Antibiot. 34:* 1619–1621.

160. Tanaka, H., Kuroda, A., Marusawa, H., Hatanaka, H., Kino, T., Goto, T., Hashimoto, M., Taga, T. 1987. Structure of FK-506: A novel immunosuppressant isolated from streptomyces. *J. Am. Chem. Soc. 109:* 5031–5033.

161. Thomson, A. W. 1989. FK-506—How much potential? *Immunol. Today. 10:* 6–9.

162. Tocci, M. J., Matkovich, D. A., Collier, K. A., Kwok, P., Dumont, F., Lin, S., Degudicibus, S., Siekierka, J. J., Chin, J. and Hutchinson, N. I. 1989. The immunosuppressant FK-506 selectively inhibits expression of early T-cell activation genes. *J. Immunol. 143:* 718–726.

163. Tosato, G., Pike, S. E., Koski, I. R., and Blaese, R. M. 1982. Selective inhibition of immunoregulatory cell functions by cyclosporin A. *J. Immunol. 128:* 1986–1991.

164. Traber, R., Hofmann H. and Kobel, H. 1989. Cyclosporins—new analogues by precursor directed biosynthesis. *J. Antibiot. 42:* 591–597.

165. Tropschug, M., Barthelmess, I. B. and Neupert, W. 1989. Sensitivity to Cyclosporin A is mediated by Cyclophilin in *Neurospora crassa* and *Saccharomyces cerevisiae. Nature 342:* 953–955.

166. Tropschug, M., Nicholson, D. W., Hartl, F. U., Köhler, H., Pfanner, N., Wachter, E. and Neupert, W. 1988. Cyclosporin A-binding protein (Cyclophilin) of *Neurospora crassa. J. Biol. Chem. 263:* 14433–14400.

167. Tropschug, M., Wachter, E., Mayer, S., Schönbrunner, E. R. and Schmid, F. 1990. Isolation and sequence of an FK-506-binding protein from *Neurospora crassa* which catalyses protein folding. *Nature 346:* 674–677.

168. Tsiji, R. F., Yamamoto, M., Nakamura, A., Kataoka, T., Magae, J., Nagai, K. and Yamasaki, M. 1990. Selective immunosuppression of Prodigiosin 25-C and FK-506 in the murine immune system. *J. Antibiot. 43:* 1293–1301.

169. Ullmann, K. S., Northrop, J. P., Verweij, C. L. and Crabtree, G. R. 1990. Transmission of signals from the T lymphocyte antigen receptor to the genes responsible for cell proliferation and immune function: the missing link. *Annu. Rev. Immunol.* *8:* 421–452.
170. von Wartburg, A. and Traber, R., 1989. Cyclosporins, fungal metabolites with immunosuppressive activities. *Progr. Med. Chem. 25:* 1–33.
171. Walliser, P., Benzie, C. R. and Kay, J. E. 1989. Inhibition of murine B-lymphocyte proliferation by the novel immunosuppressive drug FK-506. *Immunol. 68:* 434–435.
172. Waring, P. and Mülebacher, A. (1990). Fungal warfare in the medicine chest. *New Scientist 128:* 41–44.
173. Wenger, R. M. 1983. Synthesis of Cyclosporine and analogues: Structure, activity, relationships of new cyclosporine derivatives. *Transplant. Proc. 15,* Suppl. 1.:2230–2241.
174. Yewdell, J. W. and Bennink, J. R. 1989. Brefeldin a specifically inhibits presentation of protein antigens to cytotoxic T-lymphocytes. *Science 244:* 1072–1075.
175. Yoshimura, N., Matsui, S., Hamashima, T. and Oka, T. 1989a. Effect of a new immunosuppressive agent, FK-506, on human lymphocyte responses *in vitro.* I. Inhibition of expression of alloantigen-activated suppressor cells, as well as induction of alloreactivity. *Transplantation 47:* 356–359.
176. Yoshimura, N., Matsui, S., Hamashima, T. and Oka, T. 1989b. Effect of a new immunosuppressive agent, FK-506, on human lymphocyte responses *in vitro.* II. Inhibition of the production of IL-2 and gamma-IFN, but not cell-stimulating factor 2^1. *Transplantation 47:* 356–359.
177. Yuh, D. D., Zurcher, R. P., Carmichael, P. G. and Morris R. E. 1989. Efficacity of Didemnin B in suppressing allograft rejection in mice and rats. *Transplant. Proc. 21:* 1141–1143.

2

The Biosynthesis and Enzymology of an Immunosuppressant, Immunomycin, Produced by *Streptomyces hygroscopicus* var. *ascomyceticus*

K. M. Byrne, A. Shafiee, J. B. Nielsen, B. Arison, R. L. Monaghan, and L. Kaplan
Merck Sharp and Dohme Research Laboratories
Rahway, NJ 07065

INTRODUCTION

In the course of screening fermentation broths for compounds exhibiting immunosuppressive activity, extracts of culture MA6475, *Streptomyces hygroscopicus* var. *ascomyceticus* (ATCC 14891), were found to be active in an IL-2 dependent T-cell proliferation assay [1]. Culture MA6475 was reported in 1962 to produce ascomycin, an antifungal antibiotic of unknown structure [2,3]. Isolation of the active immunosuppressant material from fermentations of MA6475 yielded a mixture of two immunosuppressant compounds: immunomycin, the major component, and a minor component designated L-683,795 (Figure 2.1). Structural elucidation of the compounds showed them to be identical to two compounds recently elucidated by researchers at Fujisawa Pharmaceutical Co., Ltd., designated FK520 and FK523, respectively [4,5].

The most extensively studied member of this group of immunosuppressants is FK506 (Figure 2.1). It was discovered in 1984 by researchers at Fujisawa Pharmaceutical Co. FK506 is produced by *Streptomyces tsukubaensis* and belongs to the macrolide class of natural products. A high level of efficacy has been seen in early clinical trials of FK506 used for controlling rejection in organ transplant patients [6]. The *in vitro* immunosuppressive potency of FK506 is 10 to 100 times that of cyclosporin A [7,8]. Reports of the *in vitro* biological activity of immunomycin indicate it to be approximately half as potent as FK506 [4].

Rapamycin is a natural product structurally related to FK506. Although a triene-macrolide somewhat larger than FK506, it also contains pipecolic acid and a substituted cyclohexane moiety. Some aspects of the biosynthesis of rapamycin were recently disclosed [9,10]. To date, the biosynthesis of the FK506-group of compounds

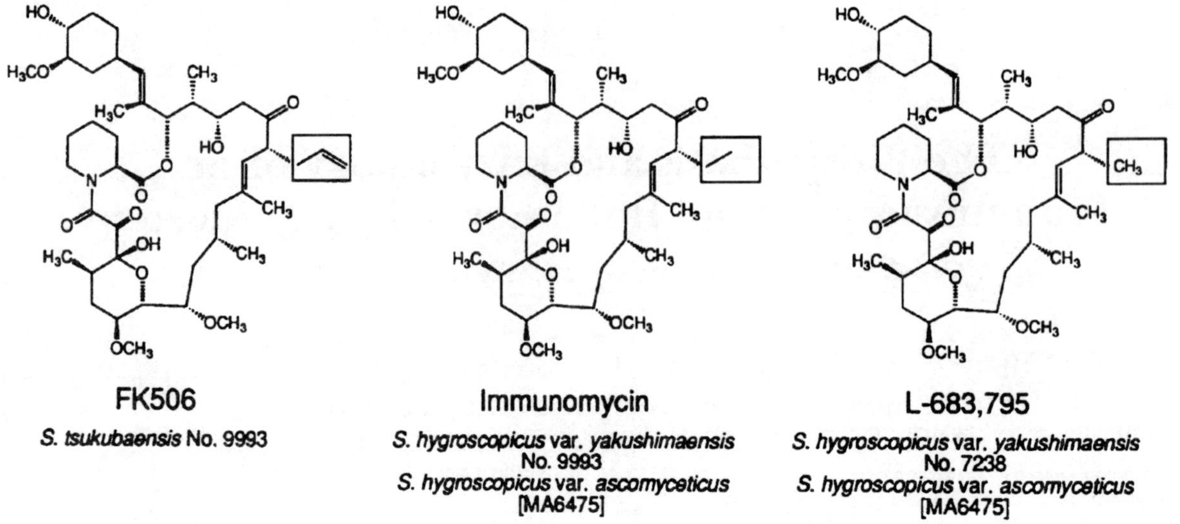

Figure 2.1 / Structures of FK506, immunomycin, L-683,795. Structural variation occurs within the boxed-in area.

has not been reported, nor have there been any reports on the enzymology of FK506 biosynthesis. There is an extensive literature on enzymes that activate amino acids for non-ribosomal peptide antibiotic synthesis [11]. The best characterized are those involved in purely peptide products like gramicidin and tyrocidin. However, there are a number of hybrid structures which arise from both polyketide and amino acid or carboxylic acid precursors in which it seems likely that the activation of amino or carboxylic acid precursor is accomplished by an analogous reaction. Briefly, based on what is known for peptide synthesis, this reaction involves ATP-dependent adenylation of the amino acid to form the amino acyl adenylate, and transfer to an essential thiol group on the activating enzyme to form an acid-stable thioester. Both these steps occur even in the absence of the subsequent acceptor for the activated amino acid, a great advantage when a molecule of clearly hybrid origins like the immunomycin complex is being investigated. The adenylation reaction may be assayed by amino acid-dependent pyrophosphate-ATP exchange, while binding to the enzyme in thioester linkage can be assayed by binding of labelled amino acid to protein in acid-stable form. Based on the assumption that the two activities described here might be relevant for pipecolate activation in immunomycin biosynthesis, cell-free extracts producing immunomycin were examined for pipecolate-dependent PP-ATP exchange and binding.

In the studies reported here, ^{13}C-, ^{14}C- and ^{15}N-labelled substrates were used to discern the biosynthetic pathway of immunomycin and the related compounds FK506 and L-683,795. Two enzymes in the biosynthetic pathway of immunomycin, S-adenosyl-L-methionine: 31-O-desmethylimmunomycin O-methyltransferase (DIMT)[1] and pipecolate activating enzyme were isolated and characterized.

Abbreviations: DIMT, S-adenosyl-L-methionine:31-O-desmethylimmunomycin O-methyltransferase; DTT, dithiothreitol; SAM, S-adenosyl-L-methionine; PMSF, phenylmethylsulfonyl fluoride; and PAGE, polyacrylamide gel electophoresis; IEP, phosphoenolpyruvate; MOPS, morpholinoethanesulfonic acid.

MATERIALS AND METHODS

ISOTOPES

Sodium acetate-1-^{13}C, 90 atom % ^{13}C and sodium propionate-1-^{13}C, 90 atom % ^{13}C were obtained from MSD Isotopes, Division of Merck & Co., Inc. and used as substrates in the MA6492 fermentations. Sodium butyrate-4-^{13}C, 90.7 atom % ^{13}C was also obtained from MSD Isotopes. All other ^{13}C-labelled substrates were 99 atom % ^{13}C. Acetic acid-1-^{13}C, acetic acid-2-^{13}C and acetic acid-1,2-^{13}C$_2$, propionic acid-1-^{13}C, methyl-^{13}C-L-methionine were purchased from Sigma Chemical Co. DL-Lysine-1-^{13}C and sodium butyrate-1-^{13}C were supplied by Cambridge Isotope Laboratory. L-Lysine HCl [UL-^{14}C] and shikimic acid-^{14}C [UL] were obtained from ICN Biomedicals, and NEN Research Products, respectively. [Carboxyl-^{14}C]-DL-Pipecolic acid (7.1 mCi/mmole) was prepared by Drug Metabolism at Merck in 1976. It was repurified by preparative TLC before use.

ORGANISMS AND CULTURE CONDITIONS

Culture MA6475 was obtained from the American Type Culture Collection (No. 14891). Culture MA6678 was obtained following N-methyl-N-nitroso-guanidine mutagenesis of culture MA6475. Both cultures were grown either from a well grown slant on ISP Medium 4 (Difco Laboratories) at 27°C, or from 1 ml of a frozen vegetative mycelium in seed medium diluted 1:1 with 20% sterile glycerol. Seed cultures were grown in 250 ml baffled shake flasks at 27°C and 220 rpm for 48 hours in 50 ml of basal medium containing: glucose 2%, yeast extract 2%, Hy-Case SF 2%, KNO$_3$ 0.2%, CaCl$_2$•2H$_2$O 0.0002%, ZnSO$_4$•7H$_2$O 0.0001%, MnSO$_4$•H$_2$O 0.00005%, FeSO$_4$•7H$_2$O 0.0025%, MgSO$_4$•7H$_2$O 0.05%, and NaCl 0.05%; the pH of the medium was adjusted to 7.0 before autoclaving. A 1.0 ml aliquot was used to inoculate 30 ml of GYG medium consisting of glucose 2.2%, glycerol 2.5%, corn steep liquor 0.1%, yeast extract 1.5%, L-tyrosine 0.4%, CaCO$_3$ 0.025%, lactic acid 0.17%, morpholinoethanesulfonic acid (MOPS) 1.0%, in a 250 ml non-baffled shake flask. The pH of the medium was adjusted to 6.8 with 6M NaOH before autoclaving. The pH of aqueous solutions of labelled substrates were adjusted to 6.8 and filter sterilized. Washed cell incubations with labelled precursors were carried out for 18 hours. Addition of precursors to fermentation broths were made in three equal portions between 24 and 36 hours.

Culture MA6492, *S. tsukubaensis*, was obtained from the Fermentation Research Institute, Agency of Industrial Science and Technology, Tsukuba Gun, Iburaki Prefecture, Japan, culture number FERM BP-927, and grown as previously described [12]. Fermentations were harvested at 90 hours. Addition of ^{13}C-labelled precursors to complex medium fermentations were made in 3-5 stages from 43 to 73 hours to maximize incorporation.

WASHED CELL INCUBATIONS

Washed cell incubations were carried out using a modification of the method described by Nielsen & Kaplan [13]. Cultures were harvested at the beginning of product formation (24-30 hours), washed twice, and the cells resuspended back to the original volume in incubation buffer containing glucose at 1.0%, MOPS at 0.25%, and the appropriate ^{13}C-labelled precursor. A volume of 30 ml resuspended cells was placed in a non-baffled shake flask (or 2.5 ml in a 25 × 150 mm tube) and shaken at 27°C and 220 rpm for 18-24 hours.

ISOLATION AND PURIFICATION OF LABELLED PRODUCTS

Isolation of the ^{13}C-labelled products was accomplished by extracting the whole broth with an equal volume of MeOH. After centrifugation to remove media and cellular debris, the MeOH from the supernatant was evaporated under reduced pressure. Extraction of the remaining aqueous mixture with EtOAc, followed by drying over Na$_2$SO$_4$ and evaporation to dryness afforded the desired materials in crude form. After washing with hexane the resulting material was purified by semi-preparative HPLC. Analytical HPLC was done using a Whatman Partisil 5 ODS-3 analytical column equilibrated at 60°C in a mobile phase of acetonitrile/0.1% H$_3$PO$_4$ (65:35) and run at 1.0 ml/minute. Compounds of interest were detected at 205nm. Retention times for L-683,795 and immunomycin were 8.8 and 10.1 min, respectively. Semi-preparative HPLC was accomplished using a Rainin Dynamax 10mm ×25cm column equilibrated at 50°C in acetonitrile/0.1% H$_3$PO$_4$ (50:50), and a flow rate of 4.0 ml/minute.

Compounds labelled with ^{14}C were purified through the EtOAc step described above, and then either assayed directly or adsorbed onto a Waters C-18 Sep-Pak cartridge. Incorporation into the molecule of interest was determined by fractionating an aliquot of the sample on the analytical HPLC system and counting the collected peak in a liquid scintillation counter.

DIMT ENZYME ASSAY AND PRODUCT IDENTIFICATION

The assay was carried out in a 1.0 ml mixture containing 0.025 mM substrate (31-desmethylimmunomycin), 1mM MgSO$_4$ and enzyme in 50 mM phosphate buffer, pH 7.5. Reaction was initiated by the addition of 1 nmole of ^{14}C-methyl labelled S-adenosyl-L-methionine (SAM) at 46 mCi/mmole and incubated at 34°C for 20 minutes. The reaction was terminated by the addition of ethyl acetate and the product was analyzed by silica gel TLC (CHCl$_3$:MeOH, 9/1). Areas containing radioactivity were scraped from the plate, eluted with EtOAc, and then analyzed and quantitated by analytical HPLC.

ISOLATION OF DIMT All procedures were carried out at 4°C. Washed mycelium was suspended in 50mM phosphate buffer, pH 7.5 containing 1mM phenylmethylsulfonyl fluoride (PMSF) and 1mg/ml of lysozyme, stirred overnight and centrifuged at 15,000 rpm for 30 min. The pellet was sonicated for 2 minutes at 30 second intervals and the cell homogenate centrifuged as above. The supernatant was brought to 1% streptomycin sulfate, stirred for 30 minutes and centrifuged at 105K × g. The resulting supernatant, designated as the crude extract, was made to contain 0.1mM SAM and 10% ethanol. Further purification was accomplished by precipitation from 60% ammonium sulfate, dialysis against 50mM phosphate buffer, pH 7.5, containing 1mM PMSF, 0.1mM SAM and 10% ethanol (buffer L). The DIMT activity was completely eluted with buffer L containing 0.3M KCl, dialyzed against buffer L, applied on the preparative MonoQ HR10/10-FPLC column, and eluted with a linear gradient from zero to 1.0M KCl in buffer L. Fractions showing the DIMT activity were pooled and chromatographed on an analytical MonoQ HR5/5-FPLC column developed as above except that the KCl concentration was 0.5M. Active fractions were pooled and subjected to chromatofocusing on a MonoP HR 5/20 column after dialysis against 20 mM bis-tris buffer at pH 7.0. This column was then developed with polybuffer 74 (Pharmacia) with pH interval of 7–4. The peak DIMT active fraction was applied on a Superose-12 column, eluted with buffer L containing 150mM NaCl and the peak DIMT activity was finally purified using 12.5% native polyacrylamide electrophoresis (native-PAGE). Following electrophoresis, cuts were made across the face of the gel and each gel cut was eluted and examined for enzyme activity.

PP-ATP EXCHANGE ASSAY

PP-ATP exchange was measured by a modification of Lee and Lipmann's method [14]. The reaction mixture consisted of 2.5 mM ATP, 4 mM MgSO$_4$, 0.5 mM sodium pyrophosphate, 0.1 mM EDTA, 1 mM DTT, 75 mM HEPES pH 7.3, 30,000 dpm [^{32}P] pyrophosphate, 2–5 µl of enzyme, and 10 mM L-pipecolic acid were incubated in a final volume of 50 µl for 10 min at 30°C. The reaction was terminated by the addition of 0.3 ml of a 1% suspension of activated charcoal (Merck) in 3% perchloric acid and 10 mM sodium pyrophosphate. The charcoal was collected by filtration through glass-fiber filters, washed with 6 ml cold water and counted in 5 ml Scintiverse (Fisher). Binding of ^{14}C-pipecolate was measured essentially as described by Keller [15].

PIPECOLATE ACTIVATING ENZYME EXTRACTION AND ISOLATION Mycelium was washed three times in 0.5 M KCl to remove extracellular and loosely bound proteases. Freshly packed mycelium was resuspended in 3 volumes of extraction buffer containing 0.025 M HEPES pH 7.3, 20% glycerol, 0.5 mM EDTA, 0.2 mM dithiothreitol (DTT), 0.5% Triton X-100, 0.1 mM PMSF and 0.1 mM diisopropyl fluoride, and disrupted by sonication at 4°C for 2 min. The extracts were brought to 65% saturation with ammonium chloride. The pellet was stored overnight at 4°C without desalting. It was then dissolved in purification buffer containing 0.025 M HEPES pH 7.3, 10% glycerol, 0.5 mM EDTA, 0.2 mM DTT and dialyzed. The desalted extract was applied to microgranular DEAE cellulose, eluted by 0.2–0.3 M KCl applied as a linear gradient from 0.0 to 0.6 M KCl, and precipitated with 70% ammonium sulfate. The washed pellet was dissolved in purification buffer containing 0.1M KCl and applied to a Sephacryl S300 column (2 cm I.D. × 100 cm). The bulk of the included protein, which eluted as a broad peak, was concentrated and applied to a MonoQ HR 10/10. Elution of the enzyme was achieved by a shallow KCl gradient between 0.2 and 0.45 M.

Native molecular weight was determined by gel filtration on Superose 12 (Pharmacia) and TSK-SWP (7.5 × 750 mm) in 50 mM HEPES pH 7.3, 150 mM KCl, 0.5 mM DTT at 4°C, after calibration of each with bovine thyroglobulin (600kDa), ferritin (450kDa), catalase (240kDa), aldolase (160 kDa), and bovine serum albumin (67 kDa). The addition of at least 100 mM KCl gave nearly linear calibration curves on the two columns employed.

Preparative SDS electrophoresis was performed using 1.5 mm thick 6% Laemmli gels, while one-millimeter-thick 6 or 8% SDS gels from Novex (Encinatas, CA) were used for analytical purposes. The isoelectric point was determined on Phast (Pharmacia) isoelectric focussing gels, pH 3–9 and 4.5–6.0. Fingerprinting with V8 protease was performed as described in [16].

NMR AND MS

The ^{13}C NMR spectra were obtained in CDCl$_3$ on a Varian XL-400 spectrometer operating at 100 MHz. Chemical shifts are given in ppm relative to the central CDCl$_3$ triplet set at 77.0 ppm. Analysis of ^{15}N enrichments was performed using a Finnigan-MAT model 90 mass spectrometer operating in a FAB mode and ionization achieved using a cesium ion gun.

RESULTS AND DISCUSSION

ACETATE, PROPIONATE AND METHIONINE INCORPORATION

The ^{13}C NMR enrichment values obtained for immunomycin, L-683,795, and FK506 produced in the presence of singly and doubly labelled acetate, 1-^{13}C-propionate, and ^{13}C-methyl-L-methionine are summarized in Table 2.1. Immunomycin and L-683,795 produced in the presence of 1-^{13}C-propionate led to enrichment values greater than 10-fold at carbons 10, 16, 18, 24, and 26. L-683,795 exhibited an additional enrichment site at carbon 20. These results suggest that immunomycin is composed of 5 propionates, while L-683,795 contains 6 propionates. The additional propionate in L-683,795 occurs in the region where the FK506 family of molecules vary. Like immunomycin, FK506 labelled from 1-^{13}C-propionate showed enrichment

Table 2.1 / Summary of incorporation the of ^{13}C-acetate, ^{13}C-propionate and ^{13}C-methyl-methionine precursors into immunomycin, L-683,795 and FK506

Precursor	Carbon Atom	Enrichment Ratios		
		Immunomycin	L-683,795	FK506
1-^{13}C-Acetate	8	4.5	2.0	2.0
	20	3.0	2.0	2.0
	22	4.0	2.0	2.0
	35	2.5	0.0	0.0
1,2-^{13}C-Acetate	8	6.8	3.0	—
	9	4.9	3.0	—
	20	2.5	0.0	—
	21	2.5	0.0	—
	22	4.0	3.0	—
	23	5.1	3.0	—
	35	2.8	0.0	—
	36	2.5	0.0	—
1-^{13}C-Propionate	10	11.6	15.0	3.0
	16	11.7	20.0	4.0
	18	11.7	20.0	4.0
	20	0.0	20.0	0.0
	24	14.0	20.0	4.0
	26	13.0	20.0	3.0
	35	0.0	0.0	3.0
^{13}C-Methyl-Methionine	13-OCH$_3$	>40	45 ± 5	—
	15-OCH$_3$	>40	45 ± 5	—
	31-OCH$_3$	>40	45 ± 5	—

at carbons 10, 16, 18, 24, and 26. In addition, carbon 35 of FK506 exhibited the highest enrichment from 1-^{13}C-propionate. This labelling shows the same lactone ring pattern as immunomycin, but identifies the origin of the allyl side-chain of FK506 as arising from propionate.

The labelling patterns obtained using the variously labelled ^{13}C-acetates indicated both direct incorporation from acetate (presumably through the biologically activated form of acetate, malonyl-CoA), and the expected incorporation due to metabolism of acetate through the Krebs cycle to propionate (via its biologically activated form, methylmalonyl-CoA). In the cases where singly labelled acetate substrates served as precursors, variations in the enrichment levels and comparison to the propionate labelling pattern indicated whether incorporation was direct or via scrambling. L-683,795 and FK506 produced in the presence of 1-^{13}C-acetate exhibited identical enrichment values of 2-fold natural abundance, at the same carbon atoms: 8, 20, and 22 (Table 2.1). Immunomycin co-produced as the major component along with L-683,795 under the same conditions showed enrichment at four carbons: 8, 20, 22, and 35. The enrichments at carbons 8 and 22, however, were at least 50% higher than at carbons 20 and 35.

The sites of acetate incorporation were confirmed by the enrichment pattern observed for immunomycin and L-683,795 produced in the presence of doubly labelled acetate. When acetate was incorporated as a unit the characteristic doublet appeared, and the corresponding coupling constants were determined. The only carbons exhibiting significant one bond coupling in immunomycin were carbons 8–9, 22–23, and to a lesser extent carbons 35–36, and 20–21. This result, coupled with the smaller enrichment ratios for carbons 35–36 and 20–21, suggested that a more direct precursor than acetate may exist for carbons 20, 21, 35, and 36 in immunomycin. The results of acetate incorporation into L-683,795 were similar to that obtained for immunomycin, except L-683,795 showed significant coupling only at carbons 8–9, and 22–23. As expected, no couplings for carbons 20, 21, and 35 were observed. This is in agreement with the $1-^{13}C$-propionate data indicating that these carbons in L-683,795 are derived from propionate.

The direction of the polyketide chain was clearly discernable from the $1-^{13}C$-propionate labelling, as each enriched site was α, in a clockwise direction, to a carbon bearing a branched methyl group. The labelling pattern obtained from $1-^{13}C$- and $1,2-^{13}C$-acetate also indicated a clockwise direction. These taken together suggest the tail of the chain would be at the propionate bearing the C7 cyclohexyl branch and the head of the chain would arise from the terminal acetate unit at carbons 8 and 9, and serve in linking the polyketide to the pipecolate moiety.

The carbon spectra of immunomycin and L-683,795 isolated from fermentations of *S. hygroscopicus* var. *ascomyceticus* supplemented with methyl-^{13}C-L-methionine exhibited identical enrichment profiles (Table 2.1). The three methoxy carbons were highly enriched in both products, and virtually no other sites of enrichment were observed.

BUTYRATE AND VALINE ADDITIONS The lower incorporation levels of acetate into carbons 20, 21, 35, and 36 than into carbons 8, 9, 22, and 23 indicated the involvement of a more immediate precursor in the biosynthesis of the four former carbons. Carbon spectra were obtained on samples of immunomycin isolated from washed cell incubations of *S. hygroscopicus* var. *ascomyceticus* supplemented with either $1-^{13}C$- or $4-^{13}C$-butyrate (Table 2.2). $1-^{13}C$-butyrate feeding led to a 20-fold enrichment at carbon 20 of immunomycin, and minor enrichments at the carbons derived from carbon 1 of propionate, as expected from metabolism of butyrate to propionate [17]. Labelling with $4-^{13}C$-butyrate confirmed the orientation of the incorporated butyrate unit. The carbon NMR spectrum showed a 13-fold enrichment at carbon 36 of immunomycin, and minor enrichments at the anticipated branched methyl groups arising from carbon 3 of propionate.

Valine was first shown to act as precursor of butyrate units in streptomycete metabolism by Omura while investigating the biosynthesis of the aglycone region of the antibiotics leucomycin and tylosin [17]. Valine was also implicated to be the isobutyrate precursor in virginiamycin M_1 and related A2315A [18,19], and the butyrate precursor in monensin biosynthesis produced by *Streptomyces cinnamonensis* [20]. Valine dehydrogenase, the enzyme involved in the first step of metabolism of valine to butyrate, has been isolated and partially characterized from both *Streptomyces aureofaciens* and *Streptomyces coelicolor* A.3(2) [21,22]. The final step in butyrate production from valine, the enzymatic conversion of isobutyryl-CoA to butyryl-CoA, was recently investigated in *S. cinnamonensis*, and the stereochemistry involved in the conversion process established [23,24]. Incorporation of valine into immunomycin was examined using $2-^{13}C$-DL-valine in incubation with washed

Table 2.2 / Incorporation of ^{13}C-labelled precursors butyrate, valine, lysine and erythrose into immunomycin

Enriched Carbon Atom	Chemical Shift δ (ppm)	Enrichment Ratios				
		[1-^{13}C]-Butyrate	[4-^{13}C]-Butyrate	[2-^{13}C]-DL-Valine	[1-^{13}C]-DL-Lysine	[1-^{13}C]-D-Erythrose
1	169.0				80	
8	164.6			2.0		
10	97.0	2.0		4.0		
12	32.7					2.0
14	72.9					2.0
16	32.9	2.0		3.0		
18	48.6	1.5		2.5		
20	122.4	20.0		40.0		
24	70.0	2.0		3.0		
26	77.0	2.0		3.0		
31	84.1					9.5
32	73.2					2.0
36	11.7		13.0			
11-CH$_3$	16.2		1.5			
17-CH$_3$	20.5		1.5			

cells of *S. hygroscopicus* var. *ascomyceticus* (Table 2.2). A 40-fold enrichment was observed at carbon 20, with the same minor enrichments as seen in the butyrate labelled sample of immunomycin. Thus, in the case of immunomycin, valine catabolism to butyrate provides the four carbon segment from carbon 20 to carbon 36. In FK506 the corresponding five carbon segment arises from a propionate coupled to an acetate unit. The equivalent segment in L-683,795 (carbon 20 to carbon 35) arises from one propionate unit.

LYSINE ADDITIONS The biosynthetic origin of the pipecolate moiety was explored using ^{14}C- and ^{13}C-lysine. In an isotope competition experiment, varying concentrations of unlabelled DL-pipecolate were added to washed cell incubations of MA6678 supplemented with labelled DL-^{14}C-lysine. The results (Table 2.3) indicate a relatively high (2.13%) incorporation of lysine into immunomycin in the absence of exogenous pipecolate (e.g., incorporation of ^{14}C-valine into immunomycin was 1.02% under similar conditions). The specific activity measurements of the labelled immunomycin clearly indicate how effectively pipecolate competes with labelled lysine for incorporation into immunomycin. Thus, with labelled lysine added at 0.5 mM, 42% of the label is competed out with pipecolate added at only 0.3 mM. Pipecolic acid, or an activated form of pipecolate, is strongly indicated as an intermediate in the incorporation pathway from lysine to immunomycin.

Carbon-13 labelled lysine was used to confirm the precursor role of lysine in immunomycin biosynthesis and establish the site of lysine incorporation. A washed cell incubation of MA6678 was supplemented with 1-^{13}C-DL-lysine. The carbon NMR of the isolated immunomycin product showed an 80-fold enrichment of the resonance corresponding to carbon 1 of immunomycin.

Table 2.3 / Labelling of immunomycin with L-^{14}C-lysine, and competition with DL-pipecolic acid

Pipecolate Added (mM/L)	Immunomycin (mM/L)	Immunomycin (mCi/L)	Immunomycin Spec. Act. (mCi/mM)	Incorporation %
0.0	0.056	0.747	13.33	2.13
0.1	0.058	0.611	10.53	1.74
0.3	0.054	0.417	7.72	1.19
1.0	0.056	0.261	4.66	0.74
3.0	0.062	0.179	2.74	0.51
10.0	0.068	0.105	1.54	0.45

(Table 2.3). This is the expected site of enrichment if lysine is metabolized directly to pipecolic acid. The route of conversion from lysine to pipecolate varies among organisms. L-lysine catabolism in the genus *Pseudomonas* is known to occur by at least three different pathways [25]. Kingston's group reported that in the case of virginiamycin biosynthesis in *Streptomyces virginiae*, cyclization of lysine to 4-oxopipecolic acid occurs with retention of the e-amino nitrogen atom, presumably by α-deamination of lysine and cyclization to 1-piperideine-2-carboxylate [26]. Alternatively, Broquist recently revised the proposed primary biosynthetic pathway of lysine to pipecolic acid in the fungus *Rhizoctania leguminicola* from one proceeding by α-deamination and cyclization to 1-piperideine-2-carboxylate, to one proceeding by elimination of the e-amino group and cyclization to 1-piperideine-6-carboxylate [27].

The deamination step in lysine catabolism to pipecolate in MA6678 was investigated by examining which lysine nitrogen atom is retained in the conversion to pipecolate. DL-lysine-α-^{15}N HCl, 99% ^{15}N, and DL-lysine-e-^{15}N HCl, 99% ^{15}N were incubated in separate washed cell preparations of MA6678. The resulting immunomycin material was purified and analyzed by mass spectrometry to determine which of the ^{15}N atoms was incorporated into the pipecolate moiety of immunomycin. The increase in the intensity (I) of the m/e + 1 fragment ions resulting from incorporation of ^{15}N vs ^{14}N was calculated by comparison to the I(m/e + 1) of fragment ions from unenriched immunomycin material. The comparison was done for the three main fragment ions of m/e 564, 756, and 774, all of which contain the pipecolate moiety. The percent increase of these ions is given in Table 2.4.

The mean increase for immunomycin prepared with e-^{15}N-lysine as substrate is 49%, indicating a 33% incorporation of substrate into immunomycin product. The 33% enrichment level offered the possibility of detecting coupling between the ^{15}N nuclei and adjacent ^{13}C nuclei in the carbon NMR spectrum. A 25-hour carbon NMR accumulation on a 5 mg sample afforded sufficient signal to noise ratio that, with spectrum enhancement, the coupling of the ^{15}N to natural abundance ^{13}C nuclei at carbons 2 and 6 of immunomycin could be observed. Coupling constants of 8.0 Hz and 6.5 Hz were detected at C2 and C6, respectively. These NMR results confirm the data obtained by mass spectrometry. We conclude from these results that the e-amino group of lysine is retained and the α-amino group is lost in the metabolism of lysine to pipecolate in cultures of MA6678, consistent with the pipecolate pathway found in *S. virginiae* [26] and *Pseudomonas* species [25].

Table 2.4 / Fragment ion intensities of immunomycin labelled with either α-, or ε-^{15}N-DL-lysine.

Fragment Ion	I(m/e + 1)			
	Immunomycin (Reference)	Immunomycin from α-^{15}N-lys	Immunomycin from ε-^{15}N-lys	Net increase for (ε-^{15}N)
564	34%	36%	86%	52%
756	49%	51%	97%	48%
774	48%	49%	95%	47%

INCORPORATION OF SHIKIMIC ACID AND D-ERYTHROSE The origin of the cyclohexane moiety in the structurally related rapamycin was hypothesized to originate from shikimic acid [9]. We investigated the biosynthesis of this region using uniformly labelled ^{14}C-shikimic acid and 1-^{13}C-D-erythrose. When washed cells of MA6678 were incubated with [(G)^{14}C]-shikimic acid of specific activity of 420 mCi/mm, incorporation into immunomycin of 4.2% was obtained. It was desired to establish the site of shikimate incorporation into immunomycin using a ^{13}C-labelled substrate. The lack of commercially available ^{13}C-labelled shikimic acid was overcome by using a direct precursor of shikimate, ^{13}C-labelled D-erythrose.

The carbon NMR of immunomycin isolated from an incubation of washed cells of MA6678 with 1-^{13}C-D-erythrose showed a high level of enrichment at carbon 31, and minor enrichments at carbons 12, 14 and 32 (Table 2.2). The preponderance of labelling at carbon 31 not only confirms the ^{14}C-shikimic acid labelling and places the site of labelling in the cyclohexane moiety, but it also suggests the orientation in which erythrose-4-phosphate and phosphoenolpyruvate (PEP) condense to form the shikimic acid precursor that gives rise to the cyclohexane moiety and the adjoining carbon 28 of immunomycin. Figure 2.2 indicates two possible ways in which the shikimic acid (or activated analogue) formed from condensation of erythrose-4-phosphate and PEP could be metabolized to form the cyclohexane ring of immuno-

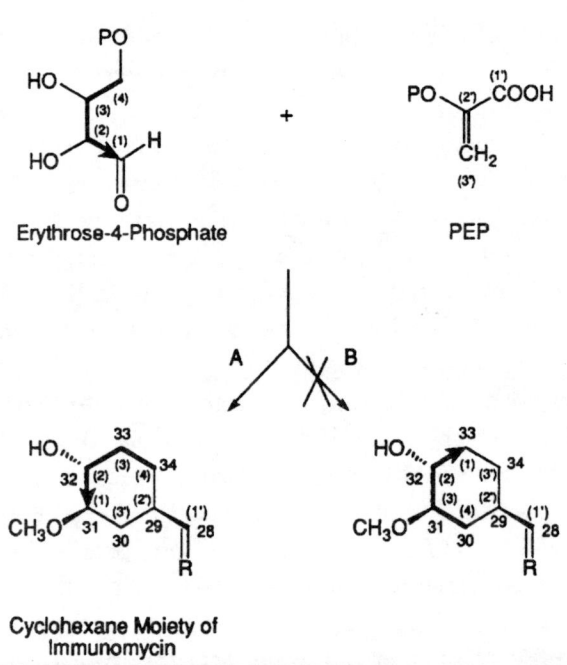

Figure 2.2 / Proposed biosynthesis of the cyclohexyl moiety of immunomycin from erythrose-4-phosphate and phosphoenolpyruvate. R denotes the remaining portion of the immunomycin, FK506 or L-683,795 molecule.

mycin. Pathway B can be ruled out since it would result in enrichment of carbon 33 of immunomycin from 1-^{13}C-erythrose. On the other hand, pathway A predicts the experimental finding that 1-^{13}C-erythrose enriches carbon 31 of immunomycin. Our proposed biosynthetic scheme for

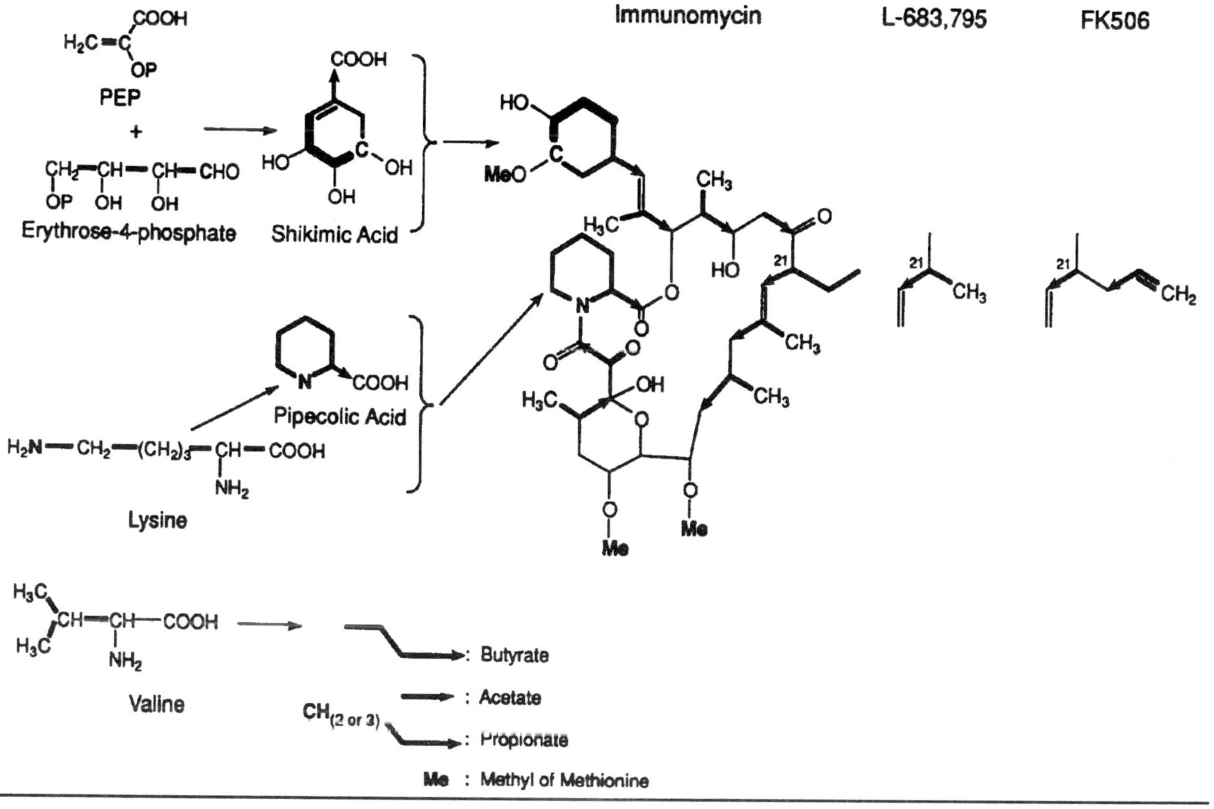

Figure 2.3 / Summary of the biosynthesis of immunomycin, L-683,795, and FK506 based upon precursor incorporation results.

immunomycin, L-683,795 and FK506 based upon all incorporation data summarized in Figure 2.3.

ISOLATION, PURIFICATION, AND CHARACTERIZATION OF DIMT The purification steps and corresponding activities of the DIMT enzyme are shown in Table 2.5. The crude extract showed an absolute requirement for SAM and magnesium ions, while EDTA inhibited the activity (Fig. 2.4). The enzyme activity was also inhibited by ammonium sulfate, but recovery of activity occurred after complete dialysis. Optimal reaction temperature and pH for activity in the crude extract were 35°C and 8.5, respectively. The DIMT activity was isolated to a high degree of purity after a series of purification steps which included ammonium sulfate precipitation, DEAE-Sepharose, preparative and analytical MonoQ, chromatofocusing and Superose-12 column chromatographies (Table 2.5). Final purification of DIMT was achieved when a highly purified fraction from Superose-12 column was subjected to native-polyacrylamide gel electrophoresis (PAGE). The fraction eluted from the native-PAGE after electrophoresis was enzymatically active and showed a single band on SDS-PAGE (Figure 2.5). Thus, DIMT isolated from *S. hygroscopicus* var. *ascomyceticus* exhibits an apparent molecular weight of 32,000, a pI of 4.4, and an absolute requirement for SAM and magnesium ions.

These activity requirements of DIMT agree with those reported in recent years for several methyltransferases participating in the biosynthesis of other therapeutically important antibiotics. Two methyltransferases involved in the biosynthesis of tylosin were shown to have both SAM and

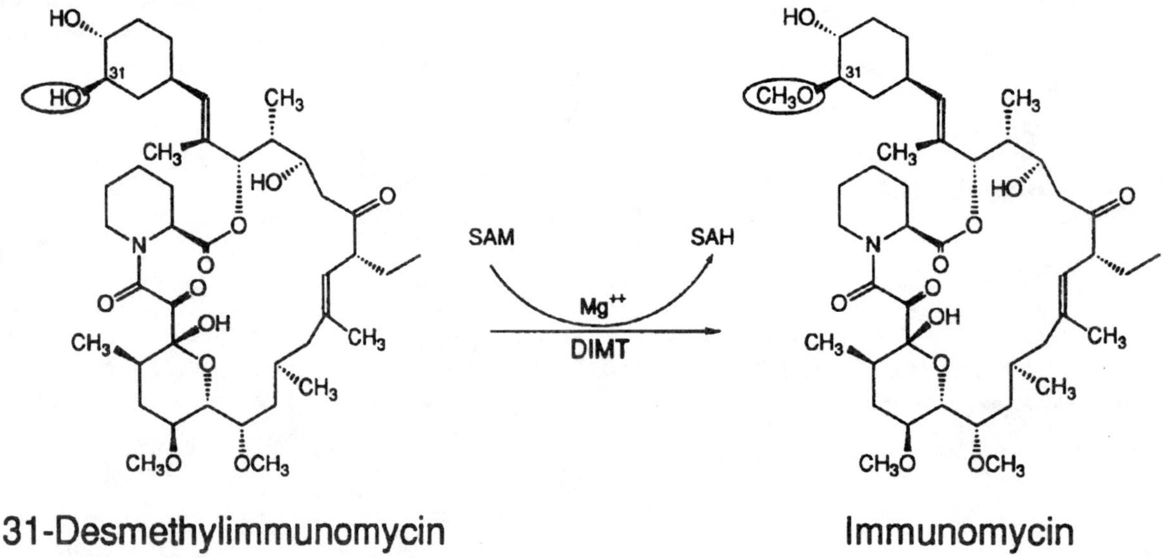

Figure 2.4 / Reaction catalyzed by 31-O-desmethylimmunomycin O-methyltransferase (DIMT).

Table 2.5 / Purification of 31-O-desmethylimmunomycin O-methyltransferease (DIMT)

Fraction	Total Protein (mg)	Sp. Act. (pmole/hr/mg)	Total Act. (pmole/hr)	Fold Purification	% Recovery
Crude extract	1150	340	391000	—	100
30–60% $(NH_4)_2SO_4$	500	243*	121500	0	31
First monoQ	33.1	8627	286277	25	73
Second monoQ	17.27	15625	269852	46	69
Superose-12	5.68	19909	113088	59	29
Native-PAGE	0.2	47600	9520	140	2.5

*Reflects possible inhibition by ammonium sulfate

metal ion requirements for activity [28]. Similarly, methyltransferases from the avermectin producer, *Streptomyces avermitilis*, and from the erythromycin producer, *Saccharopolyspora erythreae*, have been characterized and shown to have a high degree of specificity and requirements for metal ions and SAM [29,30]. Fawaz and Jones [31] reported the isolation and characterization of a methyltransferase from *Streptomyces antibioticus*. This enzyme catalyzes the transfer of the methyl group from SAM to 3-hydroxyanthranilic acid in the biosynthesis of actinomycin. It shows no metal ion requirement for activity, however. In view of the fact that methyltransferases partici-

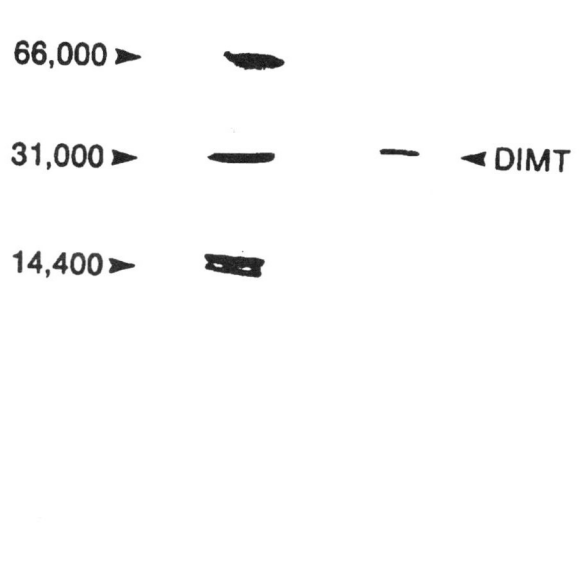

Figure 2.5 / SDS-PAGE of DIMT. Sample was eluted from a native-PAGE gel slice following recovery from the Superose-12 column. The DIMT band in the right lane is shown by an arrow on the right. Molecular weight markers, bovine serum albumin (66 kDa), carbonic anhydrase (31 kDa), and lysozyme (14.4 kda), were run in the left lane.

pate rather extensively in the biosynthesis of many important antibiotics, and may play a regulatory role in several [28], the availability of purified DIMT should facilitate the study of the biochemistry and genetics of immunomycin biosynthesis.

ISOLATION, PURIFICATION, AND CHARACTERIZATION OF THE PIPECOLATE ACTIVATING ENZYME Sonicated extracts of cells harvested at 60–68 hours were examined for the ability to catalyze a PP-ATP exchange reaction dependent on pipecolic acid, after desalting to remove amino acids. DL-pipecolic and L-pipecolate both showed some stimulation of the fairly high endogenous rate. The stimulation was dose dependent up to about 5mM in the case of L-pipecolate and 10 mM for DL-pipecolate, suggesting that the response was due to the L form but was not inhibited by the D isomer. The exchange activity was used to follow purification. After a concentrative step with ammonium sulfate and selective elution from DEAE cellulose the exchange activity was resolved into two main peaks, one independent of added amino acids and one almost completely dependent on L-pipecolate. This latter enzyme fraction had appreciable binding capacity for ^{14}C-pipecolate. Binding to a TCA precipitable form required ATP and Mg^{++}, was very rapid and was abolished by 1 mM N-ethylmaleimide or dichlorodithionitrobenzoate, sulfhydryl inhibitors. The product was stable to formic acid but unstable to performic acid, releasing the radioactivity in a form co-migrating with pipecolic acid on TLC. These are properties consistent with an enzyme bound thioester [11]. Further purification was achieved by a shallow gradient on MonoQ, followed by final purification by electrophoresis. This gave a preparation exhibiting a family of bands migrating at about 170 kDa in SDS gels (Figure 2.6, lane 5). Fingerprinting by V8 protease of 5 bands resolved by extended electrophoresis from the 170 kDa region showed that they all are closely related, suggesting that they arise from proteolysis. Heterogeneity is even more pronounced if serine protease inhibitors are not included in the extraction buffer, or if 0.2mM PMSF is not included in buffers for all steps prior to MonoQ chromatography. We conclude that the multiplicity of bands is due to proteolytic "fraying" of a single monomer of about 170 kDa, despite the inclusion of a collection of serine protease inhibitors in the early steps.

Gel filtration of the pipecolate activating enzyme purified to the MonoQ stage showed that the enzyme behaves as a dimer, migrating with a size of about 300kDa. The isoelectric point of the purified enzyme, as determined on Phast isoelectric focussing gels, was about 5.2. At this pH the enzyme was quite unstable ($t_{1/2}$, 30 minutes).

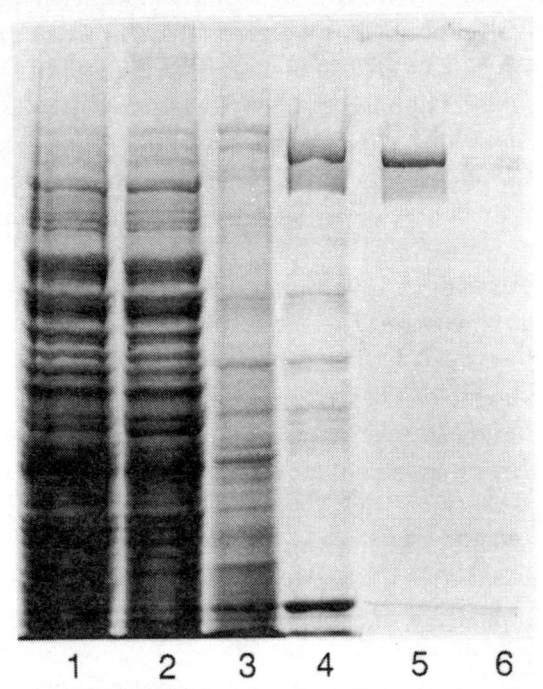

Figure 2.6 / Purification of the pipecolate activating enzyme. Lane 1. Crude extract. Lane 2. Ammonium sulfate 30–65%. Lane 3. Pool from DE-52 chromatography. Lane 4. Pool from MonoQ chromatography. Lane 5. Concentrated eluant from electrophoretic purification. Lane 6. One mg *E. coli* RNA polymerase standard, showing bands at 165, 155 and 39 kDa. The gel was stained in Coomassie Blue R.

SPECIFICITY OF THE PIPECOLATE ACTIVATING ENZYME

Analogs of pipecolic acid were tested for their ability to stimulate the PP-ATP exchange reaction exhibited by the purified pipecolate activating enzyme (Table 2.6). All protein amino acids except L-proline were completely inactive, as was D-proline. The prolyl analog of FK506 has been reported as a minor component in fermentation broths of *S. tsukubaensis* [32].

Table 2.6 / Substrate specificity of the pipecolate activating enzyme

Substrate	% of L-Pipecolate activity*
L-proline	19
L-4-hydroxyproline	12
cis,trans-4-methyl proline	178
cis-4-chloro-L-proline	33
cis-5-methyl-D,L-proline	25
piperazic acid	24
5-hydroxypipecolate	7
thiazolidine-4-carboxylic acid (thioproline)	4

*100% = initial rate of PP-ATP exchange with L-pipecolate at 10 mM (1.3mg/mL), which is a saturating level. The analogues were all tested at 2.5 mg/mL.

SUMMARY

Labelling of immunomycin, L-683,795 and FK506 with carbon-13 and carbon-14 precursors permitted establishment of the biosynthetic precursors of this family of compounds. The origin of the polyketide chain was discerned using carbon-13 labelled acetate, propionate and butyrate. The common portion of the macrolide ring of immunomycin, L-683,795 and FK506 contains 5 propionates and 2 acetates. The variable region of the ring, encompassing carbons 20 through 37, arises from differences in polyketide assembly between carbons 19 to 22. L-683,795 contains a propionate at this site, immunomycin contains a butyrate, and FK506 contains a propionate plus one acetate unit. In the case of immunomycin, 2-^{13}C-DL-valine enriched the resonance of carbon 20 by 40-fold, indicating that the butyrate unit arose from valine catabolism. The origin of the four carbon segment between carbons 12 and 15 still has to be established. Labelling with ^{14}C-shikimic acid suggested that the tail end of the polyketide chain begins with a seven carbon unit of shikimic acid origin. Labelling with 4-^{13}C-D-erythrose confirmed the ^{14}C-shikimic acid result, and also provided evidence as to the orientation in which erythrose-4-phosphate and phospoenolpyruvate condense to form the shikimic acid precursor that eventually goes on to constitute the cyclohexane moiety and the adjoining carbon 28 of immunomycin. Carbon 28 serves as the attachment site of the substituted cyclohexane ring to the first propionate unit. The head end of the polyketide chain is attached by an amide linkage to a pipecolic acid moiety. Carbon-13 and nitrogen-15 labelled DL-lysine were used to prove that the pipecolic acid moiety is produced by deamination and cyclization of lysine. In this conversion, the α-amino group of lysine is lost, while the ϵ-amino group was shown to be retained in the product.

Incubation with ^{13}C-methyl-L-methionine showed that the three O-methyl groups at carbons 13, 15, and 31 of immunomycin originate from the methyl group of methionine. The enzyme responsible for O-methylation at carbon 31 was isolated and characterized. It is a 32,000 molecular weight protein with an isoelectric point of 4.4. The enzyme responsible for pipecolate activation in immunomycin biosynthesis was also isolated and characterized. It is a monomer of 170kDa with an apparent native size of about 300kDa, suggesting that it associates as a dimer, and has an isoelectric point of 5.2. Cis,trans-4-methyl-proline and L-pipecolate served as the best substrates for this enzyme, while L-proline was the only protein acid to be activated.

We have taken a multi-pronged approach to describing the biosynthesis of the FK506 family of compounds. Feeding studies with stable and radioactive isotopes have defined the origin of the common and variable polyketide portions and the two cyclic substituents. Furthermore, the primary metabolic precursors for the cyclic substituents have been elucidated and for one a start has been made on the enzymology of its incorporation. In addition, an O-methyl transferase specific for one of the three methylations from methionine has been isolated and characterized.

The authors wish to acknowledge D. Vesey, M.J. Hsu, and C. Schreiber for their technical assistance in fermentation and enzyme isolation, J. Smith for ^{15}N mass spectral data, T. Chen for providing 31-O-desmethylimmunomycin, C. Ruby for the 31-O-desmethylimmunomycin producing mutant of MA6475, M. Nallin-Omstead and R. Borris for assistance in labelling and purification of FK506.

REFERENCES

1. Dumont. F. J., K. M. Byrne, N. H. Sigal, L. Kaplan, R. L. Monaghan, and G. Garrity. 1989. Novel immunosuppressive agent. European Patent Office publication 0323865.
2. Arai, T., Y. Koyama, T. Suenaga, and H. Honda. 1962. Ascomycin, an antifungal antibiotic. *J. Antibiotics* (Series A) 15: 231–232.
3. Arai, T. 1966. Ascomycin and process for its production. U.S. Patent: 3,244,592.
4. Hatanaka, H., M. Iwami, T. Kino, T. Goto, and M. Okuhara. 1988. FR-900520 and FR-920523, novel immunosuppressants isolated from a *Streptomyces*. I. Taxonomy of the producing strain. *J. Antibiotics* 41: 1586–1591.
5. Hatanaka, H., T. Kino, S. Miyata, N. Inamura, A. Kuroda, T. Goto, H. Tanaka, and M. Okuhara. 1988. FR-900520 and FR-900523, novel immunosuppressants isolated from a *Streptomyces*. II. Fermentation, isolation and physico-chemical and biological characteristics. *J. Antibiotics* 41: 1592–1601.
6. Starzl, T. E., J. Fung, R. Venkataramman, S. Todo, A. J. Demetris, and A. Jain. 1989. FK506 for liver, kidney, and pancreas transplantation. *The Lancet, II*: 1000–1004.
7. Sawada, S., G. Suzuki, Y. Kawase, and F. Takaku. 1987. Novel immunosuppressive agent, FK506. *In vitro* effects on the cloned T cell activation. *J. Immunol.* 139: 1797–1803.
8. Dumont, F. J., M. R. Melino, M. J. Staruch, S. L. Koprak, P. A. Fischer, and N.H. Sigal. 1990. The immunosuppressive macrolides FK-506 and rapamycin act as reciprocal antagonists in murine T cells. *J. Immunol.* 144: 1418–1424.

9. Paiva, N. L. 1988 Biosynthesis of rapamycin by *Streptomyces hygroscopicus*. Massachusetts Institute of Technology. *Dissertation Abstracts*.
10. Paiva, N. L., A. L. Demain, and M. L. Roberts. 1991. Incorporation of acetate, propionate and methionine into rapamycin by *Streptomyces hygroscopicus. J. Nat. Prod.* 54: 167–177.
11. Kleinkauf, H. and H. von Dohren. 1987. Biosynthesis of Peptide Antibiotics. *Ann. Rev. Microbiol.* 41:259–289.
12. Kino, T., H. Hatanaka, M. Hashimoto, M. Nishiyama, T. Goto, M. Okuhara, M. Kohsaka, H. Aoki, and H. Imanaka. 1987. FK-506, a novel immunosuppressant isolated from a *Streptomyces*. I. Fermentation, isolation, and physico-chemical and biological characteristics. *J. Antibiotics* 40: 1249–1255.
13. Nielsen, J. B. K. and L. Kaplan. 1989. A resting cell system for efrotomycin biosynthesis. *J. Antibiotics* 42: 944–951.
14. Lee, S. and F. Lipmann. 1975. Tyrocidine synthetase system. *Methods Enzymol.* 43:585–602.
15. Keller, U. K. 1987. Actinomycin synthetases. *J. Biol. Chem.* 262: 5852–5856.
16. Cleveland, D. W., S. G. Fischer, M. W. Kirschner, and U. K. Laemmli. 1977. Peptide mapping by limited proteolysis in sodium dodecyl sulfate and analysis by gel electrophoresis. *J. Biol. Chem.* 252: 1102–1106.
17. Omura, S., K. Tsuzuki, Y. Tanaka, H. Sakakibara, M. Aizawa, and G. Lukacs. 1983. Valine as a precursor of n-butyrate unit in the biosynthesis of macrolide aglycone. *J. Antibiotics* 36: 614–616.
18. Kingston, D. G. I., M. X. Kolpak, J. W. LeFevre, and I. Borup-Grochtmann. 1983. Biosynthesis of antibiotics of the virginiamycin family. 3. Biosynthesis of virginiamycin M_1. *J. Am. Chem. Soc.* 105: 5106–5110.
19. Kingston, D. G. I. and J. W. LeFevre. 1984. Biosynthesis of antibiotics of the virginiamycin family. 4. Biosynthesis of A2315A. *J. Org. Chem.* 49: 2588–2593.
20. Pospisil, S., P. Sedmera, M. Havranek, V. Krumphanzl, and Z. Vanek. 1983. Biosynthesis of Monensins A and B. *J. Antibiotics* 36: 617–619.
21. Vancurova, I., A. Vancura, J. Volc, J. Neuzil, M. Flieger, G. Basarova, and V. Behal. 1988. Isolation and characterization of valine dehydrogenase from *Streptomyces aureofaciens. J. Bacteriol.* 170: 5192–5196.
22. Navarrete, R., J. A. Vara, and R. Hutchinson. 1990. Purification of an inducible L-valine dehydrogenase of *Streptomyces coelicolor* A3(2). *J. Gen. Microbiol.* 136: 273–281.
23. Gani, D., D. O'Hagan, K. Reynolds, and J. A. Robinson. 1985. Biosynthesis of the polyether antibiotic monensin-A: stereochemical aspects of the incorporation and metabolism of isobutyrate. *J. Chem. Soc.,* Chem. Commun., pp. 1002–1004.
24. Reynolds, K., D. O'Hagan, D. Gani, and J. A. Robinson. 1988. Butyrate metabolism in *Streptomycetes*. Characterization of an intramolecular vicinal interchange rearrangement linking isobutyrate and butyrate in *Streptomyces cinnamonensis. J. Chem. Soc.,* Perkin Trans. I, pp. 3195–3207.
25. Fothergill, J. C. and J. R. Guest. 1977. Catabolism of L-Lysine by *Pseudomonas aeruginosa. J. Gen. Microbiol.* 99: 139–155.
26. Reed, J. W., M. B. Purvis, D. G. I. Kingston, A. Biot, and F. Gossele. 1989. Biosynthesis of the virginiamycin family. 7. Stereo- and regiochemical studies on the formation of the 3-hydroxypicolinic acid and pipecolic acid units. *J. Org. Chem.* 54: 1161–1165.
27. Wickwire, B. M., C. M. Harris, T. M. Harris, and H. P. Broquist. 1990. Pipecolic acid biosynthesis in *Rhizoctonia leguminicola*. I. The lysine, saccharopine, 1-piperideine-6-carboxylic acid pathway. *J. Biol. Chem.* 265: 14742–14747.
28. Kreuzman, A. K., J. R. Turner, and Wu-Kuang Yeh. 1988. Two distinct methyltransferases catalyzing penultimate and terminal reaction of macrolide antibiotic (tylosin) biosynthesis. *J. Biol. Chem.* 263: 15626–15633.
29. Schulman, M. D., D. Valentino, M. Nallin, and L. Kaplan. 1986. Avermectin B2 O-methyltransferase activity in *Streptomyces avermitilis* mutant that produces increased amounts of the avermectins. *Antimicrobial Agents and Chemother.* 29: 620–624.
30. Corcoran, J. W. 1975. S-Adenosylmethionine: erythromycin C O-methyltransferase. *Methods Enzymol.* 43: 487–498.
31. Fawaz, F. and G. H. Jones. 1988. Actinomycin synthesis in *Streptomyces antibioticus. J. Biol. Chem.* 263: 4602–4606.

32. Hatanaka, H., T. Kino, M. Asano, T. Goto, H. Tanaka, and M. Okuhara. 1988. FK 506 related compounds produced by *Streptomyces tsukubaensis* No. 9993. *J. Antibiot.* 42: 620–622.

3

WS 9482, an Immunomodulator of Microbial Origin

Shizue Izumi, Tsutomu Kaizu, Michio Yamashita, Masanori Okamoto, Masakuni Okuhara
Exploratory Research Laboratories, Fujisawa Pharmaceutical Co., Ltd., 5-2-3, Tokodai, Tsukuba, Ibaraki, Japan

Class II major histocompatibility molecules (Ia) on the antigen presenting cells (APCs) are attractive screening targets for immunomodulators because APCs are the first to be involved in an immune response. The search for compounds with activities influencing the portion of Ia expressing peritoneal macrophages led to the discovery that the fermentation product, WS 9482, isolated from Kitasatosporia kifunese. It is a potent inducer of Ia antigen expression. The effects of WS 9482 on animal models of autoimmune diseases and organ transplantation have been assessed. Type II collagen (CII) arthritis in rats could be prevented by sustained administration of WS 9482 and the antibody levels to CII were lower in the treated rats. Survival of skin allografts was prolonged from 6–7 days in the controls to 11–13 days in rats treated with this agent. These results support the potential of WS 9482 in the treatment of autoimmune diseases and allograft rejection.

INTRODUCTION

Ia molecules on the APCs have important functions in immune regulation (1,2,3). These molecules bind fragments of protein antigen (or peptides) and present these bound peptides to T cells. APC-T cell interactions which result from the formation of a complex among the T cell receptor (TCR), the antigen molecule and the Ia molecule of the APC, can induce an immune response (1,2,3). It is therefore reasonable to assume that manipulation of Ia molecule levels expressed on APCs must modulate the immune system and such modulation presents some interesting therapeutic possibilities.

In this connection, it has been suggested that IFN-γ, a potent inducer of Ia expression, may be beneficial in the treatment of rheumatoid arthritis (4). IFN-γ like substances with Ia inducing activity could therefore also be expected to have a therapeutic potential in autoimmune diseases.

Aiming at developing drugs that effectively restore impaired immune response, we have undertaken a comprehensive program to discover agents which positively modulate Ia molecule expression on APCs. In this paper we describe the primary screening method for inducers of Ia antigen expression, and the discovery of WS 9482 and its chemical and biological properties.

Abbreviations used in this paper:

- APC, antigen presenting cell; CII, type II collagen;
- PEC, peritoneal exudate cells; ME, mercaptoethanol;
- OPD, o-phenylene diamine; HRP, horse radish peroxidase;
- CIA, collagen-induced arthritis; IFA, Freund's incomplete adjuvant; DTH, delayed-type hypersensitivity;
- FCA, Freund's complete adjuvant; AA, adjuvant arthritis.

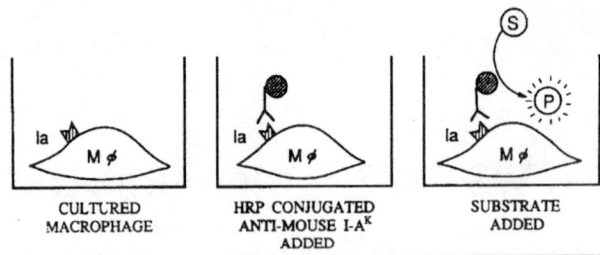

Figure 3.1 / Determination of Ia antigen expression

SCREENING

The screening was carried out using culture filtrates of soil isolates and of microbes from our culture collections. The methods used for the screening work are as follows.

PREPARATION OF PERITONEAL EXUDATE CELLS (PEC)

C3H/HeJ mice were injected i.p. with 2 ml of 3% thioglycollate medium. Four days after the injection, TG-PEC were harvested by peritoneal lavage with 3 ml of ice cold DMEM containing 20 units of heparin/ml. They were collected by centrifugation at 1000 rpm for 5 min, and resuspended at 2×10^6 cells/ml in DMEM supplemented with 10% FCS and 50μM of 2ME.

DETERMINATION OF IA ANTIGEN INDUCTION OF PEC (MACROPHAGES)

Serial 2-fold dilutions of test samples were prepared in 96-well microtiter plates in a volume of 50μl, which were seeded with 50μl of PEC (1×10^5 cells/well). The samples were then incubated for 3 days at 37°C in a humidified CO_2 incubator. The plates were aspirated and 100μl of HRP conjugated anti-mouse 1-A^k monoclonal antibody diluted in DMEM containing 5% FCS was added and left for 2 hr at room temperature. After incubation and washing (3 times), OPD containing 0.018% H_2O_2 was added to the wells. After further incubation for 30 min, 50μl of 1N HCl was added and color development was measured by using a multichannel spectrophotometer. A schematic representation of the assay method is illustrated in Fig. 3.1.

FERMENTATION

Ia inducing activity was detected in a culture filtrate of *Kitasatosporia kifunese* No. 9482 (Fig. 3.2), a strain from our culture collection. The seed culture was prepared in a medium containing potato starch 1%, sucrose 1%, glucose 1%, cottonseed flour 1%, soybean flour 0.5%, (pH 6.5) by

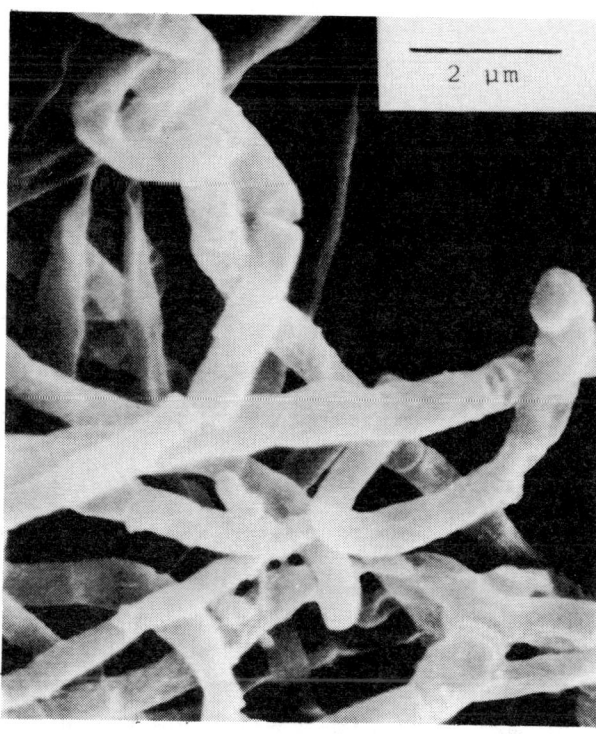

Figure 3.2

shaking for 3 days at 30°C. The culture was transferred into a 200-liter jar fermenter with 150 liters of production medium containing sucrose 4%, soybean powder 1.5%, wheat germ 0.5%, peanut powder 1%, $FeSO_4 \cdot 7H_2O$ 10µg/ml (pH 6.5). The culture was agitated (200rpm) at 30°C for 7 days with an air flow rate of 150 liter/min. The peak level of the activity was 57µg/ml.

ISOLATION AND STRUCTURE

The isolation and purification procedures used to obtain the active compound are outlined in Fig. 3.3. The culture broth was filtered to remove the cells. The supernatant (220L) was subjected to column chromatography on Dowex 1 × 2 (OH⁻ form). The objective compound was eluted with 0.5N H_2SO_4. The eluted products were charged onto a column of activated carbon and selectively eluted with 20% aq.MeOH. Further purification of the active compound was achieved by rechromato-

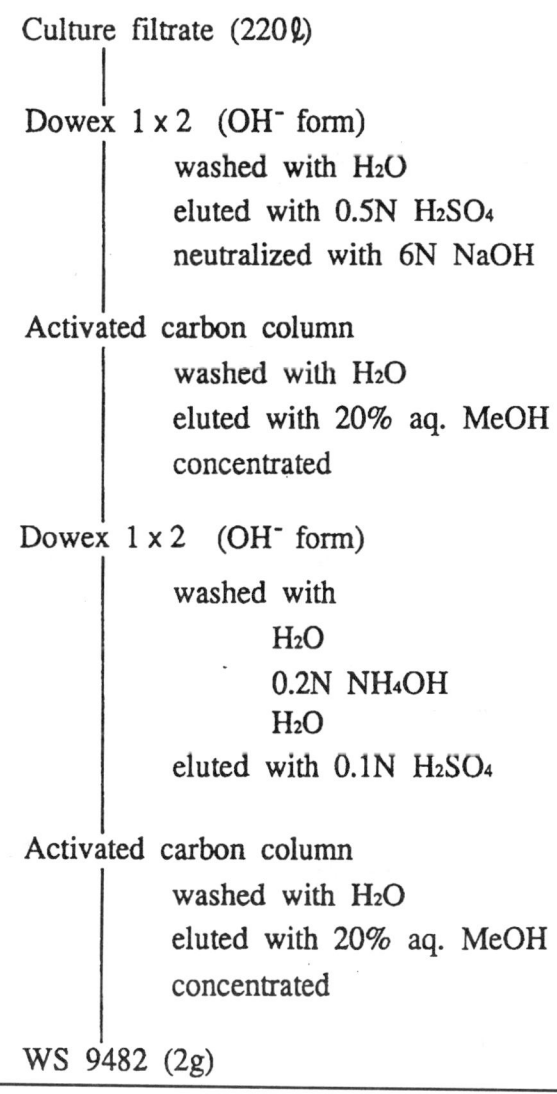

Figure 3.3 / Purification procedure for WS 9482

graphy on Dowex 1 × 2 (OH⁻ form), and then on activated carbon using 0.1N H_2SO_4 and 20% aq.MeOH as eluants, respectively. The active eluate was concentrated and lyophilized to give 2 g of WS 9482. The physico-chemical properties of WS 9482 are shown in Table 3.1. WS 9482 was obtained as a white powder. The molecular formula was established to be $C_8H_{12}N_2O_6$ from the results of the FAB-MS and ^{13}C NMR data. WS 9482 had been presumed to be identical to kifunensine because the WS 9482 producing strain, *Kitasatosporia kifunense* No. 9482 was a producer of the

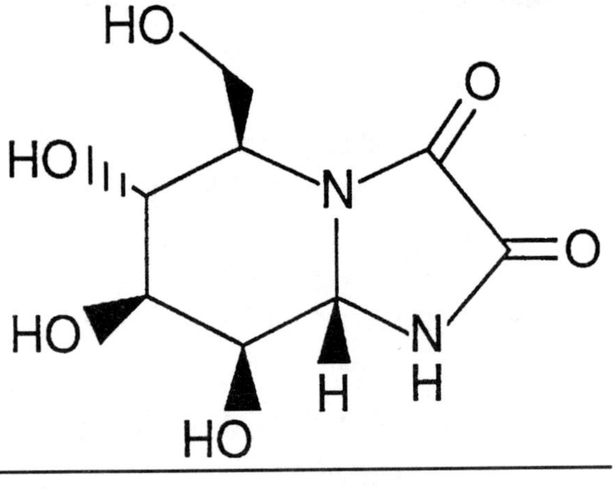

Figure 3.4 / Structure of WS 9482

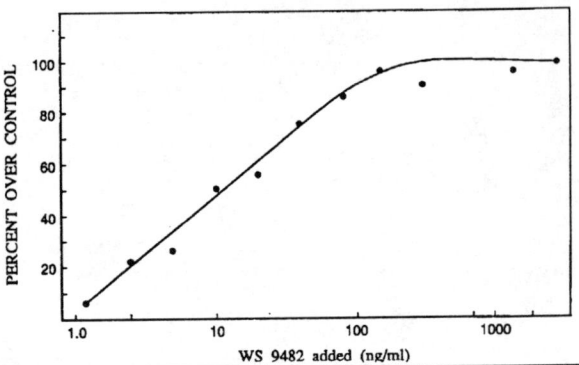

Figure 3.5 / Dose-response for Ia induction by WS 9482 in vivo

kifunensine, which was originally isolated in our laboratories in 1984(5). The structure of WS 9482 was confirmed by comparison of its IR, ^{1}H-and ^{13}C-NMR data with those of kifunensine. The two compounds were found to be identical (Fig. 3.4).

BIOLOGICAL ACTIVITIES

Ia INDUCING ACTIVITY IN VITRO AND IN VIVO

Experiments were performed to test the effects of various concentrations of WS 9482 on the induction of macrophage Ia expression in vitro. The amount of Ia molecule expressed on macrophages was detected with HRP conjugated anti-mouse I-A^k monoclonal antibody. As can be seen in Fig. 3.5, a clear dose response curve was obtained at concentrations up to 100ng/ml, which showed a 2-fold increase over the control.

WS 9482 also induced Ia antigen expression on antigen presenting cells in vivo. The dose response curve obtained 24 hours after subcutaneous injection with 100 mg/kg of WS 9482 showed a 3 to 4-fold increase of I-A^k antigen over the control (Fig. 3.6). To determine the kinetics of the 9482-induced response, mice were injected subcutaneously with 100 mg/kg of WS 9482, and peritoneal exudate cells were harvested on days 1, 2, and 3 after injection. Maximum increase of 3-fold over the vehicle control was found on day 1. The expression of Ia declined rapidly to low levels on days 2 and 3 (Fig. 3.7).

Table 3.1 / Physico-chemical properties of WS 9482

Appearance	White powder
Molecular formula	$C_8H_{12}N_2O_6$
FAB-MS (m/z)	233 (M + H)$^+$
$[\alpha]_D^{23}$ (c 0.1, H$_2$O)	+40°
IR ν_{max}^{KBr} cm^{-1}	3400~2800, 1730, 1710, 1450, 1400
	1100, 1010

IMMUNOMODULATING EFFECTS OF WS 9482 IN VIVO

To determine if WS 9482 would modulate the immune response in vivo, we measured its effect on delayed-type hypersensitivity (DTH) reactions to sheep red blood cells (SRBC). BALB/c mice

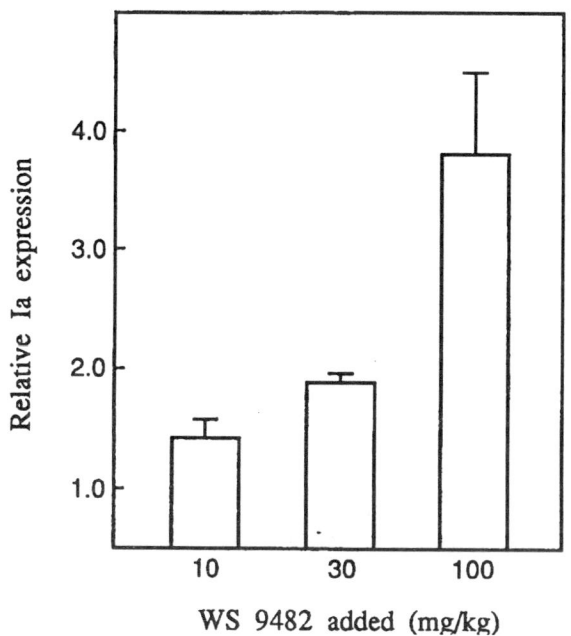

Figure 3.6 / Dose-response for IA induction by WS 9482 in vivo

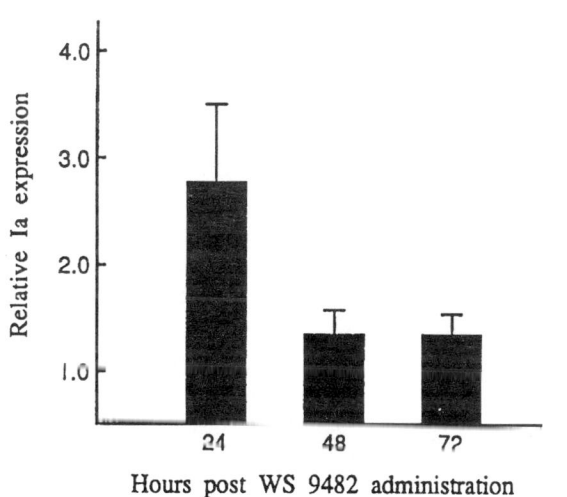

Figure 3.7 / Kinetics of Ia induction in vivo

Table 3.2 / Effect of WS 9482 on DTH reactions to SRBC in BALB/c mice

Drug effect was calculated as percent changes in footpad swelling in relation to animals receiving vehicle only.

Dose (mg/kg)	Footpad swelling ($\times 10^{-2}$mm)	Inhibition (%)
—	18.4 ± 8.9	—
vehicle[a]	89.5 ± 15.9	—
0.1	87.5 ± 6.0	3
1.0	74.3 ± 12.1[b]	21
10	59.8 ± 12.9[c]	42
100	26.2 ± 19.0[d]	89

ED_{50}:18mg/kg

[a] 0.5% methylcellulose solution
[b] significantly different from vehicle control group, $P<0.05$
[c] same as footnote b, $P<0.01$
[d] same as footnote b, $P<0.001$

were challenged 6 days after immunization with SRBC by injection of 10^8 SRBC in the right hind footpad. Twenty-four hours after the challenge the increase in the thickness of footpad was measured. The drug was administered subcutaneously on day 0, 1, 2, 5 and 6. WS 9482 showed a strong and dose-dependent suppressive effect on DTH reactions to SRBC (Table 3.2), and its ED_{50} (the dose causing 50% effect) was 18 mg/kg. Oral administration of WS 9482 with similar protocol timing also decreased DTH reactions to SRBC. The ED_{50} was 22mg/kg (data not shown).

EFFECT OF WS 9482 ON THE INDUCTION OF COLLAGEN-INDUCED ARTHRITIS (CIA)

As WS 9482 was found to have suppressive properties on DTH reactions to SRBC, we then examined the effects of the drug on the development of CIA, which is a standard animal model of human rheumatoid arthritis (RA). Lewis rats in groups of 10 were immunized intradermally with an emulsion of bovine CII and Freund's incomplete ad-

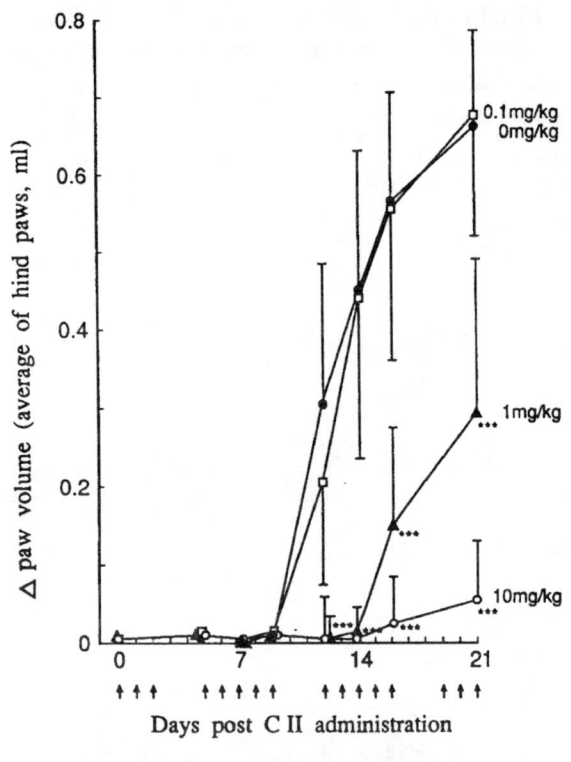

*** Significantly different from vehicle group, P<0.001

Figure 3.8 / Effect of WS 9482 on the induction of CIA in rats

juvant (IFA). The rats were examined over the course of 3 weeks to record the edema volume of the hind paws. The drug was administered subcutaneously every day from days 0 to 21 except on weekends. One and 10mg/kg WS 9482 significantly reduced the paw edema, but 0.1mg/kg had scarcely any effect (Fig. 3.8).

EFFECT ON ANTICOLLAGEN ANTIBODY TITER

Because a critical level of antibody to CII is considered to be important to the development of arthritis, serum anti-CII antibody titers were measured on day 21 after CII immunization. The results shown in Table 3.3 indicate that treatment of rats

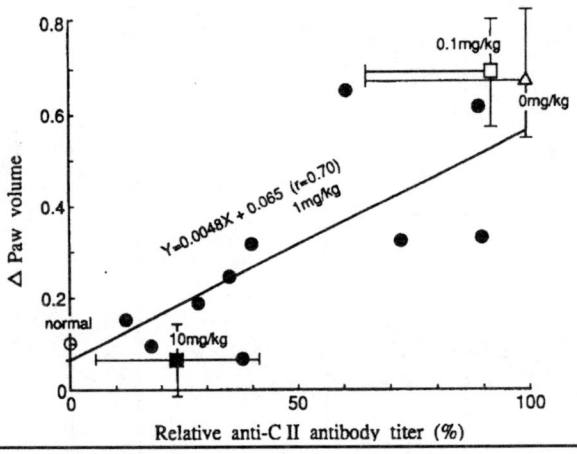

Figure 3.9 / Correlation of antibody levels to CII and increment of paw edema

with 1.0 and 10mg/kg WS 9482 caused significant decreases in the level of serum CII antibody. A comparison of antibody titer and arthritis severity (Δpaw volume) for individual rats revealed a positive correlation (Fig. 3.9), and suggested that the drug might exert its beneficial effects on arthritis by decreasing anti-CII antibody.

Table 3.3 / Effect of WS 9482 on anti-CII antibody titer

Dose (mg/kg)	Antibody to C II (unit/ml)	Relative anti-C II Ab titer
vehicle	0.346 ± 0.120	100 ± 35
0.1	0.326 ± 0.100	94 ± 29
1.0	0.165 ± 0.100[a]	48 ± 29
10	0.076 ± 0.061[b]	23 ± 18

[a]Significantly different from vehicle group, P<0.01
[b]Same as footnote a, P<0.001

Table 3.4 / Effect of WS 9482 on DTH reactions in Lewis rats immunized with CII

CII immunization	Treatment	Incidence of arthritis	Increase in ear thickness (mm)		
			24h	48h	72h
−	Vehicle	0/3	0.09 ± 0.04[c]	0.03 ± 0.01[b]	0.08 ± 0.01[a]
+	Vehicle	4/4	0.62 ± 0.06	0.72 ± 0.09	0.71 ± 0.14
+	WS9482	0/4	0.15 ± 0.01[c]	0.27 ± 0.04[b]	0.36 ± 0.04

[a] $P<0.05$ versus arthritic control rats, by Student's t-test
[b] same as footnote a, $P<0.01$
[c] same as footnote a, $P<0.001$

EFFECT OF WS 9482 ON DTH REACTIONS TO CII

DTH to CII was measured in immunized rats which were challenged with CII. Lewis rats were immu-nized intradermally with bovine CII emulsified with IFA. Three weeks later, the animals were challenged with bovine CII in the right ear and with the vehicle in the left ear as a control. The change in ear thickness was measured 24, 48, and 72 hours after challenge. When the animals were treated with the drug (3mg/kg/day) for 19 consecutive days (except weekends) before challenge with the antigen, the development of DTH reaction to CII was significantly inhibited. However, the suppressive effect began to fade after 48 hours, as evidenced by the decreasing differences in ear thickness (Table 3.4).

EFFECT OF WS 9482 ON THE INDUCTION OF ADJUVANT-INDUCED ARTHRITIS (AA)

The effect of WS 9482 for rat adjuvant arthritis, another experimental animal model for human RA, was tested. Lewis rats in groups of 10 were inoculated into the right paw with FCA and were treated with the drug at 1.0 or 10mg/kg, administered subcutaneously for 13 consecutive days except weekends, starting on the day of inoculation. The arthritis which developed in the left paw (a secondary lesion) was examined over the course of 2 weeks. The results on days 12 and 14 are shown in Table 3.5. Treatment with 10mg/kg of WS 9482 significantly reduced the footpad swelling due to the arthritis.

Table 3.5. Effect of WS 9482 on adjuvant arthritis in rats

Dose (mg/kg)	Increment of paw edema (ml)	
	Day 12	Day 14
vehicle[a]	0.98 ± 0.52	1.26 ± 0.47
1.0	0.95 ± 0.35 (3.1)[b]	1.04 ± 0.29 (17.2)
10	0.32 ± 0.48[c] (66.8)	0.55 ± 0.72[d] (56.3)

[a] 0.5% methylcellulose solution
[b] The values in parentheses represent the percent of inhibition
[c] Significantly different from vehicle control group, $P<0.01$
[d] Same as footnote c, $P<0.05$

EFFECT OF WS 9482 ON SKIN ALLOGRAFT SURVIVAL IN RATS

Since WS 9482 was demonstrated to suppress the development of both CII and adjuvant arthritis, we tested its effect on a rat skin allograft model. Ear skin grafts of F344 rats were transplanted on the lateral thorax of WKA recipients. Survival of the transplanted skin was assessed by visual inspection of the grafts. The time of graft rejection was defined as the day after transplantation on which more than 90% of the graft epithelium necrotized. Group survival was expressed as median survival time(MST). In this system, F344 skin transplanted to untreated WKA rats survived 6.4±0.2 days. WS 9482 was administered subcutaneously in doses of 3, 10 or 30 mg/kg, 5 days a week starting on the day of grafting. As shown in Table 3.6, graft survival time was prolonged significantly in every case. A regimen of 10mg/kg resulted in an almost two-fold prolongation of allograft survival over the untreated control. However, with 30mg/kg, four of 5 recipients died on day 11. The cause of death was probably drug toxicity, although one of the animals retained a healthy graft for 24 days.

Table 3.6 Effect of WS 9482 on skin allograft survival in rats

Drug	Dose (mg/kg)	Graft survival time (days)	Median (range)
Vehicle[b]		6 6 6 7 7	6.0 (6–7)
WS9482	3	6 6 8 8	7.0 (6–8)
	10	11 12 13 13	12.5 (11–13)[c]
	30	(10)[a](10) (10) (10) 24	24.0 (24)

[a]The numbers in parentheses represent the day when recipient died with active graft.
[b]0.5% methylcellulose solution
[c]$P<0.001$ as compared with vehicle treated group (Mann-Whitney's U-test)

EFFECTS OF WS 9482 ON THE SURFACE PHENOTYPE OF LYMPH NODE POPULATIONS

We studied the effects of WS 9482 treatment on lymphocyte subsets in the lymph nodes of arthritic rats. Cells were obtained from the lymph nodes of Lewis rats which had been immunized 16 days earlier with CII/FCA, and subsets of lymphocytes were monitored by FACS analysis. In CIA rats, a profound enlargement of the lymph nodes was observed, along with a nearly 20-fold increase in absolute number of lymph node cells. An increase in lymphocytes occurred in both T and B cell populations, although the B cell population showed the more drastic increase. Among T cells, both $CD4^+$ T cells and $CD8^+$ T cells were affected. When the drug was administered for 12 consecutive days (except weekends), absolute cell numbers and lymphocyte subsets decreased in comparison to the numbers present in the lymph nodes of the arthritic rats, including B cells, $CD4^+$ T cells and $CD8^+$ T cells. The ratios of $CD4^+$ to $CD8^+$ (helper to suppressor) T cell populations in the untreated arthritic rats were nearly the same as that in the normal control, and were 3.2 and 2.9, respectively. Treatment with WS 9482 reduced this ratio to 2.0 (Table 3.7).

EFFECT OF WS 9482 ON RAT MIXED-LYMPHOCYTE CULTURES (MLC)

MLC were performed in round-bottomed 96-well microtiter plates. F344 rat lymph node cells (LNC) were x-irradiated at 150 rad, and distributed at 2.5×10^5 cells/well as stimulators. WKA rat LNC were distributed at 5×10^5 cells/well as responders. Final volume was brought to 200µl/well and cultures were incubated for 4 days at 37°C in a humidified CO_2 incubator. The proliferative activity of the stimulated cultures was determined by measuring thymidine incorporation. The dose-response effect of WS 9482 was investigated by adding various concentrations of the compound to

Table 3.7. Effect of WS 9482 on cellular composition of rat inguinal lymph nodes immunized with CII

CII immunization	Treatment	Total cells recovered ($\times 10^7$)	No. of positive cells (10^7)				$CD4^+/CD8^+$
			B cells	T cells	$CD4^+$	$CD8^+$	
−	Vehicle	0.6 ± 0.1^c	0.17 ± 0.04^b	0.46 ± 0.08^c	0.34 ± 0.09^b	0.11 ± 0.02^b	2.9
+	Vehicle	10.3 ± 1.1	4.24 ± 0.52	6.02 ± 0.67	4.60 ± 0.56	1.42 ± 0.22	3.2
+	WS9482	3.3 ± 0.9^b	1.45 ± 0.42^a	1.57 ± 0.41^b	0.94 ± 0.14^b	0.47 ± 0.07^a	2.0

^{a}P<0.05 versus arthritic control rats, by Student's t-test
^{b}same as footnote a, P<0.01
^{c}same as footnote a, P<0.001

MLC. As shown in Fig. 3.10, WS 9482 did not inhibit the lymphocyte proliferation over a range of 1.5 to 400 ng/ml, but surprisingly, augmented it. Siegel reported that addition of IFN-γ to MLC resulted in augmentation of proliferation(6). WS 9482 showed a similar pattern of activity to IFN-γ, causing maximum stimulation at 0.5 to 1.0 μg/ml.

DISCUSSION

Our recent discovery of FK506, a potent immunosuppressant of proven effectiveness in clinical transplants, is a good example of the way in which increasing knowledge about the functioning of the immune system has facilitated the search for compounds with immunomodulating activities (7). This agent was discovered by using an assay of mixed-lymphocyte reaction(MLR), which is considered to be a relevant in vitro model for allograft rejection.

To establish another rational approach for immunomodulators, we have focused our attention on the expression of Ia receptor on APCs. It is widely accepted that activated T cells cause much of the tissue damage in graft rejection and autoimmune disease (8, 9, 10), and T cell activation is triggered by means of interactions among an Ia receptor, antigen peptide and T cell receptor. Activities

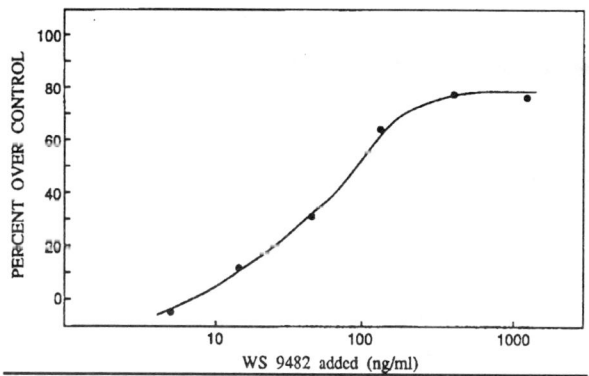

Figure 3.10 / Effect of WS 9482 on the proliferative response in rat one-way mixed lymphocyte reactions (WKA anti-F344)

which can modify the level of Ia antigen expression may regulate T cell activity, so we have searched for compounds with such activity in the fermentation broths of microorganisms.

An in vitro test system for Ia antigen expression has been developed in 96-well plates. This system could be applied to large-scale screening, although this cellular assay took 3 days to complete the evaluation of the activity. Using this screening procedure, we carried out an extensive screening program, and selected a strain of actinomycete in our culture collection. The active principle was identified as FR900494, kifunensine, which had been isolated as a immunomodulator by Iwami et al (5). This compound has been demonstrated to improve the immunosuppressive condition which was caused by immunosuppressive factors both in vitro and in vivo.

WS 9482 induced a dose-dependent expression of Ia antigen in vitro, with a maximum increase of 2-fold compared to the saline control. This agent also induced Ia antigen expression on peritoneal exudate cells in vivo.

WS 9482 was found to prevent the development of DTH reactions to SRBC. This result shows the drug has a suppressive effect on T-cell mediated immune responses in vivo. Several compounds have been reported to induce macrophage Ia expression, including bacillus Calmette Guérin (BCG) (11), lipopolysaccharide (12), and MVE-2 (11), which are high molecular weight microbial or synthetic products. WS 9482 is the first low molecular weight compound with Ia inducing as well as immunosuppressive activities.

WS 9482 also has been shown to suppress immune responses to antigen. We therefore examined the effect of this agent on the progress of the autoimmune diseases, CIA and AA, and allograft rejection in rats. The results indicated that WS 9482 efficiently inhibited some of the manifestations of autoimmune diseases, such as paw edema, antibody titer and DTH reactions to CII. WS 9482 treatment was shown to have a significant effect on the humoral immune response to CII. The sera in which the drug was most effective in the reduction of CIA severity, evaluated by size of paw edema, showed lower levels of anti-CII antibody 21 days after immunization. The suppression of disease was also accompanied by a reduction in DTH reactions.

Our working hypothesis that inducing activities of Ia expression on APC would have immunomodulatory ability led to the discovery of WS 9482, which has been demonstrated to have potent suppressive effects on the development of autoimmune diseases and allograft rejection in animal models. However, we can only speculate on whether the effect of WS 9482 on Ia antigen expression is related to its immunosuppressive activity. The injection of WS 9482 in mice caused a maximum increase of Ia antigen on day 1, and this was followed by a rapid disappearance after that. The suppressive effects of WS 9482 on DTH reactions to CII declined shortly after discontinuation of the drug administration. These results may suggest that the Ia inducing activity of WS 9482 is associated with its suppressive effects on the immune response.

There is some information in the literature which may explain the mechanism of WS 9482-induced immunosuppressive effects (Fig. 3.11). It becomes apparent that monocytes play a critical role in immunoregulation by regulating suppressor T cell activity (13, 14), so one explanation is that the primary effect of WS 9482 is on the APC (macrophage), which leads to the generation of suppressor T cells. In this respect it is interesting that WS 9482 reduced the ratio of $CD4^+$ to $CD8^+$ (helper to suppressor) T cell populations in the treatment of arthritic rats.

It is still to be confirmed whether WS 9482 preferentially acts on APC to induce Ia expression, but it is also conceivable that in addition to the Ia molecule, WS 9482 also affects the expression of other adhesion receptors on immunocompetent cells. CD8, an immunoglobulin superfamily adhesion receptor, could be another possible target molecule for the action of the drug. It was recently shown that IFN-γ preferentially augmented $CD8^+$ cell activation (6). This lymphokine has been shown to have some therapeutic effects in the treatment of autoimmune diseases in humans and animals (4, 15). Taken together, an alternative explanation would be that WS 9482, like IFN-γ, acted directly on $CD8^+$ T cells and activated them to render suppressor inducer T cells, which exerted suppressive effects on the immune response in vivo. These explanations are still speculative and further studies on the mechanism of WS 9482-mediated immunosuppressive activity are in progress.

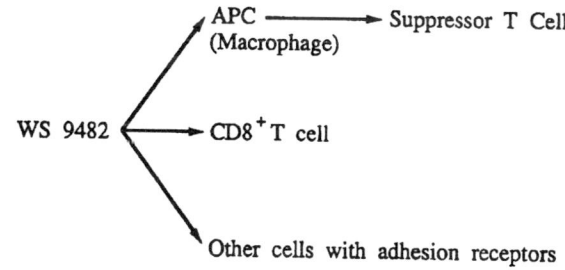

Figure 3.11 / Speculative mechanisms of WS 9482-mediated immunosuppression

ACKNOWLEDGMENTS

We are grateful for the technical expertise of Reiko Yasuda and the skilled secretarial assistance of Chieko Ohkawa in helping prepare this manuscript.

REFERENCES

1. Unanue, E. R.; & P. M. Allen: The basis for the immunoregulatory role of macrophages and other accessory cells. *Science 236:* 551–557, 1987
2. Sette, A.; S. Buus, S. Colon, J. A. Smith, C. Miles, & H. M. Grey: Structural characteristics of an antigen required for its interaction with Ia and recognition by T cells. *Nature 328:* 395–399, 1987
3. Grey, H. M.; A. S. Sette, & S. Buus: How T cells see antigen. *Sci. Amer. 261:* 38–46, 1989.
4. Pernice, W.; L. Schuchmann, J. Dippell, J. Suschke, P. Vogel, H. Truckenbrodt, F. Schindera, Ch. Humburg, & J. Brozoska: Therapy for systemic juvenile rheumatoid arthritis with γ-interferon: A pilot study of nine patients. *Arthritis Rheum. 28:* 841–845, 1985.
5. Iwami, M; O. Nakayama, H. Terano, M. Kohsaka, H. Aoki & H. Imanaka: A new immunomodulator, FR900494: Taxonomy, fermentation, isolation, and physico-chemical and biological characteristics. *J. Antibiotics 40:* 612–622, 1987.
6. Siegel, J. P.: Effects of interferon-γ on the activation of human T lymphocytes. *Cell. Immunol. 111:* 461–472, 1988.
7. Kino, T.; H. Hatanaka, M. Hashimoto, M. Nishiyama, T. Goto, M. Okuhara, M. Kohsaka, H. Aoki & H. Imanaka: FK506, a novel immunosuppressant isolated from a *Streptomyces* I. Fermentation, isolation and physico-chemical and biological characteristics.
8. Rosenberg, A. S.; T. Mizuochi, S. O. Sharrow, & A. Singer: Phenotype, Specificity, and function of T cell subsets and T cell interactions involved in skin allograft rejection. *J. Exp. Med. 165:* 1296–1315, 1987.
9. Eden, V. W.; J. Holoshitz, Z. Nevo, A. Frenkel, A. Klajman, & I. R. Cohen: Arthritis induced by a T-lymphocyte clone that responds to *Mycobacterium tuberculosis* and to cartilage proteoglycan. *Pro. Natl. Acad. Sci. USA 82:* 5117–5120, 1985.
10. Cohen, I. R.; J. Holoshitz, W. V. Eden, & A. Frenkel: T lymphocyte clones illuminate pathogenesis and affect therapy of experimental arthritis. *Arthritis Rheum. 28:* 841–845, 1985.
11. Strassmann, G.; T. A. Springer, S. J. Haskill, C. C. Miraglia, L. L. Lanier, & D. O. Adams: Antigens associated with the activation of murine mononuclear phagocytes in vivo: Differential expression of lymphocyte function-associated antigen in the several stages of development. *Cell. Immunol. 94:* 265–275, 1985
12. Wentworth, P. A.; & H. K. Ziegler: Induction of macrophage Ia expression by lipopolysaccharide and *Listeria monocytogenes* in congenitally athymic nude mice. *J. Immunol. 138:* 3167–3173, 1987.
13. Usui, M; I. Aoki, G. H. Sunshine, & M. E. Dorf: A role for macrophages in suppressor cell induction. *J. Immunol. 132:* 1728–1734, 1984.
14. Baxevanis, C. N.; G. L. Reclos, & M. Papamichail: Decreased HLA-DR antigen expression on monocytes causes impaired suppressor cell activity in multiple sclerosis. *J. Immunol. 144:* 4166–4171, 1990.
15. Jacob, C. O.; J. Holoshiz, P. V. D. Meide, S. Strober, & H. O. McDevitt: Heterogeneous effects of IFN-γ in adjuvant arthritis. *J. Immunol. 142:* 1500–1505, 1989.

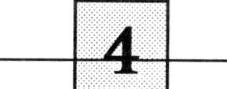

Trichosporin, Ion Channel Forming Peptides, and ISP-I, an Immunosuppressive Amino Acid

Tetsuro Fujita
Faculty of Pharmaceutical Sciences, Kyoto University, Sakyo-ku, Kyoto 606, Japan

*Two groups of fungal metabolites have been studied. The trichosporins were isolated from **Trichoderma polysporum**. Their primary structures were determined and the secondary structure of trichosporin-B-V was inferred. These peptides have uncoupling activities toward rat liver mitochondria and cause Ca^{2+}-dependent catecholamine release from adrenal medullary chromaffin cells. The activities are likely to result from their ion channel formation in lipid bilayers.*

*A second class of secondary metabolites were studied, namely a novel immunosuppressant, ISP-I, (2S,3R,4R)-(E)-2-amino-3,4-dihydroxy-2-hydroxymethyl-14-oxoeicos-6-enoic acid (myriocin = thermozymocidin), was isolated from the culture broth of **Isaria sinclairii**. The compound was found to suppress the mouse allogeneic mixed lymphocyte reaction at 10 times lower concentration than Cyclosporin A. The structure-activity relationship of ISP-I derivatives was also discussed.*

Key words: *Trichoderma polysporum*, trichosporin-Bs, peptaibol, *Isaria sinclairii*, ISP-I, immunosuppressant

INTRODUCTION

We have been studying fungal secondary metabolites with potential pharmacological applications. Results regarding two types are presented; the trichosporins and an immunosuppressant ISP-I (myriocin = thermozymocidin).

Trichosporins: The newer Japanese methods of producing the edible mushroom, *Lentinus edodes*, through giving improved yields, result in enhanc-

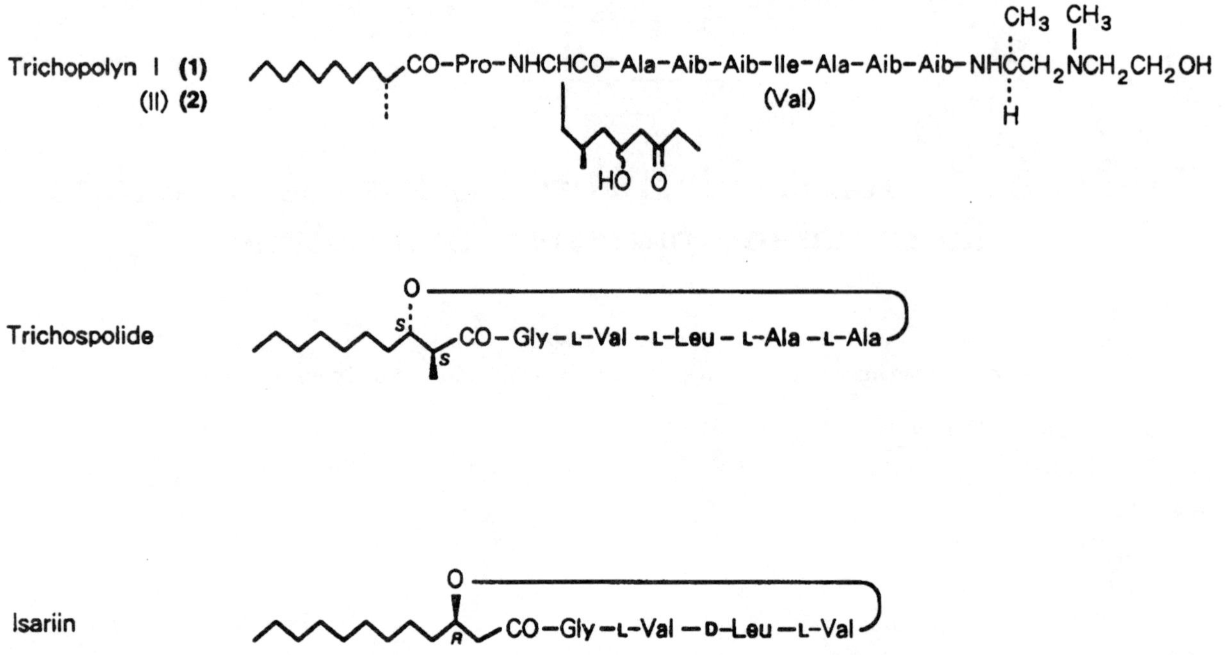

Trichopolyn I (1) / (II) (2): ~~~~CH(—)CO-Pro-NHCHCO-Ala-Aib-Aib-Ile-Ala-Aib-Aib-NHC(CH₃)(H)CH₂N(CH₃)CH₂CH₂OH
(Val side chain: CH(CH₃)CH₂C(OH)=O ... HO O)

Trichospolide: ~~~~CH(S)—CH(S)—CO-Gly-L-Val-L-Leu-L-Ala-L-Ala (cyclic via O)

Isariin: ~~~~CH(R)(O)—CO-Gly-L-Val-D-Leu-L-Val (cyclic)

Figure 4.1

ed fungal damage to mushrooms by *Trichoderma, Gliocladium* and *Hypocrea* species. In particular *Trichoderma polysporum* (link ex Pers.) Rifai (Strain TMI 60146) [1] has strong inhibitory action against *L. edodes*. We have studied antagonistic compounds from this species including the trichopolyns (1,2) [2] and the milder active trichosporin-B (TS-B)s (3–13) [3,4]. Both antibiotic peptides belong to peptaibols [5]. These peptaibols contain a high proportion of unusual amino acids, α-aminoisobutyric acid (Aib) and isovaline (Iva) and are defined as having: N-terminal amino acids "protected" by an acyl group and C-terminal ones "linked" to an amino alcohol (Fig. 4.1). The structures and activity of TS-Bs were further clarified, and are presented.

We have also studied a novel immunosuppressant, ISP-I, isolated from the culture of *Isaria sinclairii* (Berkley) Sacc. (ATCC No.24400). In this paper, we also report on the structures and biological activities of TS-Bs and ISP-I.

METHOD

Trichosporin-Bs cultures, growth media and cultural conditions have been described [1]. Separation of the trichopolyns and trichosporins was carried out by extraction of the culture broth with ethyl acetate: trichopolyns are soluble in this solvent, while trichosporins remain in the aqueous layer (Scheme 4.1).

RESULTS AND DISCUSSION

1. Trichosporins

The water soluble fraction contains the trichosporins. Reversed-phase high performance liquid chromatography (HPLC) of this fraction yielded many peaks (Fig. 4.2). The earlier and later groups of peaks on the HPLC chromatogram were named as trichosporin-A (TS-A) and trichosporin-B (TS-B), respectively. Preparative HPLC on the ODS

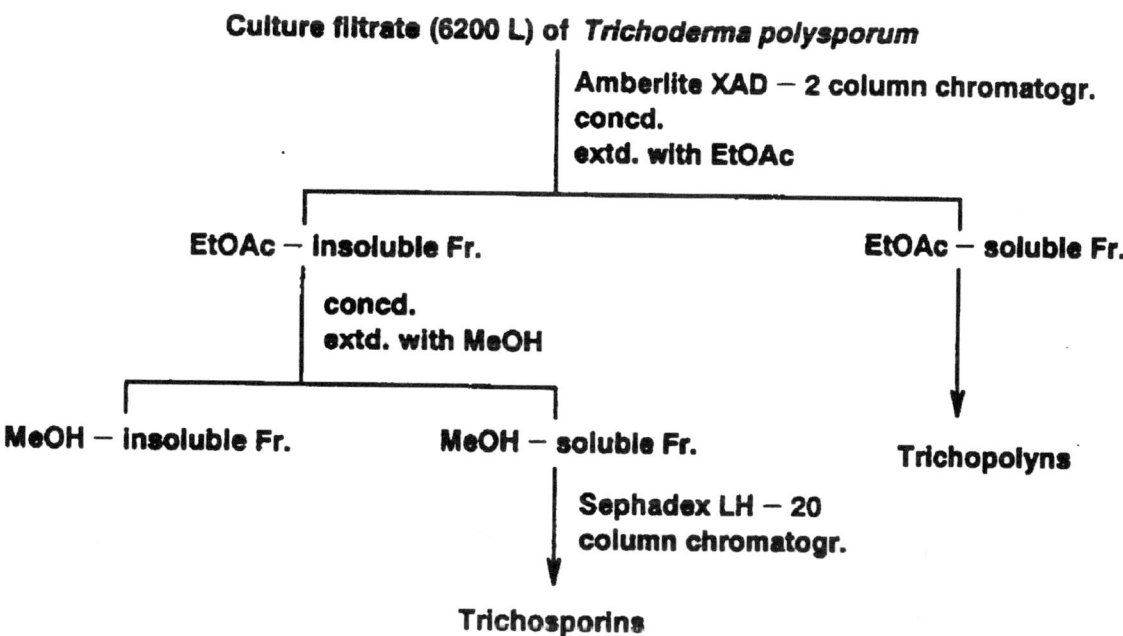

Scheme 4.1

column gave only TS-B-V as a pure compound. The other fractions were apparently impure as indicated from analysis of their amino acids and ^{1}H-nuclear magnetic resonance (NMR) spectra. Further fractionation, using a different type column, separated ten peptides (Fig. 4.3).

TS-B-V failed to ninhydrin reagent or diazomethane, while the ^{1}H-NMR spectra showed the existence of an acetyl group and a phenylalaninol. This suggests that the N-terminus is protected by an acetyl group while the C-terminal residue is linked with a phenylalaninol. The amino acid composition of TS-B-V by analysis of its acid hydrolysate, and its ^{1}H- and ^{13}C-NMR spectra showed the ratio of Aib (8), Ala (3), Gly (1), Gln (3), Ile (1), Leu (1), Pro (1), Val (1), respectively. The phenylalaninol and the optically active amino acids have L-configuration. The amino acid sequence was determined by fast atom bombardment mass spectrometry (FAB MS), fast atom bombardment mass spectrometry/mass spectrometry (FAB MS/MS) and two-dimensional NMR.

A positive ion FAB MS spectrum of TS-B-V shows the formation of two acylium ion series, A and B. The fragment ion peaks observed in each series could be assigned as shown in Fig. 4.4 [3,4]. The presence of two acylium ion series is characteristic of peptaibols which contain a very labile Aib-Pro peptide bond [6]. Therefore, the entire molecule of TS-B-V can be obtained by linking

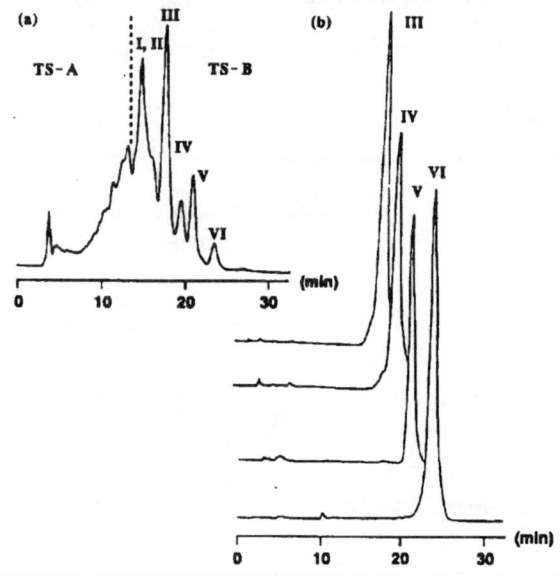

Figure 4.2

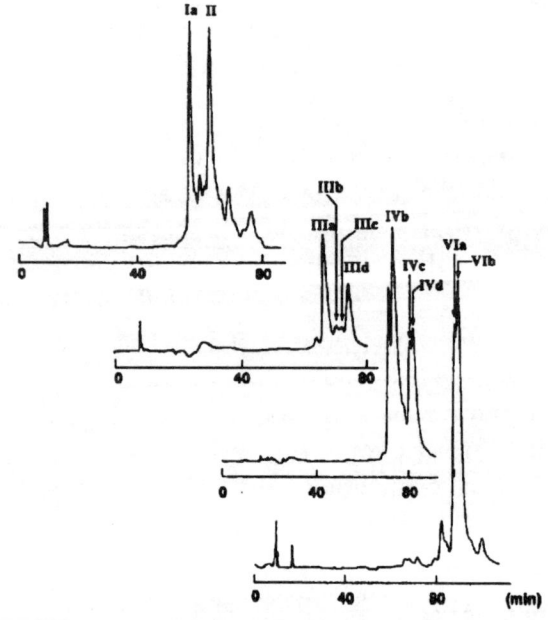

Figure 4.3

these two series. An (M+Na)⁺ cation peak could be observed at a mass charge ratio of m/z 1973. the molecular weight of 1949 (nominal mass) is consistent with the value obtained by adding all the components of TS-B-V. FAB MS/MS clearly showed the fragment peaks of series B, and clarified the sequences of some ambiguous portion such as Gln-Aib (Fig. 4.5).

The location of Ile and Leu could be determined by the sequential assignment procedures outlined by Wagner and Wüthrich [7]. The Ile NH has NH-NH cross-peaks with the Aib^8 and Aib^{10} NHs (Fig. 4.6) while the Leu NH with the Gly^{11} and Aib^{13} NHs in the nuclear Overhauser enhancement spectroscopy (NOESY) spectrum. Consequently, Ile and Leu were found to be located at positions 9 and 12, respectively.

The primary structure of TS-B-V was deduced as 11 (Table 4.1). Having defined the basic structure of TS-B-V, ten remaining peptide sequences were elucidated (see Table 4.1).

In order to examine the secondary structure of TS-B-V, circular dichroism (CD) and NMR spectroscopy were used. The CD spectra (Fig. 4.7) of this peptide were measured in methanol at various temperatures. The peptide showed negative absorptions at 207 and 221nm, indicating that this peptide has a helical conformation. Furthermore, the magnitude at both absorption bands increases as the temperature is lowered. This suggests that the rigidity of the peptide is temperature-dependent.

One and two-dimensional NMR spectra of trichosporin-B-V in CD_3OH yielded considerable information on its secondary structure. Table 4.2 shows the inter-residue NOE patterns [8] obtained from the NOESY spectra and $^3J_{NH-\alpha H}$ values. These inter-residue NOE patterns suggest that the helical structure continues from the N-terminal to the C-terminal. Furthermore, the $^3J_{NH-\alpha H}$ values, half of which are close to 6 hertz, are proximate to those of α-helices [9]. Thus, it is deduced that TS-B-V has a α-helically extended structure in CH_3OH. However, ^1H-^2D exchange rates of the amide protons

Figure 4.4

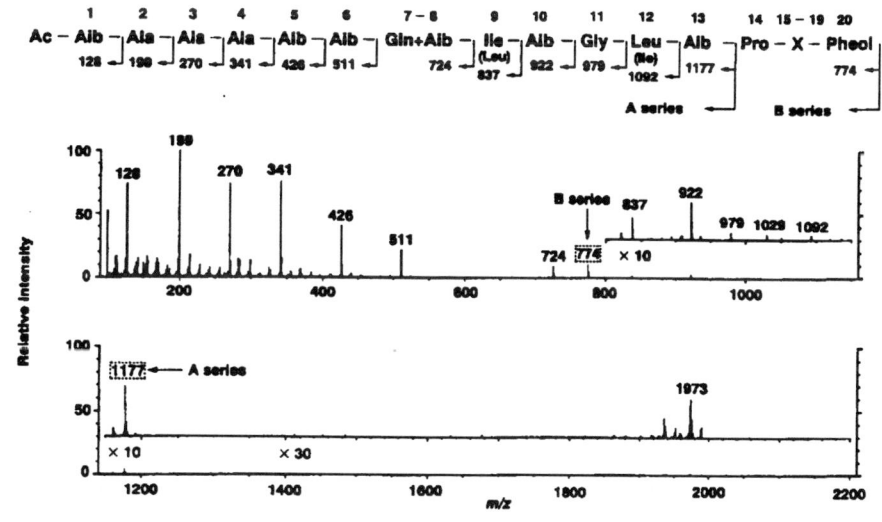

Figure 4.5

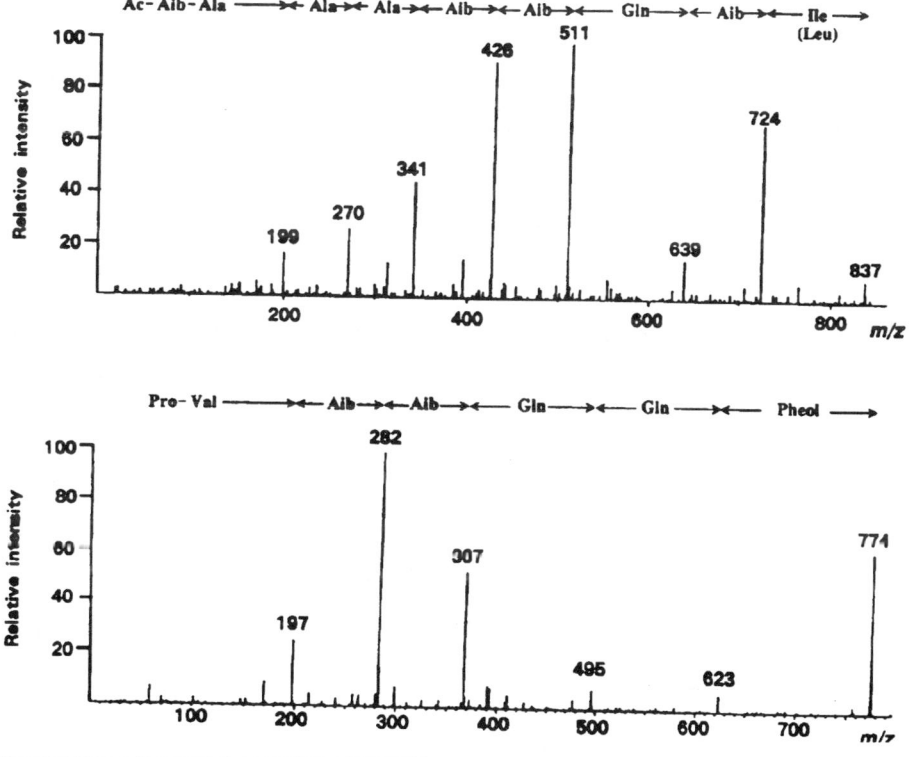

64 Fujita

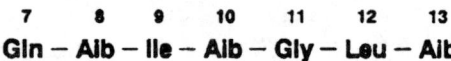

Figure 4.6

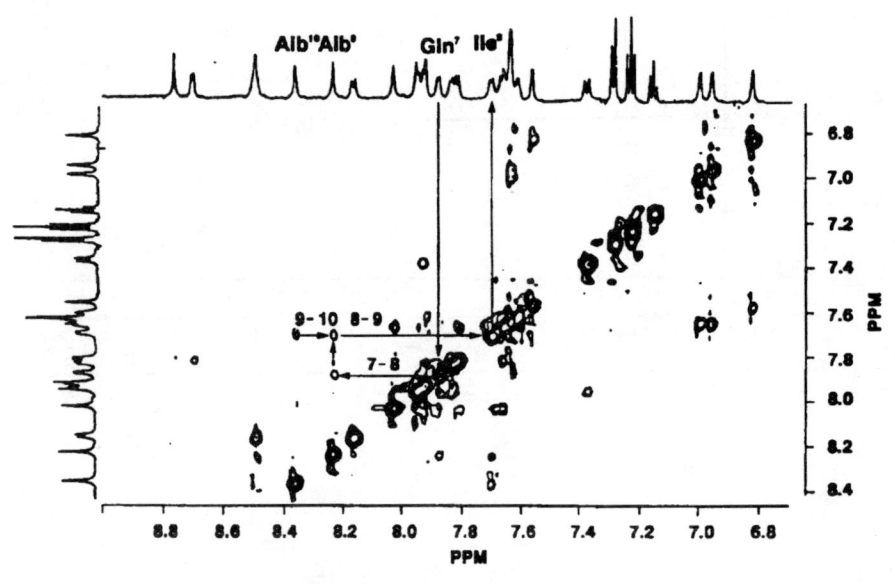

Table 4.1 / Primary structures of trichosporin-Bs

Trichosporin	1	2	3	4	5	6	7	8	9	10	11	12	13	14	15	16	17	18	19	20
B-Ia (3)	Ac-	Aib-	Ala-	Ser-	Ala-	Aib-	Aib-	Gln-	Aib-	Leu-	Aib-	Gly-	Leu-	Aib-	Pro-	Val-	Aib-	Aib-	Gln-	Gln-	Pheol
B-IIIa (4)	Ac-	Aib-	Ala-	Ala-	Ala-	Aib-	Aib-	Gln-	Aib-	Leu-	Aib-	Gly-	Leu-	Aib-	Pro-	Val-	Aib-	Aib-	Gln-	Gln-	Pheol
B-IIIb (5)	Ac-	Aib-	Ala-	Ala-	Ala-	Aib-	Aib-	Gln-	Aib-	Ile-	Aib-	Gly-	Leu-	Aib-	Pro-	Val-	Aib-	Ala-	Gln-	Gln-	Pheol
B-IIIc (6)	Ac-	Aib-	Ala-	Ala-	Ala-	Aib-	Aib-	Gln-	Aib-	Ile-	Aib-	Gly-	Leu-	Aib-	Pro-	Val-	Aib-	Aib-	Gln-	Gln-	Pheol
B-IIId (7)	Ac-	Aib-	Ala-	Ala-	Ala-	Aib-	Aib-	Gln-	Aib-	Val-	Aib-	Gly-	Leu-	Aib-	Pro-	Val-	Aib-	Aib-	Gln-	Gln-	Pheol
B-IVb (8)	Ac-	Aib-	Ala-	Ala-	Ala-	Aib-	Aib-	Gln-	Aib-	Leu-	Aib-	Gly-	Leu-	Aib-	Pro-	Val-	Aib-	Iva-	Gln-	Gln-	Pheol
B-IVc (9)	Ac-	Aib-	Ala-	Aib-	Ala-	Aib-	Aib-	Gln-	Aib-	Val-	Aib-	Gly-	Leu-	Aib-	Pro-	Val-	Aib-	Aib-	Gln-	Gln-	Pheol
B-IVd (10)	Ac-	Aib-	Ala-	Ala-	Ala-	Aib-	Aib-	Gln-	Aib-	Val-	Aib-	Gly-	Leu-	Aib-	Pro-	Val-	Aib-	Iva-	Gln-	Gln-	Pheol
B-V (11)	Ac-	Aib-	Ala-	Ala-	Ala-	Aib-	Aib-	Gln-	Aib-	Ile-	Aib-	Gly-	Leu-	Aib-	Pro-	Val-	Aib-	Aib-	Gln-	Gln-	Pheol
B-VIa (12)	Ac-	Aib-	Ala-	Aib-	Ala-	Aib-	Aib-	Gln-	Aib-	Ile-	Aib-	Gly-	Leu-	Aib-	Pro-	Val-	Aib-	Aib-	Gln-	Gln-	Pheol
B-VIb (13)	Ac-	Aib-	Ala-	Ala-	Ala-	Aib-	Aib-	Gln-	Aib-	Ile-	Aib-	Gly-	Leu-	Aib-	Pro-	Val-	Aib-	Iva-	Gln-	Gln-	Pheol

Table 4.2 / Summary of inter-residue NOEs and $^3J_{NH-C\alpha H}$ values at −5 and 26 °C (400 MHz)

Residue	0[a]	1	2	3	4	5	6	7	8	9	10	11	12	13	14[a]	15	16	17	18	19	20
NOE type[b]																					
$N_i N_{i+1}$			○	●	○	○	○	○	○	○	○	○	○	○	●			●	●	○	○
$N_i N_{i+2}$				○	●		●		○		○			●		○			●		
$\alpha_i N_{i+1}$	○			○				○		○		○	○		○				●		
$\alpha_i N_{i+3}$	○			●	●			○					○		●	○					
$\alpha_i N_{i+4}$					○										●	●					
$\alpha_i \beta_{i+3}$								○								○					
$\beta_i N_{i+1}$								○		○			○					○		●	
$^3J_{NH-C\alpha H}$ (Hz)																					
−5 °C			[c]	6.3	[c]			[c]		5.9		[c]	7.6		[c]			4.3	6.2	9.4	
26 °C			4.1	6.4	6.0			4.5		6.0		6.1	7.5		8.0			5.3	7.3	9.2	

[a] Residue o represents acetyl group. The δ protons of Pro-14 are considered to be equivalent to amide protons.
[b] ● represents the NOEs observed only when NOESY spectra were measured at −5°C.
[c] These values are ambiguous because of overlapping of the signals.

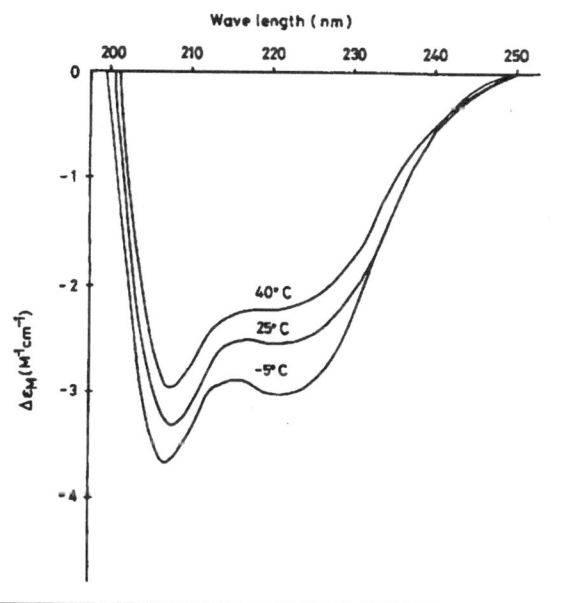

Figure 4.7

showed that the Aib¹ and Ala⁹ NHs are not hydrogen-bonded; the NH signals disappeared within 15 min. On the other hand, the half-life time of the Ala³, Ala⁴, Aib⁵ NH signals are over 2 h, and then, the others over 12 h (Fig. 4.8). If TS-B-V assumes a α-helix, the Ala³ NH should behave like the Aib¹ and Ala² NHs. In addition, the Aib¹ and Ala² NH chemical shifts were largely temperature-dependent, but the Ala³ NH did not shift. These facts suggest that the Ala³ NH is participating in hydrogen-bonding to some extent.

Thus, it can be deduced that the helical structure of the N-terminal pentapeptide is somewhat flexible and there exists two possible hydrogen-bonding patterns as illustrated in Fig. 4.9; the first turn is a mixture of a 3_{10}-helix and a α-helix. Except for the N-terminal, TS-B-V can be considered to take a α helical structure, but, based on inspection of molecular models, the axis of the α-helix should be bent at Pro¹⁴ owing to steric hindrance.

TS-Bs have uncoupling activity in rat liver mitochondria [10] and Ca^{2+}-dependent catecholamine release activity from adrenal medullary

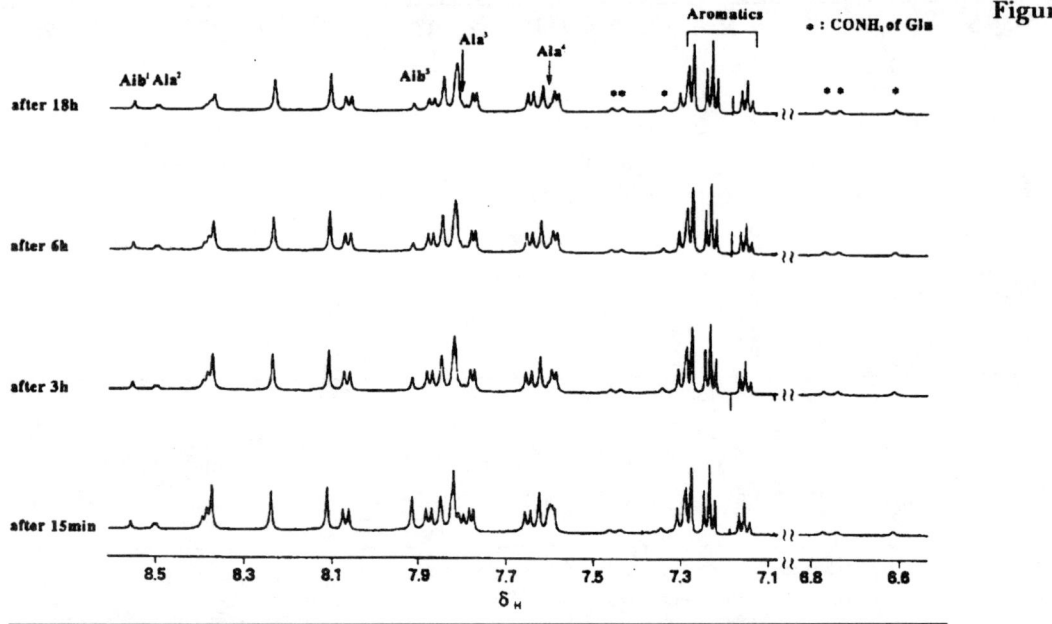

Figure 4.8

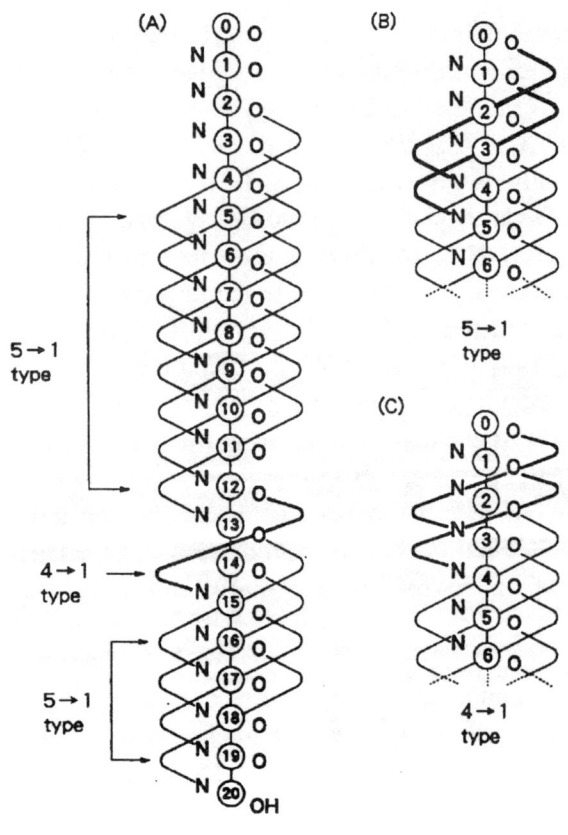

Figure 4.9

chromaffin cells as does alamethicin [11] which forms voltage dependent ion channel in artificial lipid bilayers. The activities of TS-Bs are assumed to be related to the formation of ion channels in biological membranes. Therefore, TS-Bs should become useful tools for the study of biomembrane functions.

Since the pure peptide is formed in extremely low yield, we tried to synthesize it by the solution-phase method in order to obtain a sufficient amount for examination of the biological activities and for structural confirmation [12].

The synthetic scheme for TS-B-V was designed as Scheme 4.2. The synthetic TS-B-V was identified by FAB MS, and ^{1}H- NMR spectra (Fig. 4.10). The melting point [267–270°C] and $[\alpha]_D$ [-17.7° (C = 0.3, MeOH)] of the synthetic TS-B-V were in good agreement with those of the natural substrate [268–271°C and -16.3° (C = 0.67, MeOH)].

The effects of TS-B-III fraction, which is a mixture of four peptides (TS-B-IIIa, IIIb, IIIc and IIId), on the release of catecholamines from bovine adrenal medullary chromaffin cells was examined. Incubation of the cells with TS-B-III (3-20 μM) caused an increase in catecholamine re-

Figure 4.10

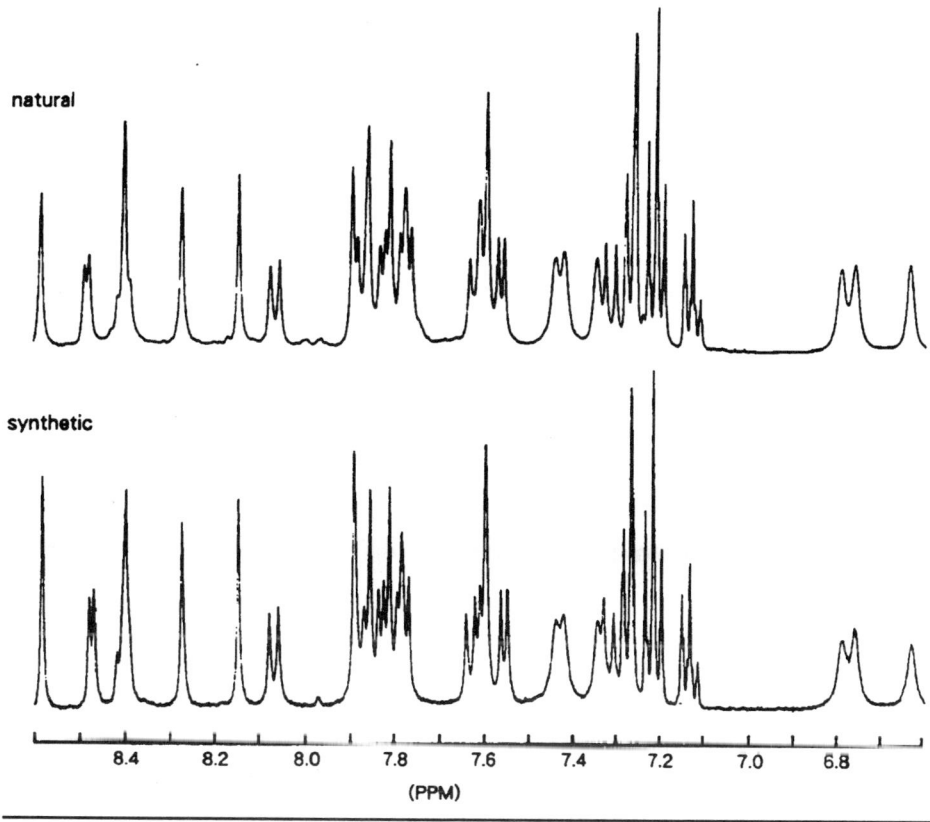

Scheme 4.2

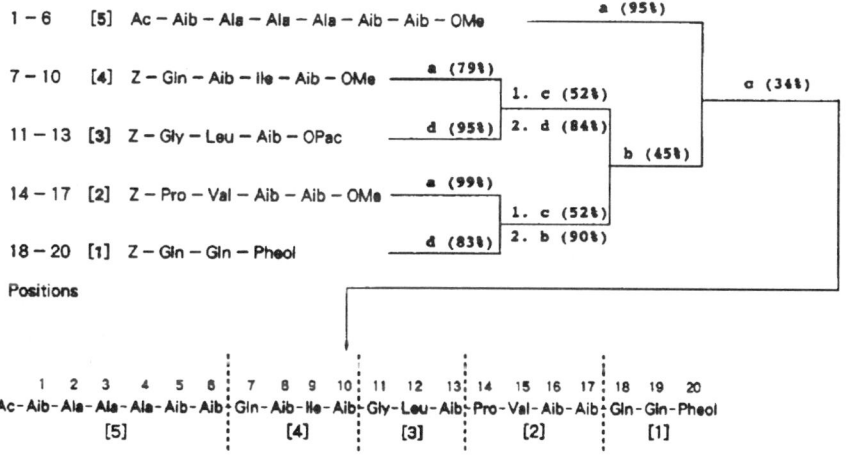

Reagents: a, 1 N NaOH; b, 10% Pd-on-charcoal; c, dicyclohexylcarbodiimide – 1 – hydroxybenzotriazole (DCC – HOBt); d, 30% HBr in acetic acid.

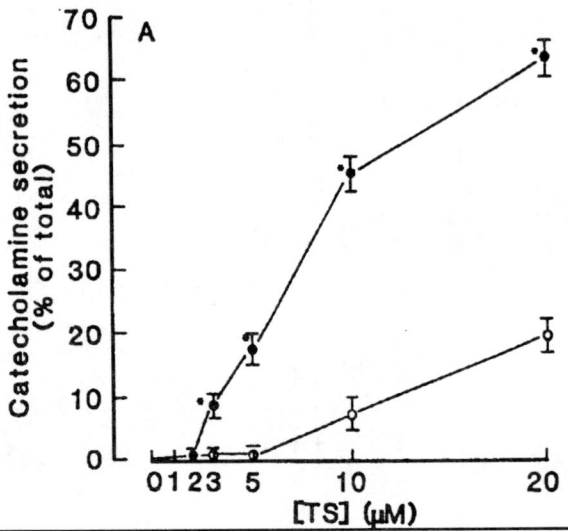

Figure 4.11

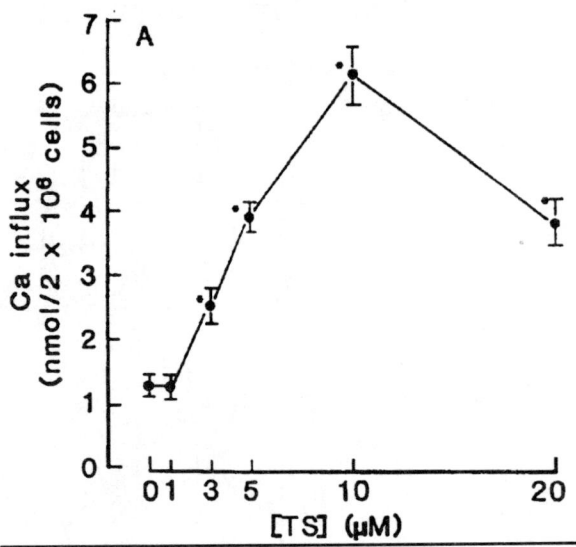

Figure 4.12

lease (Fig. 4.11). The release of catecholamines induced by low concentrations (3 and 5 µM) of TS-B-III was dependent on the presence of extracellular Ca^{2+}, whereas that induced by higher concentrations (10-20 µM) was partly independent of Ca^{2+}. Furthermore, TS-B-III increased $^{45}Ca^{2+}$ influx in a concentration-dependent manner up to 10 µM. Thus both curves for catecholamine release and $^{45}Ca^{2+}$ influx were quite similar up to 10 µM (Fig. 4.12). Near a concentration of 20 µM of TS-B-III, $^{45}Ca^{2+}$ may experience efflux due to damage of cells by high concentration of TS-B-III. These results suggest TS-B-III stimulates Ca^{2+} influx, which induces the release of catecholamines.

2. Immunosuppressant, ISP-I

Cyclosporin A [13] was initially isolated from a fungus misidentified as *Trichoderma polysporum*. The producing fungus has since been redesignated as *Tolypocladium inflatum* [14]. *Trichoderma polysporum* also produces a cyclodepsipeptide, trichospolide (Fig. 4.1). Similar cyclodepsipeptides (e.g. isariin [15]) have been obtained from other fungi, e.g. *Cordyceps* species, which are parasitic on insects. We hoped that new immunosuppressive agents may be isolated from such fungi. *Cordyceps* belongs to *Hypocreaceae*, a family of *Ascomycetes*. *Cordyceps sinensis* (Chinese name: Dong Chon Xia Cao) is parasitic on larvae of *Hepialus armoricanus*, a *Lepidoptera*. The *Cordyceps* has been used as a chinese medicine for eternal youth. *Isaria sinclairii*, the imperfect stage of *Cordyceps*, secretes compounds with considerable immunosuppressive activity in the culture broth. An immunosuppressant, ISP-I, has been isolated through the bioassay-**directed fractionation**. The activity was estimated by the mouse allogenic mixed lymphocyte reaction (MLR) [16].

The physico-chemical properties of ISP-I indicated the presence of hydroxy groups, and amino group, a carboxy group, and a trans-double bond. ^{1}H-NMR and mass spectrometry have deduced the three partial structures; an isolated hydroxymethyl group, a moiety having a proton sequence from methine on a hydroxy-bearing carbon to a methylene group, and hydrocarbon chain containing a keto group. These results and spectral analysis of the acetate indicated that ISP-I have the same basic structure as that of myriocin

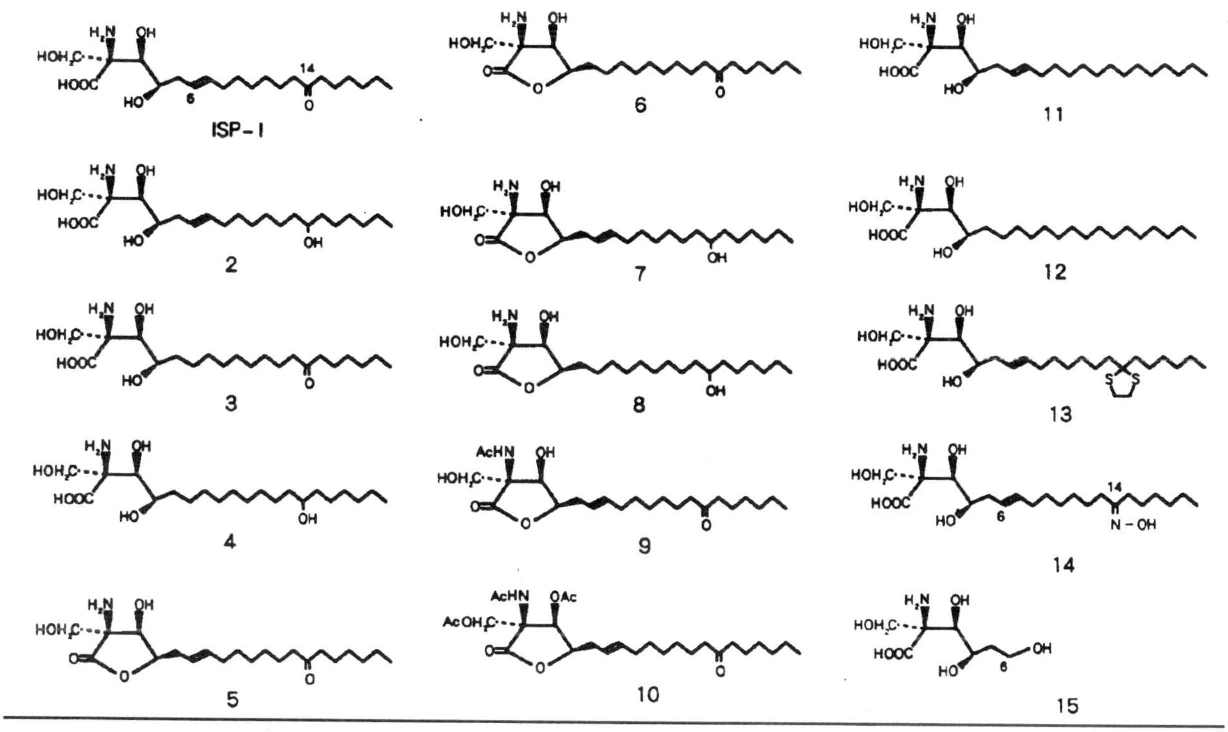

Figure 4.13

[17] and thermozymocidin [18]. Further comparison of three compounds including the absolute structure [19, 20] showed them to be identical with one another.

In order to investigate the structure-activity relationships, the twelve derivatives of ISP-I were prepared (Fig. 4.13) and their immunosuppressive activities were estimated by MLR test (Table 4.3 a, b). The MLR is known as an in vitro model of immunologic rejection on clinical organ transplantation. From these results, the relationships of structures and their immunosuppressive activities appeared as follows: 1. The hydrogenation of the double bond at C-6 decreased the immunosuppressive activity by one order of magnitude from that of ISP-I and the dihydrolactonization resulted in the more decreased activity. 2. The reduction of the ketone group at C-14 to hydroxy group, followed by lactonization does not affect the activity. The deoxygenation of the ketone group at C-14 resulted in increased activity. 3. The lactonization of ISP-I does not affect the activity. 4. The acetylation of the amino group or the complete acetyla-tion decreased the activity extremely. 5. The conversion of the ketone group to ethylenethioketal and oxime, and the cleavage by ozonolysis at double bond also resulted in further decrease of activity (1.3, 13 and 18 %, respectively).

I thank Dr. Akira Iida, Dr. Eiichi Tachikawa, Dr. Kenichiro Inoue, Mr. Satoshi Yamamoto, Mr. Masahiro Okida, Mr. Takeshi Ikumoto, Dr. Takeki Okuyama for their co-operations.

REFERENCES

1 Fujita, T., Y. Takaishi, H. Moritoki, T. Ogawa and K. Tokimoto. 1984. Fungal metabolite, I. Isolation and biological activities of hypelcins A and B (Growth inhibitor against **Lentinus edodes**) from **Hypocrea peltata**. *Chem. Pharm. Bull.,* 32: 1822–1828.

Table 4.3(a) / Effect of ISP-I and its derivatives on mouse allogeneic mixed lymphocyte reaction

compound				IC_{50} (μg/ml)
	type	C—C (6) (7)	C=X (14)	
ISP-I	COOH	C=C	C=O	0.0032
2	COOH	C=C	CHOH	0.0050
3	COOH	C—C	C=O	0.020
4	COOH	C—C	CHOH	0.022
6	γ-lactone	C—C	C=O	0.22
7	γ-lactone	C=C	CHOH	0.0020
8	γ-lactone	C—C	CHOH	0.035
9	γ-lactone (NAc)	C=C	C=O	0.22
cyclosporin A				0.0063

Table 4.3 (b). Effect of ISP-I and its derivatives on mouse allogeneic mixed lymphocyte reaction

compound				IC_{50} (μg/ml)
	type	C—C (6) (7)	C=X (14)	
ISP-I	COOH	C=C	C=O	0.0032
5	γ-lactone	C=C	C=O	0.0050
10	γ-lactone (NAc, OAc × 2)	C=C	C=O	32
cyclosporin A				0.016
ISP-I	COOH	C=C	C=O	0.0011
11	COOH	C=C	CH₂	0.00025
12	COOH	C—C	CH₂	0.013
13	COOH	C=C	C(S-S) (dithiolane)	0.25

2. Fujita, T., Y. Takaishi, A. Okamura, E. Fujita, K. Fuji, N. Hiratsuka, M. Komatsu and I. Arita. 1981. New peptide antibiotics, trichopolyns I and II, from **Trichoderma polysporum.** *J. Chem. Soc, Chem. Commun.,* 585–587.
3. Fugita, T., A. Iida, S. Uesato, Y. Takaishi, T. Shingu, M. Saito and M. Morita. 1988. Structural elucidation of trichosporin-B-Ia, IIIa, IIId, and V from **Trichoderma polysporum.** *J. Antibiot.,* 41: 814–818.
4. Iida, A., M. Okuda, S. Uesato, Y Takaishi, T. Shingu, M. Morita and T. Fujita. 1990. Fungal Metabolites. Part 3. Structural elucidation of antibiotic peptides, trichosporin-B-IIIb, c, IVb, c, d and VIa, b from **Trichoderma polysporum.** Application of fast atom bombardment mass spectrometry/ mass spectrometry to peptides containing a unique Aib-Pro peptide bond. *J. Chem. Soc, Perkin Trans. I:* 3249–3255.
5. Pandey, R. C., C. C. Cook, Jr. and K. L. Rinehart, Jr. 1977. High resolution and field desorption mass spectrometry studies and revised structures of alamethicins I and II. *J. Amer. Chem. Soc,* 99: 8469–8483.
6. Brückner, H., W. A. König, M. Greiner and G. Jung. 1979. The sequences of the membrane-modifying peptide antibiotic trichotoxin A-40. *Angew. Chem. Int. Ed. Engl.,* 18: 476–477.
7. Wagner, G. and K. Wüthrich. 1982. Sequential resonance assignments in protein ^{1}H nuclear magnetic resonance spectra. *J. Mol. Biol.,* 155: 347–366.
8. Wüthrich, K., M. Billeter and W. Braun. 1984. Polypeptide secondary structure determination by nuclear magnetic resonance observation of short proton-proton distances. *J. Mol. Biol.,* 180: 715–740.
9. Pardi, A., M. Billeter and K. Wüthrich. 1984. Calibration of the angular dependence of the amide proton-C^a proton coupling constants, $^3J_{NH\alpha}$, in a globular protein. *J. Mol. Biol.,* 180: 741–751.
10. Fujita, T., M. Okuda, A. Iida and H. Terada. Unpublished work.
11. Artalejo, A. K., C. Montiel, P. Sánchez-García, G. Uceda, J. M. Guantes and C. C. García. 1990. Alamethicin-evoked catecholamine release from cat adrenal glands. *Biochem. Biophys. Res. Commun.,* 169: 1204–1210.
12. Iida, A., S. Yoshimatsu, M. Sanekata and T. Fujita. 1990. Fungal Metabolites. Part IV. Synthesis of an antibiotic peptide, trichosporin-B-V, from **Trichoderma polysporum.** *Chem. Pharm. Bull.,* 38: 2997–3003.
13. Dreyfuss, M., E. Haerri, H. Hoffman, H. Kobel, W. Pache and H. Tscherter. 9176. Cyclosporin A and C. New metabolites from **Trichoderma polysporum.** *Eur. J. Appl. Microbiol.,* 3: 125–133.
14. Kobel, H. and R. Traber. 1982. Directed biosynthesis of cyclosporins. *Eur. J. Appl. Microbiol.,* 14: 237–240.
15. Vining, L.C. and W. A. Taber. 1962. Isariin, a new depsipeptide from **Isaria cretacea.** *Can. J. Chem.,* 40: 1579–1584.
16. Meo, T. 1981. "Immunological methods" ed. by Lefkovits, I. and B. Pernis. Academic Press Inc., New York.
17. Bagli, J.F. and D. Kluepfel. 1973. Elucidation of structure and stereochemistry of myriocin. A novel antifungal antibiotic. *J. Org.Chem.,* 38: 1253–1260.
18. Aragozzini, F., P. L. Manachini and R. Craveri. 1972. Isolation and structure determination of a new antifungal α-hydroxymethyl-α-amino acid. *Tetrahedron,* 28: 5493–5498.
19. Destro, R. and A. Colombo. 1979. Crystal structure and relative configuration of the N-acetyl-γ-lactone of the antibiotic thermozymocidin. *J. Chem. Soc, Perkin Trans. II:* 896–899.
20. Banfi, L., M. G. Beretta, L. Colombo, C. Gennari and C. Scolastico. 1983. 2-Benzoylamino-2-deoxy-2-hydroxymethyl-D-hexono-1, 4-lactones: Synthesis from D-fructose and utilization in the total synthesis of thermozymocidin (myriocin). *J. Chem. Soc, Perkin Trans. I:* 1613–1619.

The Role and Value of Microbial Screening for Agrochemicals

Sarah B Rees and Keith A Powell
ICI Agrochemicals, Jealott's Hill Research Station, BRACKNELL, Berkshire RG12 6EY, UK

Natural product screening can be of value to isolate products produced by fermentation. This offers the possibilities for chemical synthesis of natural product analogues or to stimulate the synthesis of novel chemicals based on studies of a natural product's mode of action.

This paper describes some of the problems which have to be addressed in order to develop such a screening programme and possible solutions to these problems. It concludes by recommending natural products as a route to novel agrochemicals.

INTRODUCTION

"Chemical effect" businesses are continually challenged to provide new molecules of higher added value in order to stimulate demand. In the agrochemicals business there is a drive to produce more potent yet environmentally compatible products. From the first development of agrochemicals natural products such as "Derris" have played an important role in agriculture, both as novel toxophores and also as leads to novel modes of action.

The pyrethroids provide a well known example of agrochemical development from a novel natural toxophore. Synthetic developments around the natural product have resulted in improved insecticidal potency accompanied by greatly increased photostability, indicating perhaps that increase in activity *per se* should not always be assumed to be the most important development route.

More recent developments are illustrated by two fungicides. The methoxyacrylates show broad spectrum protection as fungicides at rates of less than 1ppm [5]. The development of these compounds by ICI and BASF was stimulated as a result of the antifungal activity demonstrated by the strobilurins and oudemansins first isolated from the basidiomycete *Oudemansiella mucida* in the late 1970s by Prof Anke and his colleagues [1,2]. Another example is the pyrrolnitrin derived compound, fenpiclonil, marketed by Ciba Geigy as Beret®. The natural product pyrrolnitrin is produced by a *Pseudomonas* sp. and is thought to be responsible for the biological control of soil-borne plant pathogens by some members of this genus [1].

Natural product agrochemicals are found with activity as insecticides, fungicides, herbicides or plant growth regulators. It is interesting to note that the major historical source of insecticidal compounds is plant material while compounds with activity against plants and micro-organisms are of microbial origin. Other natural products have stimulated research as a result of their mode of action. Bialaphos, the herbicide, inhibits glutamine synthetase [3]. Phosphinothricin is the active moiety and has been developed independently as a product. Coformycin and deoxycoformycin have similarly stimulated considerable interest since both act by inhibiting adenosine deaminase to give herbicidal activity. They mimic the tetrahedral reaction intermediate or the transition state prior to the intermediate [6].

ADVANTAGES AND DISADVANTAGES OF NATURAL PRODUCTS

Of the thousands of natural products reported to have alleged insecticidal, herbicidal or fungicidal activity few are useful as agrochemical leads. The principal drawbacks are low potency or a broad spectrum, and therefore toxic, mode of action.

INAPPROPRIATE ACTIVITY

Many literature reports of, for example, phytotoxic activity are the result of tests carried out on excised coleoptiles, application to injured leaves or *in vitro* systems. While these tests may be helpful to examine specific molecules they can be very misleading as a primary screen and bear little resemblance to an intact, whole plant screen appropriate to the field use of a herbicide or plant growth regulator. Even where whole plants are used they can be irrelevant to real targets. It is essential that an early test compares activity to standard chemicals on a real target. It should not be forgotten that agrochemical activity has increased over time and hence application rates are now lower by several orders of magnitude in all pesticide areas.

TOXICITY

Natural does not mean safe. We need to be wary of broad spectrum toxicity and to realise the need to test the toxicity of natural compounds. Mycotoxins provide obvious examples but there are many others. Such compounds are unlikely to be appropriate for development because the risk to animal and environmental health is unacceptable.

THE SCREENING PROCESS

There are many alternative processes for screening natural products. This paper will outline the key issues involved in screening microorganisms for the production of biologically active metabolites. The process is multi-disciplinary requiring analytical and synthetic chemists, biologists and microbiologists, and it comprises many steps before an active ingredient can be identified and evaluated.

SCREEN INPUT

The first major issue which has to be addressed is the balance between quality and quantity. The key question is how many organisms need to be screened to generate a successful lead? In answering this question the properties of the entire screening cascade need to be considered. This can be illustrated in the simplest terms by considering how the nature of the first screen influences the number of organisms it will be reasonable to screen. For example, if the objective is to discover compounds with a specific, unusual, mode of action for which there are very few known examples, it will be appropriate to screen large numbers of organisms on a specific screen, preferably using a highly automated screening process.

Conversely a multi-component *in vivo* screen is likely to generate a much higher rate of positives per number of organisms screened; careful selection of screen input will be required to avoid duplication, rediscovery and to ensure variety. The investment in screening each organism in this instance is much higher relative to a high capacity screen. Each screening cascade must be designed for the specific objective and directed with the knowledge gained by experience of the process.

ISOLATION OF ORGANISMS TO BE SCREENED

Several approaches can be taken to organism isolation, from broad range soil isolation programmes to very specific, targeted approaches. We have found certain approaches to be particularly useful. It is often useful to go to a relevant ecological niche as a source of culture isolates, for example when screening for herbicidal compounds, plant pathogens might be a useful source. They are known to produce phytotoxic metabolites as part of the infection process, for example tabtoxin produced by *Pseudomonas tabaci* the tobacco wildfire disease [9].

Isolation of rare strains which are unlikely to have been screened before can yield novel compounds particularly if the habitat suggests that unusual biochemistry is required for growth. This approach can be pursued either by collection from exotic locations, extreme environments such as hot springs or sea bed, or by using novel isolation techniques to isolate from common sources such as soil. Such isolates may have to be treated in particular ways when they are cultured for metabolite production as they may be slow growing or have particular substrate requirements if they are to attain the appropriate physiological state to produce secondary metabolites.

Targeting particular genera may be a useful method. In our hands the wall IV group of actinomycetes has provided an example of the approach. Working with Bowen et al [4] we screened several hundred representatives of the different genera of wall IV types and noted a higher than average level of activity from members of the genus *Amycolata*.

Bowen identified a region of 16sRNA specific to the genus *Amycolata*. He cloned this region and made the appropriate 21mer probe. Concurrently he devised an isolation medium based on the numerical taxonomy database of Williams et al [8]. The probe could then be used to confirm the identity of colonies from the isolation plates as *Amycolata*. Other groups have developed different, taxon specific isolation techniques such as the use of specific phage to select interesting, or to eliminate certain, uninteresting, isolates.

A fourth approach is to employ isolation processes which select for organisms with a greater tendency to produce biologically active metabolites. For example some groups have tried probing for known secondary metabolite gene sequences such as *act* genes. In collaboration with Phillips and Wellington it has been shown that streptomycetes with multiple antibiotic resistance tend to produce more biologically active metabolites (personal communication).

CULTURING

The process of culturing isolates for secondary metabolite production is just as important as the original isolation. If the micro-organism is not subjected to conditions conducive to secondary metabolite production all efforts expended in other stages of the process are to no avail. The main considerations in the culturing phase are to induce secondary metabolism at a reasonable titre and to be able to reproduce the activity in subsequent culture. Furthermore, if high throughput is required then the use of robotics and/or disposable equipment are key elements. It is also important to ensure that no component of the method compromises screen efficacy, for example toxicity of medium ingredients in the screen must be avoided.

It is well known that different culturing conditions will induce production of different metabolites. General rules such as ammonia repression and phosphate regulation are well established. There is little knowledge about how environmental factors operate at the molecular level to affect metabolite production in a wide range of micro-organisms. The result is that much of the methodology still practised is based on experience of trial and error. We use agar surface culture which offers several advantages over liquid culture, including a greater reproducibility of metabolite production from one experiment to the next, a clean extraction process and a high titre of secondary metabolites.

After a suitable period of incubation the surface is extracted with methanol. The methanol is evaporated and the extract made up to an appropriate volume prior to application to the screen. This extraction method is effective for both cellular and extra-cellular compounds and for polar and non-polar compounds. The average yield is 2mg per plate.

The other aspect of culturing micro-organisms for secondary metabolite production that concerns microbiologists is how many media to use. We started out by using four media covering a range of C:N ratios and substrate complexity. While we did indeed note that different strains gave the most activity with different media, this was, in fact, due more to differences in titre rather than the production of completely different metabolites. We decided to use only one medium at a time enabling us to culture four times as many organisms, while changing the medium used from time to time in case there was any biosynthetic bias.

SCREENING

The design of a screening process is determined by the final target. The cascade may comprise many stages including *in vivo, in vitro,* and mode of action assays as well as tests to eliminate certain types of compound which are not desirable for a particular reason, for example toxicity.

The overall shape of the screening cascade will be determined by both the number of steps it comprises and the severity of each step. A screen which will yield only a low number of leads worthy of further investigation will have a capacity for a higher throughput than a screen capable of generating higher numbers of good quality leads. Ultimately, the numbers entering at the top of the cascade will be driven by the capacity of the effort devoted to following up the good leads.

Another important consideration is the amount of effort a particular step in the screening cascade demands. It may be appropriate to pre-screen strains prior to a step involving a labour-intensive screen. For example a simple seed germination test in agar may be a good screen for general phytotoxicity prior to a soil test since this not only demands effort to set up, but also requires 10 fold more compound.

It is inevitable in any natural product screening programme that certain compounds are going to be encountered more frequently than others. It is important that these more common metabolites are eliminated at an early stage. Some of the most commonly encountered metabolites in a typical agrochemical screening programme are cycloheximide and nigericin. Both of these compounds have broad spectrum activity as herbicides, insecticides and fungicides. We deal with these compounds by identifying extracts giving a biological activity spectrum similar to the pure compound. These extracts are then chemically assayed by tlc or hplc to determine whether the compound in question is present and if so whether the level is likely to account for all the activity seen, at which time the strain is eliminated from the screening cascade. However, on a cautionary note, it is tempting to eliminate a common metabolite producer regardless of the levels of the compound and there are good arguments to do so. In our programme we have isolated two different, novel compounds from nigericin producers.

When all the extracts have been evaluated, their activity has been confirmed, and those producing common metabolites have been eliminated it is time to select those which will be the subject of chemical fractionation leading to the isolation of the active compound. It is still impossible at this stage to determine which extracts contain the most interesting and potent compounds since any one extract may contain a high level of a weakly active compound, a low level of a highly active compound or a mixture of uninteresting compounds whose sum activity is very interesting.

The pool of good leads should exceed the fractionation capacity. The best chance of isolating novel compounds is to select for variety based on biological activity and unusual features. Compounds with certain properties may be more desirable, for example water-soluble compounds are more likely to be systemic than lipophilic compounds and these can be selected for at an early stage in the fractionation process.

FRACTIONATION/BIOASSAY

Traditionally a fractionation process begins with establishing the polarity of the active compound and any effects of pH. Purification is then carried out further by the use of various chromatographic processes. The selection of the most appropriate system will depend on the properties of the active compound, for example whether to use gas or liquid chromatography and whether to use normal or reverse phase column packings.

In tracking the active compound a major factor in the rate of progress will be the availability of a bioassay. The key features of the ideal bioassay are speed and the requirement for a small chemical sample. A bioassay to direct fractionation need not be based on the main activity for which the compound is being isolated, as long as it is known to correlate. For example we have used antifungal bioassays to resolve both insecticidal and herbicidal leads as well as fungicides.

CHARACTERISATION

The identity of a compound can often be solved before it is purified. This is of course desirable as this information will determine the importance of any further isolation. There are several tools to help the chemist resolve a structure early. Methods which yield spectral information are very valuable; for example a photodiode array will provide on-line UV spectra of fractions eluting from a column. HPLC linked to mass spectrometry can provide similar information on molecular size and fragmentation.

Such data as well as other properties on the identity of the producing micro-organism and biological activity can be matched against the Berdy Antibiotic Database to identify any known compounds. It is generally the case that the identity of an already known compound will be resolved more easily than a novel compound. In order to assign a structure to a novel compound x-ray crystallography will often be necessary and will require the isolation of sizeable quantities of compound. Before embarking on this process it may be more efficient to carry out some evaluation of the compound to determine its biological value.

EVALUATION

The biological evaluation of a natural product can really only take place once the purified compound has been isolated. This will happen months and maybe years after the original isolation and screening of the producing strain. Although in some respects this stage is parallel with the very first stages of screening a synthetic chemical, the natural product is already known to have good activity.

In addition to the value of the biological activity other considerations apply to a natural product in determining whether a more extensive research programme should be undertaken. The compound may be potent enough to be economical as a fermentation product, it may be more suitable as a lead for a chemical synthesis programme, or it may simply provide a standard for a new mode of action.

CONCLUSIONS

Natural products continue to provide leads for agrochemical products, but primarily as leads for synthesis and as models for mode of action. There are many factors to consider and to balance in such a microbial screening programme. These will be different for every programme and will be determined according to the objectives of the business. There is scope for infinite variety and for continually embracing new technology.

REFERENCES

1. Anke, T. Hecht, H.J., Schramm, G. and Steglich, W. (1979), Antibiotics from basidiomycetes IX—Oudemansin an antifungal antibiotic from *Oudemansiella mucida* (Schrades × Fr.) HOENEL (Agaricales) *J. Antibiotics, 32,* 1112–1117.
2. Anke, T., Oberwingel, F., Steglich, W. and Schramm, G. (1977), The Strobilurins—new antifungal antibiotics from the basidiomycete *Strobilurus tenacellus* (Pers × Fr.) SING. *J. Antibiotics, 30,* 806–810.
3. Bayer, E., Gingel, K. H., Magele, K., Hagenmaier, H., Jessipaw, S., Konig, W. A. and Zahner, H. (1972), *Helv. Chim. Acta., 55,* 224–239.
4. Bowen, T., Warwick, S., Challans, J. and Embley, M., (1990) The use of ribosomal RNA sequences for the identification of members of the family Pseudonocardiaceae. *Actinomycetes, 2,* 49.
5. Beautemont, K., Clough, J. M., deFraine, P. J., and Godfrey, C. R. A. (1990), Fungicidal ß-Methoxyacrylates: from Natural Products to Novel Synthetic Agricultural Fungicides. *Pesticide Science, 31,* 499–519.
6. Frieden, C., Kurtz, L. C. and Gilbert, H. (1980), Adenosine Deaminase and Adenylate Deaminase: Comparative Kinetic Studies with Transition State and Ground State Analogue Inhibitors. *Biochemistry, 19,* 5303–5309.
7. Howell, C. R. and Stipanovic, R. D. (1979) Control of *Rhizoctonia solani* on cotton seedlings with *Pseudomonas fluorescens* with an antibiotic produced by the bacterium. *Phytopathology, 69,* 480.
8. Williams, S. T., Goodfellow, M., Alderson, G., Wellington, E. M. H., Sneath, P. H. A., and Sachin, M. J., (1983) A Probability Matrix for the Identification of some Streptomycetes. *J. Gen. Microbiol., 129,* 1743.
9. Woodley, D. W. *et al., J. Biol. Chem* (1952) *197,* 409 and *198,* 807.

6

Herbicidal Amino Acids of Microbial Origin

Paul B. Lavrik, Barbara G. Isaac, Stephen W. Ayer,[1] and Richard J. Stonard

MAC-New Products Division, Monsanto Company, 700 Chesterfield Village Parkway, Chesterfield, Missouri 63198 and Institute for Marine Biosciences, National Research Council, Halifax, Nova Scotia, Canada

A screening system comprised of aquatic and terrestrial whole plants was established in our laboratories for the purpose of discovering secondary metabolites of micro-organisms having herbicidal properties. This system has proven to be particularly useful for the detection of phytotoxic amino acids as exemplified by the isolation of the novel natural product α-methylene-β-alanine, 1. In addition, several known amino acids not previously reported to be herbicidal have been identified by this screen. These include altemicidin, 2, β-methyltryptophan, 3, 2,5-dihydrophenylalanine, 4, amicoumacin B, 5, trans-4-hydroxy-L-proline, 6, anthglutin, 7 and cispentacin, 8. Altemicidin, 2, showed very good herbicidal activity against hemp sesbania, cocklebur, morningglory and common chickweed at rates equivalent to one pound per acre. A bacterial antimetabolite assay was established as a complement to whole plant screening. The use of this assay has resulted in the isolation of a novel histidine antimetabolite, pyrrolyl-3-alanine, 9.[2]

Key Words: Microbial, Natural Product, Herbicide, Amino Acid

---

[1] Present address: National Research Council, Institute for Marine Biosciences, Halifax, Nova Scotia, Canada B3H3Z1

[2] Compounds **4, 7** and **9** were presented at an SCI Pesticides Group Symposium, "Natural products as a source for new agricultural chemicals" (1).

INTRODUCTION

The remarkably selective toxicity of substances which inhibit the biosynthesis of amino acids has proven to be of great value to the development of several modern commercial herbicides (eg. glyphosate, phosphinothricin, sulfometuron methyl, imazapyr). Metabolic pathways inhibited by these products include those of branched chain amino acids, aromatic amino acid and glutamine biosynthesis. While over 75 enzymes are known to be involved in the biosynthesis of amino acids (30) the inhibition of only three[3], 5-enolpyruvyl-shikimate-3-phosphoric acid synthase (EPSP), acetohydroxyacid synthase (AHAS/ALS) and glutamine synthetase (GS) have found agricultural application. Phosphinothricylalanylalanine, a metabolite of *Streptomyces viridochromogenes* and *Streptomyces hygroscopicus* (2,15), inhibits GS and has served as the basis for the development of the non-selective post-emergent herbicides, glufosinate and bialophos. Each of these products has active ingredients which contain the amino acid phosphinothricin and are the only microbial natural products to realize commercial success in the weed control industry. However, over 300 microbial secondary metabolites are known which possess some degree of herbicidal activity.[4] Many of these are amino acids which have been reported to elicit phytotoxic effects when applied to whole plants (see Table 6.1). Several of the compounds listed in Table 6.1 are thought to inhibit amino acid biosynthesis.

Unlike many of the secondary metabolites found during the process of screening microbial broths, amino acids and their derivatives possess several general advantages as a resource for herbicide lead discovery. For example, they frequently possess sufficient structural simplicity so as to be amenable to synthetic derivation. Furthermore, as a result of their polar and ionic characteristics there is the potential for symplastic movement, and as discussed, they can have selective modes-of-action of proven utility in agricul-ture. As the rate of success of herbicide discovery from classical chemical screens diminishes and environmental pressures to produce safe pesticides increase, the need for alternative herbicide discovery strategies becomes imperative. The attractiveness of inhibiting the enzymes involved in plant amino acid biosynthesis coupled with the tremendous biosynthetic capabilities of micro-organisms suggests that the search for bioactive amino acids produced by micro-organisms could be a particularly fruitful approach for future herbicide discovery. Such compounds would ideally have high specificity, high unit activity and low environmental persistance. In addition to being desirous from the standpoint of finding new chemical leads for analogue synthesis programs, the discovery of novel herbicidal amino acids may also lead to knowledge of novel target enzymes. This information can result in the development of unique screening systems.

There have been several approaches reported for the design of screening systems for the discovery of herbicidal amino acids. Most notable among these are antimetabolite assays such as that employed in the discovery of oxetin, **10** (26) and phosalacine, **11** (27) and whole plant screens such as those used to uncover phosphinothricylalanylalanine, **12** (2) and homoalanosine, **13** (6). The antimetabolite approach, while directly targeted to the pathway of interest, suffers from the fact that it is not designed to detect compounds which will necessarily have activity on whole plants. Whole plant screens, while satisfying a downstream criterion, suffer from their obvious inability to select for specific classes of compounds or inhibitors of specific biochemical processes. To date

---

[3] Amitrole has been suggested to inhibit imidazoleglycerolphosphate dehydratase (42) although recently it has been shown not to inhibit histidine biosynthesis in *Arabidopsis* (10).

[4] Ayer, S. W. and Isaac, B. G., unpublished literature survey.

Table 6.1 / Non-Protein amino acids from microbial sources with activity on intact plants

Compound	Herbicidal Activity	Source	Ref.
Phosphinothricin & derivatives, 11, 12, 18	Broad spectrum—post-emergent activity vs. monocots and dicots. MOA = glutamine synthetase inhibitor	S. hygroscopicus S. viridochromogenes K. phosalacinea	2, 27, 20
Oxetin, 10	Turnip, alfalfa. MOA = glutamine synthetase inhibitor	Streptomyces sp., (OM-2317)	26
Homoalanosine, 13	Damage to buds and roots of cocklebur, ladysthumb, velvetlear, wild oats. MOA = unknown (reversed by Asp, Glu)	S. galilaeus	6
Gabaculine, 19	Pre- and post-emergence activity vs. barnyardgrass, downy brome, sicklepod, yellow foxtail, velvetleaf MOA = phytochrome synthesis inhibitor (transaminase inhibitor)	S. toyocaensis	18, 5
1,2,4-triazole-3-alanine, 20	Arabidopsis thaliana MOA = histidine biosynthesis inhibitor	Streptomyces sp., (KM-10329)	12
Azaserine, 20	Broad spectrum—causes root growth inhibition in cucumber, barley, flax. MAO - glutamine amidation inhibitor	S. fragilis	25
Ethionine, 21	Effects flowering process in duckweed, cocklebur, morningglory MOA = various (reversed by Met)	B. megaterium P. aeruginosa	40
Rhizobitoxine, & related enol-ether amino acid analogs, 22	Borad spectrum—causes chlorosis in soybean, crabgrass, clover, sorghum. MOA = ethylene production inhibitor (β-cystathionase inhibitor)	R. japonicum Streptomyces sp., (X-11,085) P. aeruginosa	29, 7, 31, 34, 28, 14
1-Amino-2-nitrocyclopentanecarboxylic acid, 23	Broad spectrum—causes chlorosis/stunting in chrysanthemum, peanuts, tobacco, pea. MOA = unknown (reversed by Leu)	A. wentii	43, 3
Tabtoxinine-β-lactam, 24	Broad spectrum—causes chlorosis in tobacco, oat, bean, soybean, tomato. MAO = glutamine synthetase inhibitor	P. syringae pv. tabaci	36, 38
Octicidin, 25 (Phaseolotoxins)	Broad spectrum—causes chlorosis/stunting in bean, tobacco. MOA = ornitine carbamoyl transf. inhibitor	P. syringae pv. phaseolicola	24, 39

there have been no reports of herbicidal microbial amino acids resulting from the use of an isolated enzyme screen. The search for herbicides based upon phytotoxins derived from plant pathogenic fungi or allelochemicals from plants has been recently detailed (17,33) and will not be treated here as a consequence of the different rationales for discovery in these cases.

In our laboratories we have chosen to employ an approach which utilizes a matrix of simple aquatic and terrestrial whole plant assays as our primary discovery tool as described in Materials and Methods. To a lesser extent we have also examined the utility of bacterial based antimetabolite screening. The objective of this paper is to present the results of these screens and to discuss their values as tools for discovering herbicidal amino acids. The isolation of active compounds realized from these screens, their structures and their biological properties are described.

MATERIALS AND METHODS

Lemna minor ASSAY

The *L. minor* assay employed a method similar to that described by Einhellig et al (4). Wild type *L. minor* stock cultures were maintained under sterile conditions in 250 mL flasks containing 100 mL of medium (22) at 22°C under 16 hr photoperiod. Assays were performed in 12-well tissue culture plates under sterile conditions. To each well was added 3.6 mL of media and 0.4 mL of sterile test sample. A single uniform rosette of 2-3 fronds was transferred to each well and the plates were incubated for seven days at 22°C with 16 hr photoperiod. After the incubation period, frond number per well and morphological effects such as chlorosis, frond cupping, reduced frond size, and failure of daughter fronds to separate from the original rosette were recorded.

Arabidopsis thaliana ASSAY

The *A. thaliana* assay was conducted in a 24-well tissue culture plate. To each well was added 1 mL of sterilized 1% water agar to which 100 µg each of methanolic stock solutions of captan and benomyl were added prior to agar solidification (to prevent fungal contamination). Subsequently, 100 µL of sterile test sample was pipetted onto the agar surface and the solvent was allowed to evaporate under sterile conditions. Seeds of *A. thaliana* cv Columbia were sprinkled into the test wells, the plates were covered with lids and placed in a growth chamber which was kept at 25°C, 13 hr photoperiod and 85% relative humidity. At the end of 3-4 days the assay was evaluated for phytotoxicity, chlorosis and germination inhibition.

ROOT ASSAYS

Seeds of velvetleaf (*Abutilon theophrastii*) and barnyardgrass (*Echinochloa crusgalli*) were incubated in plastic germination pouches (No. 82700, Northrup King Co., Golden Valley, MN) which contained a paper sheet folded to form a pouch that accommodated the seeds. The paper sheet was wetted with 10 mL of a 1:5 dilution of sterile test sample. The pouches were incubated for 5 days at 22°C in a specially made aluminum box (122 cm × 18 cm × 18 cm box with 1.5 mm spacers dividing the interior into 225 slots of approximately 4 mm width and covered with a clear plastic top to prevent excessive drying). Inhibition of growth was measured as a percent reduction relative to appropriate controls.

WHOLE PLANT ASSAYS

Whole plant assays were carried out in a conventional post-emergent assay regime. Plants of soybean (cv Asgrow 3178), wheat (cv Anza, Cal. Foundation Seed), barnyardgrass (Wildlife Nurseries, Oshkosh, WI), velvetleaf (Azlin Seed Service, Leland, MS), tomato (cv Rutgers, Hummert Seed Co., St. Louis, MO) and *Amaranthus caudatus* (cv Love Lies Bleeding, Stokes Seed, Buffalo, NY) were grown in 13 cm × 18 cm plastic flats using a potting medium based on Metro Mix 200 (Grace Horticultural Products, Cambridge, MA) with supplemental time-released fertilizer. The flats were maintained at 25°C daytime and 22°C nighttime temperatures with a 16 hr photoperiod. Under these conditions, all species evolved to their first true leaf in 10 days, at which time they were treated. Test samples were adjusted to 0.5% Tween 20 and sprayed on two flats using a track sprayer equipped with a Spraying Systems Inc. Model 4002 flat fan nozzle at 14 psi. A total of 20 ml of test sample was used per assay which resulted in a spray rate of 2200 L/Ha. After spraying, the flats were returned to the growth chambers and maintained for 7 days before final observations were made. Phytotoxicity or chlorosis effects were measured.

TRANSLOCATION ASSAY

Phloem transport was determined using castor-bean cotyledons according to the procedure described by Vreugdenhil and Koot-Gronsveld (41). Castor-bean seeds (*Ricinus communis* L. cv Zanzibariensis, Hummert's Seed Co., St. Louis, MO) were grown in 6 in. pots using a potting medium based on Metro Mix 200 (Grace Horticultural Products, Cambridge, MA). Aqueous test solution (1 mL of 100 ppm) was applied to cotyledons with intact endosperm and approximately 30-50 µL of exudate was collected over a period of 4-6 hrs. Exudates were analyzed by TLC, HPLC and/or HPLC-Thermospray/MS.

ANTIMETABOLITE ASSAY

Two bacterial (*Erwinia herbicola* 1711 and *Bacillus subtilis* B558), an algal (*Chlorella pyrenoidosa*) and a plant (*A. thaliana*) indicator organisms were employed in antimetabolite assays using conventional techniques as described by Pruess and Scannell (32). *E. herbicola* 1711 (obtained from Dr. Myron Sasser, Univ. of Delaware) and *B. subtilis* B558 (USDA-NRRL, Peoria, IL) assays were carried out in an AM-medium [0.5% glucose, 1.0% ammonium citrate, 1.0% K_2HPO_4, 0.1% $MgSO_4 \cdot 7H_2O$, trace elements 1 mL/L, 1.5% purified agar (Difco). *C. pyrenoidosa* (Carolina Biological Supply Co., Burlington, NC)] was maintained on Bold's agar plates at 25°C under 24 hr (100 µE) light. Assays were carried out in a 96-well plate. To each well was added 50 µL of a test sample solution, the solvent was allowed to evaporate overnight under a vertical hood, a paper disc was placed at the bottom of each well and 100 µL of algal suspension in Bold's medium was added to each well. Algae in plates were grown at 25°C under 24 hr light for 4 days. Samples which inhibited algal growth were retested with and without reversal agent(s). *A. thaliana* cv Columbia was grown and assayed as described. Reversal studies were carried out by applying paper discs presoaked with the reversal agent of choice on agar prior to seed addition.

RESULTS

SELECTION OF ACTIVE BROTHS

The selection of biologically active fermentation broths from the whole plant screening matrix was dependent upon the degree of phytotoxicity observed and the number of species affected. Broths were also referred for effects such as stunting, chlorosis and bleaching. Referrals from the antimetabolite assay were active against *B. subtilis* B558 or *E. herbicola* 1711 on minimal media and the activity was reversible using a mixture of selected amino acids. Antimetabolite assay referrals were also required to be active against one or more plant species.

The sterile filtrate of an active fermentation was subjected to a preliminary characterization in order to determine whether an amino acid was potentially responsible for the observed activity. This procedure entailed passage through miniature cation and anion exchange columns, extraction with organic solvents and passage through a reversed phase C_{18} cartridge. Concurrently, heat, acid and base stability were assessed. All samples generated in this procedure were bioassayed against an appropriate indicator species and the combined biological and physical data were used to generate a "filtrate fingerprint." Those filtrates having fingerprints characteristic of stable, polar, ionic molecules were advanced to purification as potentially containing a herbicidal amino acid.

STRUCTURES AND BIOLOGICAL ACTIVITY OF NEW METABOLITES; α-METHYLENE-β-ALANINE AND PYRROLYL-3-ALANINE

An unclassified *Streptomyces* sp., A12701, obtained from a Rhode Island soil sample was found to produce a moderately stable, water soluble, herbicidal metabolite. This compound was successfully isolated as shown in Figure 6.1 and was readily determined to be α-methylene-β-alanine, 1, by spectroscopic analysis (Table 6.2). While the isolation of α-methylene-β-alanine represents a first report of this free amino acid, its synthesis has been described previously (11). This synthesis was prompted by the isolation of 1 from the marine sponges *Fasciospongia cavernosa* (16) and *Spongia* cf. *zimocca* (44) as its methyl ester, N-acylated by a series of fatty acids.

The discovery of 1,2,4-triazole-3-alanine, 14, as a histidine antimetabolite (10, 12), and its structural similarity to the herbicide, amitrole, prompted our

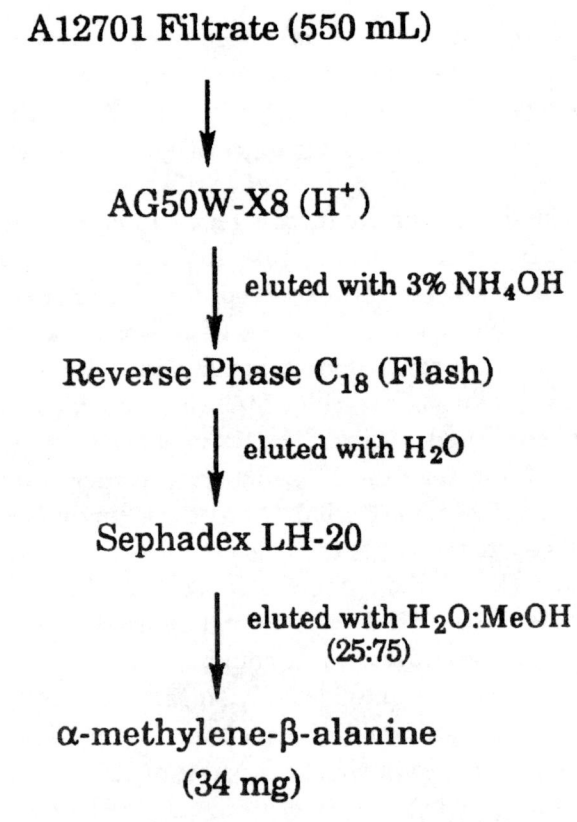

Figure 6.1 / Isolation procedure for α-methylene-β-alanine.

examination of *B. subtilis* B558 histidine antimetabolite produced by an uncharacterized *Penicillium* sp., F5412, obtained from a soil sample collected in Nebraska. The metabolite responsible for this activity was retained by cation and anion exchange (see Figure 6.2) and gave positive colorimetric reactions with ninhydrin and Ehrlich's reagents. The structure of pyrrolyl-3-alanine, 9, was proposed on the basis of spectroscopic examination (see Table 6.2). Owing to the small amount of material available, spectroscopic data alone proved to be insufficient to unambiguously distinguish between the two possible isomers, 9 and 15 (9). Confirmation of the structure as shown in 9

Table 6.2 / Selected Spectral Data for Novel Natural Products

Method	Compound 1	Compound 9	Compound 16
MS: (m/z)			
Obs.	102.0545 (M+H)$^{+1}$	155(M+H)$^{+2}$	167(M+H)$^{+2}$
Calcd.	102.0555 (for $C_4H_8NO_2$)		
1H-NMR: (δ ppm, D_2O)	6.02 (s, 1H)	6.72 (d,1H, J = 2.4 Hz)	6.65 (d, 1H, J = 2.7 Hz)
	5.62 (s, 1H)	6.65 (br s, 1H)	
	3.67 (s, 2H)	5.99 (d, 1H, J = 2.4 Hz)	5.90 (d, 1H, J = 2.7 Hz)
		3.93 (dd, 1H, J = 6.7, 5.1 Hz)	3.70 (dd, 1H, J = 11.1, 5.1 Hz)
		2.94 (m, 2H)	2.69 (dd, 1H, J = 15.3, 5.1 Hz)
			2.65 (dd, 1H, J = 15.3, 11.1 Hz)
			4.14 (d, 1H, J = 14.7 Hz)
			4.04 (d, 1H, J = 14.7 Hz)
13C-NMR: (δ ppm, D_2O)	170.8, 135.4, 124.4, 39.7	176.8, 121.5, 120.0, 117.4, 110.3, 57.9, 30.1	

^{1}FAB-MS; ^{2}Thermospray-MS

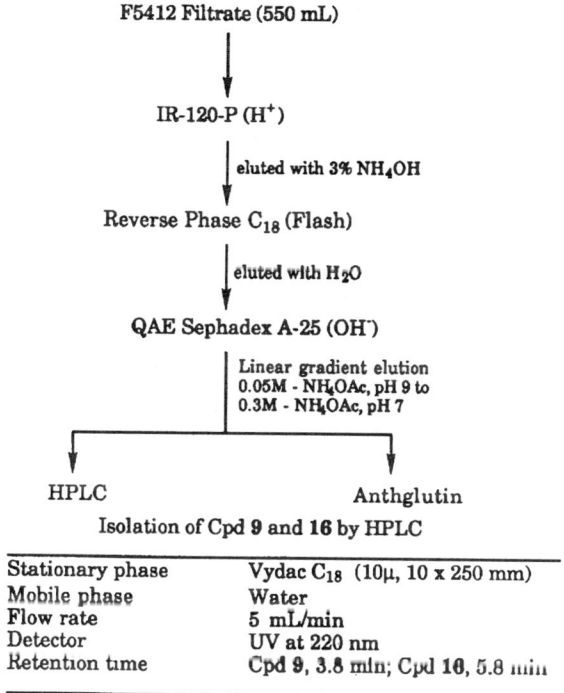

Figure 6.2 / Isolation procedure for herbicidal components from *Penicillium* sp. F5412.

was obtained by total synthesis.[5] The racemic synthetic product and pyrrolyl-3-alanine isolated from *Penicillium* sp. F5412 were identical in all respects other than optical rotation.

α-Methylene-β-alanine, 1, severely inhibited the germination of *A. thaliana* at 100 ppm and showed slight post-emergent activity against velvetleaf and tomato at an application rate of 5 pounds per acre. The antibiotic property of both natural and synthetic pyrrolyl-3-alanine, 9, toward *B. subtilis* B558 on minimal media was reversible by histidine. Compound 9 was only weakly herbicidal, retarding the growth of *L. minor* at 10 ppm and the germination of *A. thaliana* at 250 ppm.

[5]Woodard, S. S., unpublished results. 9: ^{1}H NMR (D_2O) 6.71m, 6.65 bs, 5.98 m, 3.73 dd, 2.95 m. 17: ^{1}H NMR (D_2O) 6.99 (d, 2.4Hz), 6.21 (d, 2.4Hz), 4.26 dd, 3.32 dd, 2.98 dd, 1.90 s, 1.79 s

STRUCTURES AND BIOLOGICAL ACTIVITIES OF KNOWN ANTIBIOTICS NOT PREVIOUSLY REPORTED TO BE HERBICIDAL (see Table 6.3)

In addition to pyrrolyl-3-alanine, *Penicillium* sp. F5412 also produced a known antibiotic, anthglutin, **7** (23), and a novel but inactive metabolite penicyclamine, **16**, (see Figure 6.2). Compound **16** was inactive against whole plants and as a bacterial histidine antimetabolite. Its structure was established by spectral comparison with pyrrolyl-3-alanine (see Table 6.2) and compound **17**, the synthetic acetone adduct of **9** (45). Anthglutin was active against *L. minor* at 100 ppm.

(2R*,3S*)-ß-Methyltryptophan, **3**, known metabolite of *S. flocculus* (8), was isolated from microbial isolate A7847, identified as a *Streptomyces* sp. The structure and relative stereochemistry of **3** were confirmed by synthesis. Compound **3** was inactive against *B. subtilis* B558 and *E. herbicola* 1711. When tested on whole plants it showed very good germination inhibition of *A. thaliana* seeds at 22 ppm and 40% inhibition of sweet clover when applied post-emergence at 5 lb/acre. The mode-of-action of ß-methyltryptophan was thought to involve inhibition of anthranilate synthase. 2,5-Dihydrophenylalanine, **4** (35), isolated from an unclassified *Streptomyces* sp., A5527, was weakly active against *L. minor*, reduced the germination of *A. thaliana* at 75 ppm and showed weak inhibition of the root elongation of velvetleaf and barnyard grass. The activity of compound **4** against *B. subtilis* B558 and *A. thaliana* was reversible by phenylalanine.

Bacillus subtilis, B12416, was shown to produce amicoumacin B, **5** (13). Compound **5** elicited post-emergent chlorosis on velvetleaf, tomato and soybean when applied at 1 lb/acre and severely inhibited the germination of *A. thaliana* at 25 ppm. Amicoumacin B did not translocate in the excised castor bean assay. Cispentacin, **8**, a recently reported antifungal antibiotic (21), was isolated from an unclassified *Streptomyces* sp., A11768, and its structure was confirmed by synthesis. Compound **8** showed very good post-emergent herbicidal activity against amaranth and soybean at 1 lb/acre and at the same rate, moderate activity against barnyard grass, velvetleaf and tomato. While inactive against *B. subtilis* B558 and *E. herbicola* 1711, the germination inhibition observed with **8** against *A. thaliana* at 25 ppm was reversible by valine. Translocation in the castor bean assay was not detectable.

One of the most frequently detected herbicidal antibiotics in the whole plant screen was the recently reported acaricide, altemicidin, **2** (37). The structure of **2**, as isolated from an uncharacterized *Streptomyces* sp., A5941, was confirmed by X-ray analysis. Altemicidin showed excellent translocation in the castor bean assay and the activity against *E. herbicola* 1711 was reversible by methionine. The activity against *A. thaliana* was not reversible. Very good activity was observed against hemp sesbania, cocklebur, morningglory and common chickweed when compound **2** was applied post-emergence at 1 lb/acre. *trans*-4-Hydroxy-L-proline, **6**, a constituent in plant cell wall proteins, was isolated from an unclassified *Streptomyces* sp., A17811. Compound **6** has not previously been reported as a microbial metabolite. It reduced the growth of *A. thaliana* at 1000 ppm and showed moderate inhibition of velvetleaf and barnyard grass root elongation. *trans*-4-Hydroxy-L-proline has been suggested to interfere with the post-translational modification of plant cell wall proteins (19).

DISCUSSION

The results of the microbial screening program described in this paper confirm the value of this approach for the discovery of herbicidal amino acids. In particular, the aquatic *L. minor* and agar-based *A. thaliana* assays, although not post-emergence assays, have proven to be especially suited to the detection of these compounds. To date, antimetabolite screening has not proven to be as

Table 6.3 / Selected spectral data for known amino acids with previously unreported herbicidal activity

Compound	Mass Spectra (m/z)	^{13}C - NMR (δ ppm)	Reference[5]
Cispentacin, 8	Obs.[1] 130.0868 Calcd. 130.0849 (for $C_6H_{12}NO_2$)	178.8, 58.9, 45.6, 27.4, 25.9, 19.1[4]*	21
Amicoumacin B, 5	425 (M + H⁺)[2]	177.9, 175.2, 171.0, 163.2, 141.5, 137.6, 119.6, 116.8, 109.4, 82.7, 73.1, 72.7, 52.7, 50.5, 40.8, 34.4, 30.9, 25.8, 23.8, 22.0[4]**	13
Anthglutin, 7	282 (M + H⁺)[2]	177.3, 176.3, 174.3, 152.0, 137.2, 134.4, 122.1, 116.9, 115.4, 56.8, 32.2, 28.8[4]*	23
3-(2,5-Dihydro-phenyl)-alanine, 4	168 (M + H⁺)[2]	177.4, 132.0, 127.3, 127.1, 126.9, 55.4, 41.6, 30.5, 29.1[4]*	35
ß-Methyltryptophan, 3	Obs.[1] 217.0977 Calcd. 217.0995 (for $C_{12}H_{13}N_2O_2$)	173.7, 138.6, 127.5, 123.7, 122.9, 120.1, 119.5, 115.8, 112.5, 60.0, 32.8, 13.4[4]**	8
Altemicidin, 2	Obs.[1] 377.1128 Calcd. 377.1131 (for $C_{13}H_{21}N_4O_7S$)	181.1, 175.7, 165.8, 148.8, 98.7, 77.6, 70.6, 61.9, 47.0, 44.7, 42.8, 42.4, 33.2[4]*	37
trans-4-Hydroxy-L-proline, 6	132 (M + H⁺)[3]	176.9, 72.8, 62.6, 55.7, 40.1[4]*	19

[1] FAB-MS; [2] Thermospray-MS; [3] Electrospray-MS; [4] NMR Solvent: *D_2O, **CD_3OD; [5] Previously reported isolation.

successful as the whole plant matrix in discerning amino acids having significant whole plant activity.

Two novel amino acids and several known antibiotics not previously reported to be herbicidal have been elucidated. In addition, several proprietary novel amino acids not described herein have been discovered in this screen and are currently undergoing advanced evaluation. The majority of the compounds discovered in this manner are amenable to analogue synthesis programs and several have shown a sufficient level of biological activity to warrant further investigation. Many of these substances have demonstrated excellent translocation properties. Cispentacin, dihydrophenylalanine, pyrrolyl-3-alanine and altemicidin have been implicated in the inhibition of valine, phenylalanine, histidine and methionine biosynthesis, respectively (Figs. 6.3, 6.4).

ACKNOWLEDGMENTS

The authors would like to acknowledge the contributions made to this project by M. Miller-Wideman, G. Dill, K. Crosby, D. Krupa, N. Biest, C. McGary, C. Lopez, S. Woodard, L. Letendre, T. Tran, B. Reich, N. Makkar, H. Fujiwara, T. Solsten, B. Wise and J. Kotyk.

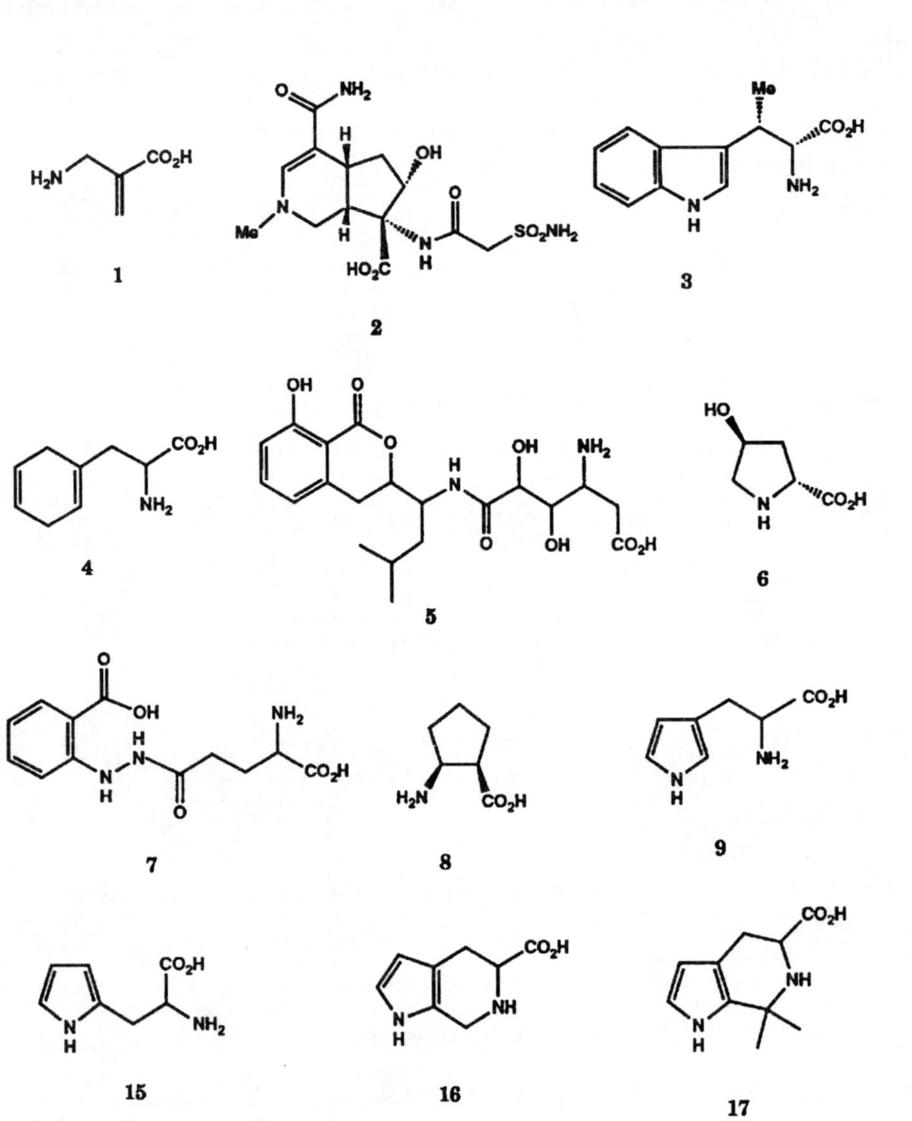

Figure 6.3 / Structures of compounds 1-9 and 15-17

REFERENCES

1. Ayer, S. W., Isaac, B. G., Krupa, D. M., Crosby, K. E., Letendre, L. J. and Stonard, R. J. 1989. Herbicidal compounds from micro-organisms. *Pest. Sci.* 27: 221–223.
2. Bayer, E., Gugel, K. H., Hagenmaier, H., Jessipow, S., Konig, W. A. and Kahner, H. 1972. Stoffwechselprodukte von Mikroorganismen-98. Phosphinothricin und Phosphinothricyl-Alanyl-Alanin. *Helv. Chim. Acta* 55: 224–239.
3. Broadbent, D. and Radley, M. E. 1966. Some effects of 1-amino-2-nitrocyclopentane-1-carboxylic acid on flowering plants. *Ann. Bot.* 30: 763–777.
4. Einhellig, F. A., Leather, G. R. and Hobbs, L. I. 1985. Use of *Lemna minor L.* as a bioassay in allelopathy. *J. Chem. Ecol.* 11: 65–72.
5. Flint, D. H. and Estreicher, H. 1983. Use of 3-aminocyclohexanecarboxylic acids for controlling the growth of unwanted plants. US Patent 4407671.

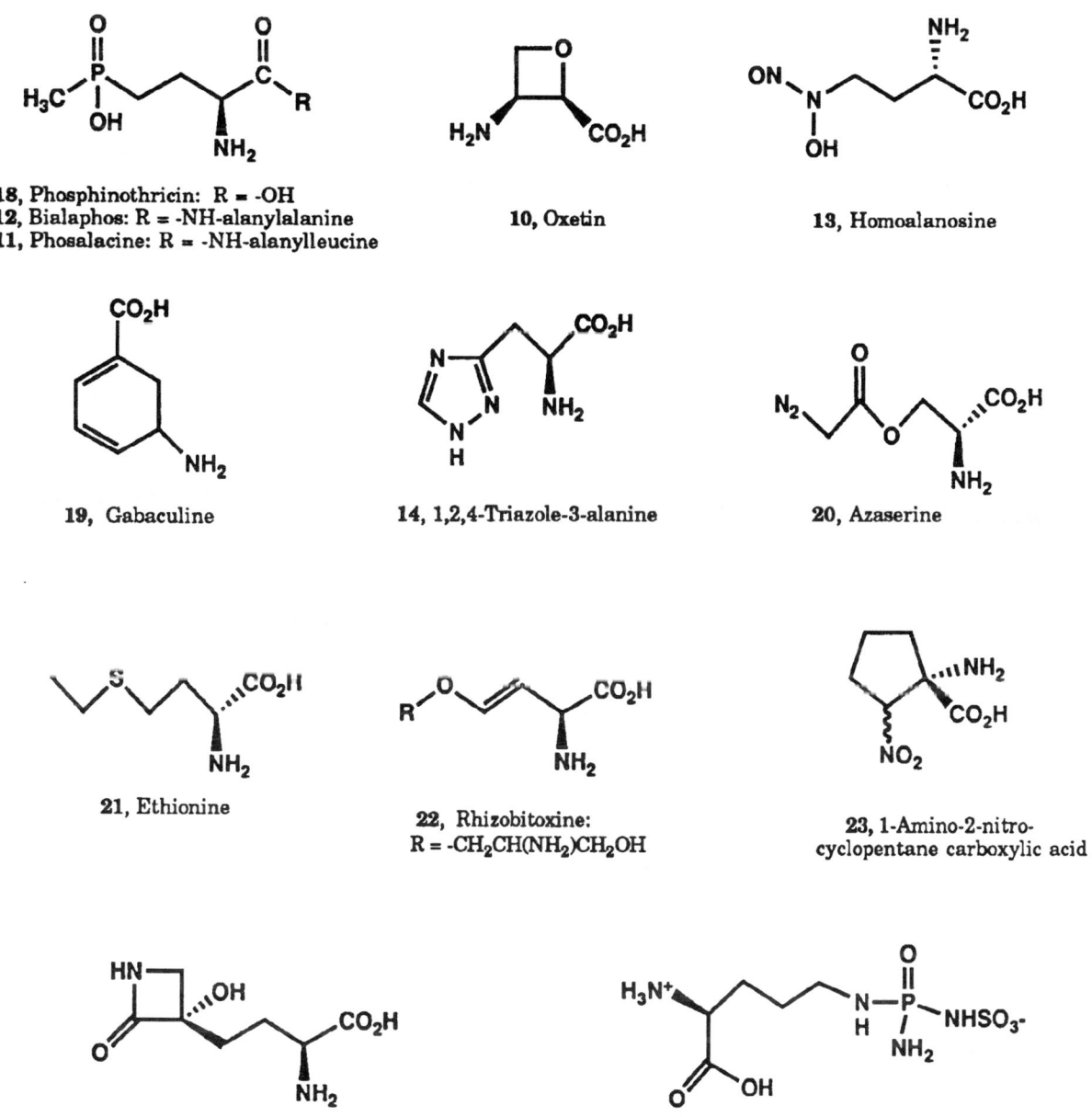

Figure 6.4 / Structures of compounds presented in Table 6.1.

6. Fushimi, S., Nishikawa, S., Mito, N., Ikemoto, M., Sasaki, M. and Seto, M. 1989. Studies on the new herbicidal antibiotic, homoalanosine. *J. Antibiotics* 42: 1370-1378.
7. Giovanelli, J., Owens, L. D. and Mudd, S. H. 1971. Mechanism of inhibition of spinach ß-cystathionase by rhizobitoxine. *Biochim. Biophys. Acta* 227: 671-684.
8. Gould, S. J., Darling, D. S. 1978. Streptonigrin Biosynthesis. 2. The isolation of ß-methyltryptophan and its intermediacy in the streptonigrin pathway. *Tet. Lett.* 35: 3207-3210.
9. Hanek, A. and Kutscher, W. 19664. Synthesis of 3-(2-pyrrolyl)-alanine. *Z. Physiol Chem.* 338: 272-275.
10. Heim, D. R. and Larrinua, I. M. 1989. Primary site of action of amitrole in *Arabidopsis thaliana* involves inhibition of root elongation but not of histidine or pigment biosynthesis. *Plant physiol.* 91: 1226-1231.
11. Holm, A. and Scheuer, P. J. 1980. Synthesis of α-methylene-ß-alanine and one of its naturally occurring α-ketoamides. *Tet. Lett.* 21: 1125-1128.
12. Imamura, N., Masatsune, M., Tianjue, Y., Oiwa, R., Tanaka, H., Omura, S. 1985. Occurrence of 1,2,4-triazole ring in actinomycetes. *J. Antibiotics* 38: 1110-1111.
13. Inouye, S. and Kondo, S. 1989. Amicoumacin and SF-2370, pharmacologically active agents of microbial origin. In: Novel Microbial Agents for Medicine and Agriculture (Demain, A. L. et al. ed.), pp. 179-194, Elsevier.
14. Johnson, H. W., Means, U. M. and Clark, F. E. 1959. Responses of seedlings to extracts of soybean nodules bearing selected strains of *Rhizobium japonicum*. *Nature* (London) 183: 308-309.
15. Kaneko, K. and Tachibana, K. 1988. Herbicides from natural products (Herbiace). *Bio. Ind.* 5: 87-93.
16. Kashman, Y., Fishelson, L. and Ne'eman, I. 1973. N-Acyl-2-methylene-ß-alanine methyl esters from the sponge *Fasciospongia cavernosa*. *Tetrahedron* 29: 3655-3657.
17. Kenfield, D., Bunkers, G., Strobel, G. A. and Sugawara, F. 1988. Potential new herbicides-phytotoxins from plant pathogens. *Weed Technology* 2: 519-524.
18. Kobayashi, K., Myazawa, S., Terahara, A., Mishiwa, H. and Kurihara, H. 1976. Gabaculine: Gamma-aminobutyrate aminotransferase inhibitor of microbial origin. *Tetrahedron Lett.* 1976: 573-540.
19. Lea, P. J. and Norris, R. D. The use of amino acid analogues in studies on plant metabolism. *Phytochemistry*. 1976. 15: 585-595.
20. Leason, M., Cunliffe, D., Parkin, D., Lea, P. J. and Miflin, B. J. 1982. Inhibition of pea leaf glutamine synthetase by methionine sulfoximine, phosphinothricin and other glutamate analogs. *Phytochemistry* 21: 855-857.
21. Masataka, K., Nishio, M., Saitoh, K., Miyaki, T., Oki, T. and Kawaguchi, H. 1989. Cispentacin, a new antifungal antibiotic. *J. Antibiotics* 42: 1749-1755.
22. McLaren, J. S., and Smith, H. 1976. The effect of abscisic acid on growth, photosynthetic rate and carbohydrate metabolism in *Lemna minor*. L. *New Phytologist* 76: 11-20.
23. Minato, S. 1979. Isolation of anthglutin, an inhibitor of γ-glutamyl transpeptidase from *Penicillium oxalicum*. *Arch. Biochem. Biophys.* 192: 235-240.
24. Moore, R. E., Niemczura, W. P., Kwok, O. C. H. and Patil, S. S. 1984. Inhibitors of ornithine carbamoyltransferase from *Pseudomonas syringae* pv. *phaseolicola*. Revised structure of Phaseolotoxin. *Tetrahedron Lett.* 25: 3931-3934.
25. Norman, A. G. 1955. Inhibition of root growth by azaserine. *Science* 121: 213-214.
26. Omura, S., Murata, M., Imanura, N., Iwai, Y., Tanaka, H., Furusaki, A. and Matsumoto, T. 1984. Oxetin, a new antimetabolite from an actinomycete. *J. Antibiotics* 37: 1324-1332.
27. Omura, S., Hinotozawa, K., Imanura, N. and Murata, M. 1984. The structure of phosalacine, a new herbicidal antibiotic containing phosphinothricin. *J. Antibiotics* 37: 939-940.
28. Owens, L. D., Lieberman, M. and Kunishi, A. 1971. Inhibition of ethylene production by rhizobitoxine. *Plant Physiol.* 48: 1-4.
29. Owens, L. D., Thompson, J. F., Pitcher, R. G., and Williams, T. 1972. Structure of Rhizobitoxine, an antimetabolite enol-ether amino-acid from *Rhizobium japonicum*. *J. Chem. Soc.*: 714.

30. Pillmore, J. B. 1989. Amino acid biosynthesis—Aladdin's cave of pesticide targets? *Brit. Crop. Protection Conference Monogr. 4*, 23–30.
31. Pruess, D. L., Scannell, J. P., Kellett, M., Ax, H. A., Janecek, J., Williams, T. H., Stempel, A. and Berger, J. 1974. Antimetabolites produced by microorganisms. X. L-2-amino-4-(2-aminoethoxy)-*trans*-3-butenoic acid. *J. Antibiotics 27:* 229–233.
32. Pruess, D. L. and Scannell, J. P. 1974. Antimetabolites from microorganisms. *Adv. Appl. Microbiol. 17:* 19–62; and references cited therein.
33. Putnam, A. R. 1988. Allelochemicals from plants as herbicides. *Weed Technology 2:* 510–518.
34. Scannell, J. P., Pruess, D. L., Demny, T. C., Sello, L. H., Williams, T. and Stempel, A. 1972. Antimetabolites produced by microorganisms. V. L-2-amino-4-methoxy-*trans*-3-butenoic acid. *J. Antibiotics 25:* 122–127.
35. Scannell, J. P., Pruess, D. L., Demny, T. C., Williams, T. and Stempel, A. 1970. L-3-(2,5-dihydrophenyl)alanine, an antimetabolite of L-phenylalanine produced by a Streptomycete. *J. Antibiotics 23:* 618–619.
36. Steward, W. W. 1971. Isolation and proof of structure of wildfire toxin. *Nature* (London) *229:* 174–178.
37. Takahashi, A., Kurasawa, S., Ikeda, D., Okami, Y. and Takeuchi, T. 1989. Altemicidin, a new acaricidal and antitumor substance. *J. Antibiotics 42:* 1556–1566.
38. Thomas, M. D., Langston-Unkefer, P. J., Uchytil, T. F. and Durbin, R. D. 1983. Inhibition of glutamine synthetase from pea by tabtoxinine-ß-lactam. *Plant Physiol. 71:* 912–915.
39. Turner, J. G. and Mitchell, R. E. 1985. Association between symptom development and inhibition or ornithine carbamoyltransferase in bean leaves treated with phaseolotoxin. *Plant Physiol. 79:* 468–473.
40. Umemura, K. and Oota, Y. 1965. Effects of nucleic acid- and protein-antimetabolites on frond and flower production in *Lemna gibba*. 1965. *Plant & Cell Physiol. 6:* 73–85.
41. Vreugdenhil, D. and Koot-Gronsveld, E. A. M. 1988. Characterization of phloem exudation from castor-bean cotyledons. *Planta 174:* 380–384.
42. Wiater, A., Hulanicka, D. and Klopotowski, T. 1971. Structural requirements for inhibition of yeast imidazoleglycerol phosphate dehyratase by triazole and anion inhibitors. *Acta Biochem. Pol. 18:* 289–97.
43. Woltz, S. S. and Littrell, R. H. 1968. Production of yellow strapleaf of chrysanthemum and similar diseases with an antimetabolite produced by *Aspergillus wentii*. *Phytopathology 58:* 1476–1480.
44. Yunker, M. B. and Scheuer, P. J. 1978. Alpha oxygenated fatty acids occurring as amides of 2-methylene-ß-alanine in a marine sponge. *Tet. Lett.* pp. 4651–4652.

| 7 |

Isolation and Insecticidal Activity of Verruculogens and Tryptoquivalines

John B Pillmoor, Brian D Bush and Duncan A Gates
Schering Agrochemicals Ltd., Chesterford Park Research Station, Saffron Walden, Essex, CB10 1XL, UK

An insecticidally active fermentation broth, identified as part of an in vivo *screening programme, was found to contain the tremorgenic mycotoxins verruculogen, fumitremorgins A and B and tryptoquivaline I as the major bioactive components. A previously unknown N^{16}-deoxy analogue of tryptoquivaline I was also isolated.*

Further biological evaluation of the isolated metabolites showed, however, only a limited spectrum of activity, mainly against the vetch aphid Megoura viciae. *The potential and problems of an* in vivo *screening approach to the discovery of agrochemically active natural products are discussed in the light of this work.*

INTRODUCTION

The potential for using natural products either directly as agrochemicals, or as leads for synthesis programmes is well established leading to increased Company involvement in screening fermentation broths and various other biological extracts for agrochemical activity. A major difference between the agrochemical and pharmaceutical industries is that the former can, if it wishes, employ true *in vivo* tests with the target organisms and does not have to rely exclusively on model biological or biochemical systems for the initial screening. However, in contrast to normal *in vivo* screening of synthetic chemicals, screening of, for example, fermentation broths can be constrained by the low quantities of broth normally available and the potentially low concentrations of active metabolites. These limitations can, nevertheless, be overcome by miniaturisation and high volume rate (2,000 litres/hectare) applications. The discovery of a crop insecticidally active fermentation broth containing a number of tremorgenic mycotoxins is reported here as an example of the successful use of an *in vivo* screening approach.

Tremorgenic mycotoxins are well recognised fungal secondary products that cause sustained tremors in animals (see [3] for review) but, at the time of our involvement with this work, had not previously been reported as possessing insecticidal activity. However, a number of tremorgenic mycotoxins has subsequently been shown to be toxic to *Spodoptera frugiperda* and *Heliothis zea* in artificial diet tests [4]. The present work shows that selected tremorgenic mycotoxins also possess other insecticidal activities.

MATERIALS AND METHODS

SPRAY APPLICATION TO *MEGOURA VICIAE*

Crude fermentation broth or processed extracts were formulated in 33% (v/v) methanol with Tween 20 and Pluronic L61 at final concentrations of 1% (v/v) and 0.1% (v/v) respectively. Application was made by overhead spray at approximately 2000 l/ha (equivalent to run-off) to *Viciae fabae* Minor plants at the two leaf stage. After drying (approximately four hours), plants were infested with *M. viciae* of mixed ages. The plants were then held in constant environment rooms at 22°C under 14 hour/day illumination. Initial symptoms were observed after 24 hours followed by visual assessment of mortality after 48 hours.

PROPHYLACTIC TESTS WITH FURTHER INSECT SPECIES

Purified components were formulated in 33% (v/v) methanol with Citowett at a final concentration of 0.05% (v/v). Application was made by dipping the appropriate plant or plant part in the treatment solution followed by infestation after drying (approximately four hours). The following species were used:

- *Nilaparvata lugens* on one-leaf *Oryzae sativa* plants
- *Nephotettix cincticeps* on one-leaf *Oryzae sativa* plants
- *Aphis fabae* on leaf discs from 2-leaf *Phaseolus vulgaris* plants.
- *Myzus persicae* on leaf discs from 2-leaf *Solanum tuberosum* Minor plants.
- *Megoura viciae* on leaf discs from 2-leaf *V. fabae* Minor plants.

In all cases, visual assessment of mortality was made after 48 hours in controlled environment rooms at 22°C.

XYLEM TRANSPORT TESTS

P. vulgaris and *V. fabae* Minor plants were excised at the fully developed primary leaf stage and placed in small vessels containing 1.3ml of the test solution. The plants were immediately infested with *A. fabae* and *M. viciae* respectively. Mortality was visually assessed after 48 hours in controlled environment rooms at 22°C.

BIOASSAYS

Caenorhabditis elegans was maintained on an *Escherichia coli* culture growing on nutrient agar [composition: NaCl (3g), peptone (Oxoid) (2.5g), agar (17g), 1ml of 1M $MgSO_4$, 1ml of 1M $CaCl_2$, 25ml of 1M KH_2PO_4 (pH6.0), 1ml of 5mg/ml cholesterol in ethanol, 973ml distilled water]. The nematodes were harvested from a 7 day old culture by washing the plate with buffer (0.05M NaCl, 0.025M K_2HPO_4 (pH6.0)). Samples (100µl) were formulated in 10% (v/v) acetone or methanol and pipetted into wells of a 96-well microplate. 100µl of *C. elegans* suspension, containing 10-20 nematodes, were then added. Mortality was assessed visually using a microscope after 24 hours at 22°C. For the *M. viciae* bioassay, 500 µl of sample were formulated in 33% (v/v) methanol and placed in a small vial. A section from 2-leaf *V. fabae* Minor plants comprising of the first leaf pair was prepared and placed in the test solution. The plant section was immediately infested with *M. viciae* of mixed ages and maintained at 22°C. Visual assessment of mortality was made after 24 or 48 hours.

RESULTS

As part of our *in vivo* screening programme a fermentation broth was found to have good activity against *M. viciae*. The first sign of activity was detachment of the aphids from the plant. They subsequently moved about in a jerky and uncoordinated manner and were mostly dead after two days.

Further analysis of the broth in a number of miniaturised bioassay tests showed activity against the nematode *C. elegans* and this bioassay was used initially to support the chemical characterisation work. However, the medium component, proline, was subsequently found to be the major compound causing activity against *C. elegans*. A miniaturised bioassay using the target organism *M. viciae* was therefore used for all the following chemical characterisation work.

The culture filtrate was initially subjected to extraction. Following the isolation scheme outlined in Figure 7.1, a single biologically active component (present at approximately 10 mg/l) was isolated. High resolution electron impact mass spectroscopy (MS) indicated a molecular formula of $C_{27}H_{33}N_3O_7$. A literature search (Chemical Abstracts) and spectroscopic analysis [ultra-violet (UV) and nuclear magnetic resonance (NMR)] revealed that all data were identical in every aspect to those described in the literature for verruculogen [5] (Figure 7.2). Retesting of the purified verruculogen in the full *in vivo* screens confirmed

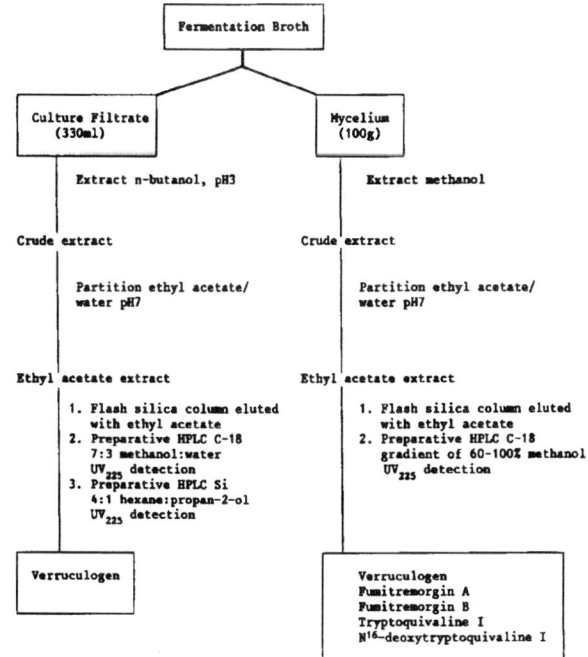

Figure 7.1 / Extraction scheme.

Table 7.1 / Activity of the Tremorgenic Mycotoxins against *Megoura viciae*

Metabolite	% Mortality at Application Rates (mg/l) of:			
	100	10	1	0.1
Verruculogen	100	100	70	20
Fumitremorgin A	100	100	100	80
Fumitremorgin B	100	50	0	0
Tryptoquivaline I	100	80	20	-
Standard (Permethrin)	-	100	80	0

Spray application was made to bean plants four hours prior to infestation with *M. viciae*. Biological activity was visually assessed after 48 hours.

good activity against *M. viciae* with activity down to 1 mg/l (Table 7.1).

In order to obtain more material, the mycelium from the fermentation broth was subsequently extracted and fractionated following the

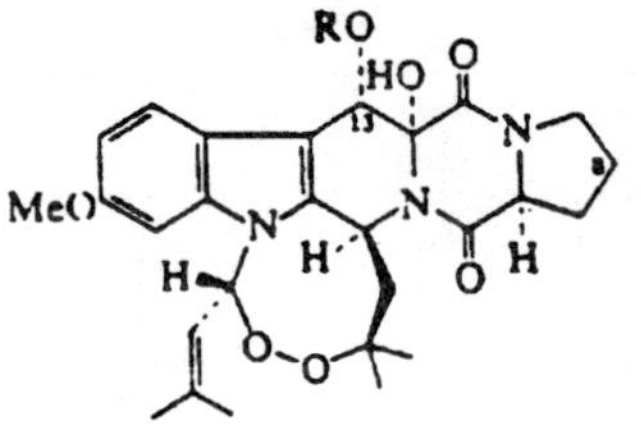

Verruculogen   R = H
Fumitremorgin A   R = CH$_2$CHC(CH$_3$)$_2$

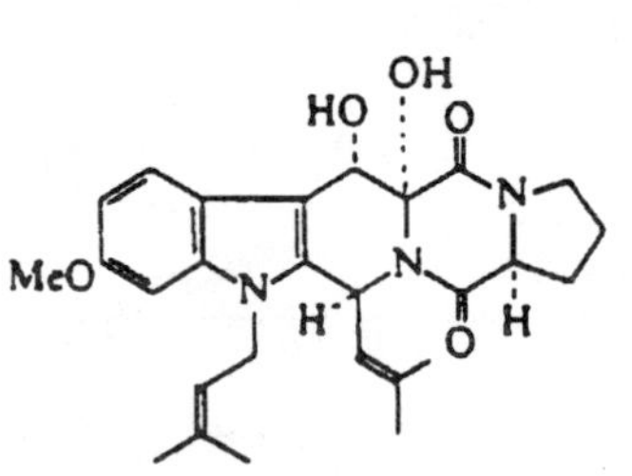

Fumitremorgin B

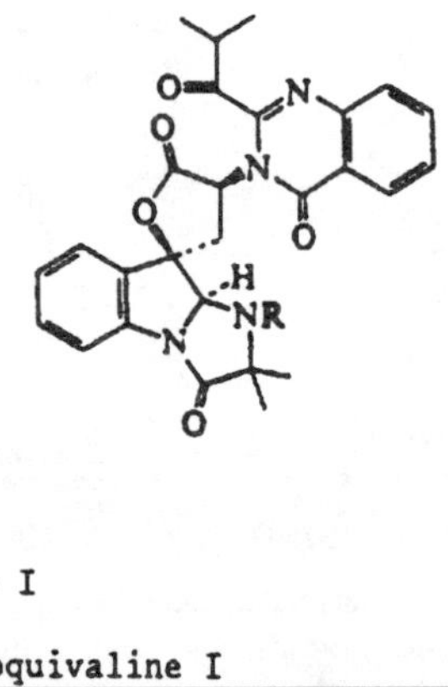

Tryptoquivaline I   R = OH
N^{16}-deoxy tryptoquivaline I   R = H

Figure 7.2 / Structures of the metabolites isolated.

procedure in Figure 7.1. Along with verruculogen a number of other biologically active components was also found to be present in the final HPLC eluate. From this eluate four components were purified and characterised. Three were established as known compounds whose MS, UV and NMR data were identical to those reported in the literature: fumitremorgin A [8], fumitremorgin B [10], and tryptoquivaline I [9] (Figure 7.2). The fourth compound had very similar spectroscopic properties to tryptoquivaline I, but with a molecular weight of 486, 16 mass units less than tryptoquivaline I. In addition, this fourth compound did not show any colour with tetrazolium blue, a reagent usually indicating the presence of hydroxylamine functional groups, in contrast to the strong colouration formed with tryptoquivaline I. These data are thus consistent with the previously unreported N^{16}-deoxy analogue of tryptoquivaline I shown in Figure 7.2. None of the other tryptoquivaline-related minor components was investigated further.

The purified fumitremorgins A and B and tryptoquivaline I were retested against *M. viciae* in the prophylactic spray application test (Table 7.1). Fumitremorgin A was found to be the most active of the compounds tested with good control evident at an application rate of 0.1mg/l. No further testing of the N^{16}-deoxy tryptoquivaline I was undertaken due to limited supplies but the other four purified compounds were then evaluated further against an extended range of insects in a prophylactic dip test (Table 7.2). Although activity of verruculogen, fumitremorgin A and tryptoquivaline I against *M. viciae* was retained in this test, the sensitivity was some 10 times lower than in the previous spray application test (Table 7.1). The lower sensitivity of the prophylactic dip test probably accounts for the apparent lack of activity of fumitremorgin B in this test. Unfortunately no significant activity was evident against any of the other species tested. Three of the compounds were

Table 7.2 / Activity of Verruculogen, Fumitremorgins A and B and Tryptoquivaline I in the further insect tests

Species (mg/l)	Rate	% Mortality				
		Verruculogen	Fumitremorgin A	Fumitremorgin B	Tryptoquivaline I	Standard
Megoura viciae	64	90	95	0	82	-
	25	39	74	0	50	-
	10	14	20	0	0	100[a]
Aphis fabae	64	23	29	0	0	-
	25	0	0	0	0	-
	10	0	0	0	0	100[a]
Myzus persicae	64	0	0	0	0	-
	10	-	-	-	-	100[a]
Nilaparvata lugens	64	0	0	0	0	100[b]
Nephotettix cincticeps	64	0	0	0	-	100[c]

Application was made by dipping plants in the treatment solution four hours prior to infestation. Mortality was visually assessed after 48 hours, apart from *N. cincticeps* which was assessed after 24 hours.

Standards: a = pirimicarb; b = ethofenprox; c = pyrethroid

shown to be active via transport in the xylem following uptake via cut stems (Table 7.3). In this test the activity against *Aphis fabae* was similar to that against *M. viciae*. Tryptoquivaline I was the most active compound in this test.

DISCUSSION

THE INSECTICIDAL ACTIVITY OF THE TREMORGENIC MYCOTOXINS

The present work has demonstrated that a number of tremorgenic mycotoxins has good activity against *M. viciae* in an *in vivo* test system (Table 7.1). All the compounds caused very distinctive, tremorgenic effects only a few hours after application. Fumitremorgin A was the most active of the compounds tested. A novel analogue, N^{16}-deoxy tryptoquivaline I, was also isolated and characterised as part of the study. Extended evaluations against a range of insect species, however, showed little activity against other, commercially more important insects (Table 7.2). Although one of the compounds investigated here (verruculogen) has previously been reported to be active against *S. frugiperda* and *H. zea* larvae in an artificial diet test [4], the present results indicate that the compounds tested have little potential as commercial products themselves. However, it was interesting to note that systemic activity against *A. fabae* was evident following uptake via cut

Table 7.3 / Systemic activity of Verruculogen, Fumitremorgin A and Tryptoquivaline I against *Megoura viciae* and *Aphis fabae*

Species	% Mortality			
	Verruculogen	Fumitremorgin A	Tryptoquivaline I	Standard (Pirimicarb)
Megoura viciae	58	30	96	100
Aphis fabae	31	20	94	100

Cuttings of the host plant were placed in 1. 3ml of a 160 mg/l treatment solution and infested immediately. Mortality was assessed visually after 48 hours.

stems (Table 7.3).

Despite the poor activity of the compounds themselves, the possibility of undertaking a synthesis programme based on these compounds as lead structures was considered. This had some attraction as the compounds are reported to exert their action through binding to the chloride channel of the GABA receptor in rat brain [6]. Similar effects might be expected on the same complex in invertebrates which is an attractive biochemical target for new insecticides [7] and may be the target of another group of natural products, the avermectins [1]. However, as any synthesis programme was perceived to be both costly and technically difficult with the additional problem of having to eliminate the well-reported toxicity of these compounds to mammals [2], this was not progressed.

ADVANTAGES AND DISADVANTAGES OF *IN VIVO* SCREENING

The present work has highlighted a number of points with respect to the discovery of agriculturally active natural products. Foremost, it has demonstrated that an *in vivo* screening approach can be used successfully to discover new, biologically active compounds in crude fermentation broths. However, like all natural product work, the true commercial value of the discovery cannot be ascertained until the structure of the active component has been elucidated and the effective application rate determined. Wider screening of either the crude fermentation broth or, preferably, a semi-purified fraction may be appropriate before committing too much effort on the chemical characterisation. However, the danger of eliminating potentially interesting samples that show only a limited spectrum of activity because the active metabolite is present at low concentrations is very real. The problem of re-discovering metabolites that have already been reported in the literature is also a recurring one. We have found that only the actual experience of handling a metabolite in-house can provide the necessary information and confidence to allow all subsequent samples containing the metabolite to be rapidly eliminated. A fine balance needs to be drawn between discarding quickly a broth found to contain a previously characterised metabolite and doing enough work to be reasonably confident that the broth does not also contain a novel analogue, or a totally different and novel compound, whose presence is masked by the known metabolite. Rapid elimination of samples containing tremorgenic mycotoxins is, however, relatively easy due to their characteristic symptomology.

The present work also highlights the dangers of relying on bioassays involving a 'non-target' indicator organism as the sole guide for the purification. In many cases it is preferable to use the true target organism in the bioassay, or at least to check that the correct active component is being purified at regular intervals during the chemical characterisation work.

Finally, it is worth mentioning the problems of obtaining sufficient material to allow full evaluation of the potential of an agrochemically active natural product that is not accessible via chemical synthesis. Obtaining the gram quantities required for growth room and initial glasshouse evaluations is generally not too great a problem. However, decisions on whether a compound is worthy of development and, hence, significant financial expenditure in the agrochemical area cannot be made until at least some field data are obtained. This might require kilogramme amounts of compound. In many cases it would not be feasible to obtain up to a kilogramme of pure compound using the initial fermentation and downstream processing protocols. It might then be difficult to justify the large expenditure required to improve the fermentation yield sufficiently to obtain the quantities of compound required unless the compound is particularly novel or the glasshouse results are particularly promising.

CONCLUSION

Our experience indicates that *in vivo* screening of natural products is a viable means of identifying new, agrochemically active compounds. Although the tremorgenic mycotoxins reported in the present study did not appear to have sufficient activity to warrant further evaluation, increasing commercial use of natural products, or synthetic derivatives thereof, will undoubtedly be part of crop protection in the future.

ACKNOWLEDGEMENTS

We would like to thank all our colleagues who have helped with this project and, in particular, Clive Pearce for the NMR work, David Martin for the mass spectrometry work and the Insecticide Biology Team in Berlin for undertaking the extended insect tests.

REFERENCES

1. Babu, J. R. 1988. Avermectins: Biological and pesticidal activities. *ACS Symp. Ser. 380* (Biol. Act. Nat. Prod.: Potential Use Agric.) (Cultler H. G., Ed): 91–108.
2. Cole, R. J. and Cox, R. H. 1981. *Handbook of Toxic Fungal Metabolites*, Chapter 8, pp. 355–509, Academic Press, New York.
3. Cysewski, S. J. 1988. Chemistry of the tremorgenic mycotoxins. In: *Mycotoxic Fungi, Mycotoxins, Mycotoxicoses*. Volume I (Wyllie T. D. and Morehouse L. G., Eds), pp. 357–364. Dekker, New York.
4. Dowd, P. F., Cole, R. J. and Vesonder, R. F. 1988. Toxicity of selected tremorgenic mycotoxins and related compounds to *Spodoptera frugiperda* and *Heliothis zea. J. Antibiot.* 41:1868–1872.
5. Fayos, J., Lokensgard, D., Clawdy, J., Cole, R. J. and Kirksey, J. W. 1974. Structure of verruculogen, a tremor producing peroxide from *Penicillium verruculosum. J. Amer. Chem. Soc.* 96:6785–6787.
6. Gant, D. B., Cole, R. J., Valdes, J. J., Eldefrawi, M. E. and A. T. Eldefrawi. 1987. Action of tremorgenic mycotoxins on $GABA_A$ receptors. *Life Sci.* 41:2207–2214.
7. Lunt, G. G., Brown, K. R., Riley, K. and Rutherford, D. M. 1988. The biochemical characterisation of insect GABA receptors. In: *Neurotox '88. Molecular basis of drug design and pesticide action* (Lunt, G. G., Ed), pp. 185–192. Excerpta Medica, Amsterdam

8. Yamazaki, M., Fujimoto H. and Kawasaki, T. 1980. Chemistry of tremorgenic metabolites. I. Fumitremorgin A from *Aspergillus fumigatus*. *Chem. Pharm. Bull.* 28: 245–254.
9. Yamazaki, M., Fujimoto H. and Okuyama E., 1978. Structure determination of six fungal metabolites, tryptoquivaline E, F, G, H, I and J from *Aspergillus fumigatus*. *Chem. Pharm. Bull.* 26: 111–117.
10. Yamazaki, M., Sasago, K. and Miyaki, K., 1974. Structure of fumitremorgin B (FTB), a tremorgenic toxin from *Aspergillus fumigatus*. *J. Chem. Soc. Chem. Commun.* 10: 409–409.

Screening of Herbicides Based on Antimicrobial Activity
Cellulose Synthesis Inhibitors

Yoshitake Tanaka, Haruo Tanaka and Satoshi Ōmura

Research Center for Biological Function, The Kitasato Institute, and School of Pharmaceutical Sciences, Kitasato Unviersity, Minato-ku, Tokyo 108, Japan

A new screening method for herbicides based on the selective antimicrobial activity against Phytophthora parasitica *which contains cellulose in the cell wall was established. Among about 20,000 soil microorganisms, the two strains WK-1875 and OM-5714 were found by the screening method to produce cellulose synthesis inhibitors. Two new compounds, phthoramycin (1) and phthoxazolin (2), were isolated from* Streptomyces *spp. WK-1875 and OM-5714, respectively. They exhibited inhibition against radish seedlings and incorporation of [^{14}C] glucose into extracellular cellulose fraction in* Acetobacter xylinum. *Phthoxazolin is the first specific inhibitor of cellulose synthesis from microorganisms while phthoramycin is not as specific.*

Key Words: phthoramycin, phthoxazolin, cellulose synthesis, inhibitor, herbicide, *Streptomyces*

INTRODUCTION

We have been responsible for screening of enzyme inhibitors based on antimicrobial activity [17], and discovered during this work the glutamine synthetase inhibitors phosalacine [7, 8] and oxetin [9], the antifolates diazaquinomycins A and B, HMPGG and AM-8402 [10], and the acyl-CoA synthetase inhibitors triacsins A and B [11].

From our experience, we learned that the screening system based on microbial growth inhibition can be useful only when the following qualifications are satisfied: (1) A similar enzyme is present in the test microorganisms. (2) The microbial enzyme is essential for the growth, or can be made essential by selecting culture media and conditions or by mutations. If these qualifications are satisfied, the screening method is more effective than using directly an enzyme system. Many false inhibitors which would otherwise be picked up by the screening with an enzyme system can be eliminated and we can now obtain more easily those useful compounds that are active *in vivo.* As an example, the glutamine synthetase inhibitors which were obtained by our screening method exhibited herbicidal activity.

Recently, we established another new screening method for herbicides (cellulose synthesis inhibitors) based on antimicrobial activity and discovered two new compounds, phthoramycin [6, 12] and phthoxazolin [13], with herbicidal activity. In this paper, we describe the rationale and method of the screening, and the isolation, structures and the biological activities of two new compounds from soil actinomycetes.

SCREENING

Screening for inhibitors of cellulose synthesis was based on the selective antimicrobial activity of actinomycete cultures against *Phytophthora parasitica,* a phytopathogenic fungus known to contain cellulose as one of the cell wall constituents [2], but no activity against common fungi such as *Candida albicans,* which do not contain cellulose in the cells.

From about 20,000 soil isolates, strains WK-1875 (Fig. 8.1) and OM-5714 (Fig. 8.2), for which the genus *Streptomyces* was assigned by taxonomical studies, were chosen as the candidate producers. These strains were thought to produce cellulose synthesis inhibitors because the culture supernatant exhibited selective antifungal activity against *P. parasitica,* as well as herbicidal activity against radish seedlings, and inhibited the incorporation of [^{14}C]glucose into the alkali-insoluble fraction of resting cells of *Acetobacter xylinum,*

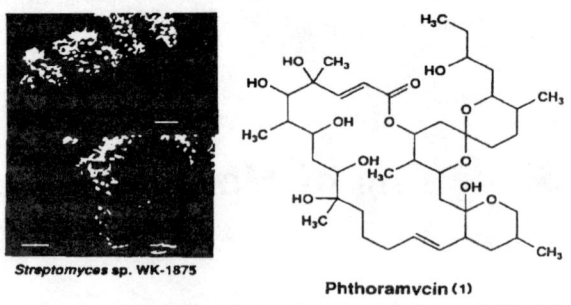

Figure 8.1 / Scanning electron micrograph of the phthoramycin producer *Streptomyces* sp. WK-1875 and the structure of phthoramycin.

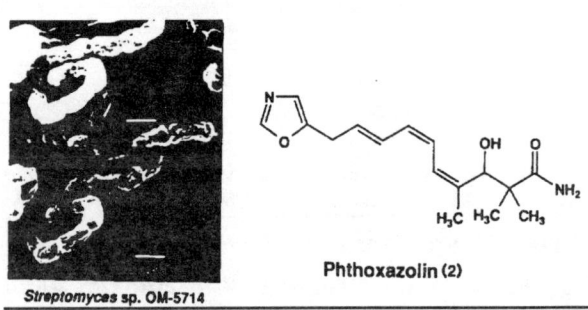

Figure 8.2 / Scanning electron micrograph of the phthoxazolin producer *Streptomyces* sp. OM-5714 and the structure of phthoxazolin.

an acetic acid bacterium known to produce extracellular cellulose from glucose [16].

ISOLATION AND STRUCTURE OF PHTHORAMYCIN

The cultured broth of *Streptomyces* sp. WK-1875 was obtained by incubation in a medium (20 liters) consisting of glycerol 2.0%, soybean powder 2.0%, NaCl 0.3%, pH 6.8, at 27°C for 3 days using a 30-liter jar fermentor. The culture supernatant (15 liters) was extracted with ethyl acetate. After evaporation of the organic layer, the oily residue (45 g) was chromatographed on a silica gel column, which was developed stepwise with chloroform-methanol (50:1-5:1). The crude powder obtained by evaporation of the active fractions was subjected to a second column chromatography on silica gel developed with benzene-acetone (10:1-2:1). The active fractions were collected and concentrated to give a white powder of pure phthoramycin (320 mg).

Phthoramycin (mp 116-119°C, $[\alpha]^{18}$-12.1° (c 0.5, methanol)) is soluble in methanol, ethyl acetate, and chloroform, but is insoluble in water and n-hexane. The molecular formula $C_{40}H_{68}O_{12}$ (MW 740.97) was established from the results of the high-resolution EI-MS of the antibiotic and FAB-MS and ^{13}C-NMR spectroscopy of the pentaacetate.

The ^{13}C-NMR spectral data of phthoramycin and its pentaacetate suggested that the antibiotic possesses a polyketide skeleton. The incorporation pattern of ^{13}C-precursors (acetate, propionate and isocaproate) into the phthoramycin molecule, in addition to extensive NMR studies, allowed deduction of the complete structure (Fig. 8.1) possessing additonal C_4- and C_{11}-alkyl chains at C-16 and C-21, respectively, in a 22-membered macrolide skeleton. Phthoramycin corresponds structurally to the demycarosyl-8-deoxy derivative of cytovaricin which has been reported by Isono et al. [15].

BIOLOGICAL ACTIVITIES OF PHTHORAMYCIN

Phthoramycin inhibits the growth of filamentous fungi: *Mucor racemosus*, and plant pathogens such as *Piricularia oryzae* and *Phytophthora parasitica* (Table 8.1). It is virtually inactive against the yeasts and bacteria tested. The antibiotic also inhibits the growth of radish seedlings (*Rhaphanus sativus* L., Table 8.2) and cellulose synthesis by resting cells of *Acetobacter xylinum*. The acute toxicity (LD_{50}) of phthoramycin in mice was 30 mg/kg (ip) and 100 mg/kg (po).

Although the antibiotic exhibited herbicidal and cellulose synthesis-inhibiting activities, it is not considered to be a specific inhibitor of cellulose synthesis because it is also active against *Piricularia oryzae* and *Mucor racemosus* which do not contain cellulose in the cell wall.

Table 8.1 / Antifungal spectrum of phthoramycin

Test organism	MIC (µg/ml)
Candida albicans KF-1	>100
Saccharomyces sake KF-26	>100
Aspergillus niger KF-103	>100
Pyricularia oryzae KF-180	6.25
Mukor racemosus IFO 4582	3.12
Phythophthora parasitica IFO 4783	1.56

Table 8.2 / Inhibition of the growth of radish (*Raphanus sativa* L.) seedlings by phthoramycin.

Compound	Amount (µg/tube)	Inhibition (%) Germination	Growth
Phthoramycin	1	0	15
	10	0	100
	100	20	100
Coumarin	1	0	25
	10	0	24
	100	0	70

Five radish seeds were incubated at 27°C for 3 days under light on inhibitor-containing wet cotton in a test tube. Percent decrease in plant height against no drug control (3–5 cm, average 3.8 cm) is shown.

ISOLATION AND STRUCTURE OF PHTHOXAZOLIN

Fermentation for phthoxazolin production by *Streptomyces* sp. OM-5714 was carried out at 27°C for 4 days with agitation and aeration in a 50-liter jar fermentor containing 30 liters of production medium (soluble starch 2.0%, glycerol 0.5%, wheat germ 1.0%, meat extract 0.3%, dried yeast cells 0.3%,

CaCO$_3$ 0.4%, and allophane 0.5%, presterile pH of 7.2). The addition of allophane, a non-crystalline aluminosilicic acid clay with phosphate-trapping activity [4], to the medium was a necessary requirement for efficient production of phthoxazolin.

The antibiotic accumulated both extra- and intracellularly, was extracted with ethyl acetate (18 liters) from the cultured broth (28 liters). After evaporation of the ethyl acetate layer, the residual oily material was applied on a silica gel column, and developed with chloroform-methanol (100:1-10:1). Active fractions were concentrated and chromatographed again on a silica gel column using mixtures of benzene-acetone (100:1-1:1) as developing solvents. The crude powder obtained from the active fractions was purified by HPLC (column: YMC D-ODS-5, Yamamura Chemicals, Kyoto, eluant: 30% acetonitrile) to give a pale yellow powder of pure phthoxazolin.

The physicochemical properties of phthoxazolin are as follows: $C_{16}H_{22}N_2O_3$, MW m/z 290.163 (Calcd. 290.163), mp 58-62°C, $[\alpha]_D^{18}$ +37.4° (c 1, CH_2Cl_2), UV λ_{max}^{MeOH} (ϵ): 275 (17.0 × 10^3). It is readily soluble in methanol, acetone, ethyl acetate and chloroform, but sparingly soluble in water and n-hexane. The ^{13}C and ^{1}H NMR, and ^{13}C-^{1}H and ^{1}H-^{1}H COSY spectroscopies besides the above physicochemical properties revealed the structure 2 (Fig. 8.2). The name "phthoxazolin" was given after its oxazole moiety and the selective activity against *Phytophthora* spp. Oxazolomycin, an antibacterial compound reported by Mori et al. [5] also, possesses the phthoxazolin moiety in the molecule.

Phthoxazolin showed selective antimicrobial activity against cellulose-containing *P. parasitica* and *P. cactorum* as shown in Table 8.3. No growth inhibition was observed at 200 µg/ml of phthoxazolin against 22 non-cellulose-containing strains of Gram-positive and Gram-negative bacteria such as *Bacillus, Staphylococcus, Escherichia, Pseudomonas, Acetobacter* spp. and yeasts and filamentous fungi including *Pyricularia, Botrytis, Microsporum* and *Trichophyton* spp. Phthoxazolin showed herb-

Table 8.3 / Antimicrobial spectrum of phthoxazolin

Test organism	Medium	MIC (µg/ml)
Staphylococcus aureus FDA 209P	A	>200
Micrococcus luteus ATCC 9341	A	>200
Bacillus subtilis PCI 219	A	>200
Escherichia coli NIHJ	A	>200
Xanthomonas oryzae KB 88	A	>200
Pseudomonas aeruginosa P 3	A	>200
Acetobacter aceti subsp. *xylinum* IFO 3288	B	>200
Phytophthora parasitica IFO 4873	C	125
Phytophthora capsici KF 278	C	31.3
Aspergillus niger ATCC 6275	D	>200
Piricularia oryzae KF 180	D	>200
Mucor racemosus IFO 4581	D	>200
Candida albicans KF 1	D	>200
Saccharomyces sake KF 26	D	>200
Trichophyton interdigitale KF-62	D	>200
Microsporum gypseum KF 65	D	>200
Penicillium chrysogenum KF 97	D	>200
Fusarium oxysporum KF 166	D	>200
Botrytis cinerea KF 184	D	>200

A: Sensitivity disc agar (Nissui), 37°C, 24 hours.
B: Composition (%): glucose 2.5, yeast extract 0.5, peptone 0.3, agar 1.5 (pH 4), 27°C, 96 hours.
C: V8 glucose agar (pH 7), 27°C, 72 hours.
D: Potato glucose agar (pH 6), 27°C, 72 hours.

icidal activity. It inhibited the growth of radish seedlings grown in laboratory test tubes under light with MIC of 25 µg/tube, and exhibited herbicidal activity against velvetleaf in a pot test as shown in Fig. 8.3. The toxicity of phthoxazolin seems to be low. It showed no or marginal effect at 100 µg/ml on the growth of B-16 melanoma cells and Vero cells incubated at 37°C in EAGLE'S MEM medium supplemented with 10% fetal bovine serum. No mice died after oral administration of phthoxazolin at 100 mg/kg.

The mode of action of phthoxazolin was studied. A microscopic observation revealed that

Table 8.4 / Inhibitory activities of phthoxazolin and coumarin against *Phytophthor parasitica*, radish seedlings and cellulose biosynthesis

Compound	Amount µg/ml (mM)[a]	Anti-*Phytophthora* activity (inhibition zone, mm)[b]	Growth inhibition of radish seedlings (%)[c]	Inhibition (%) of cellulose biosynthesis in *Acetobacter xylinum*	
				Resting cell system[d]	Cell-free system[e]
Phthoxazolin	100 (0.34)	31	100	47	69
	10 (0.03)	11	30	30	27
Coumarin	100 (0.68)	None	20	14	57
	10 (0.07)	None	0	>5	43

[a] To be read as µg per tube for growth inhibition of radish seedlings.
[b] *P. parasitica* was grown on glucose (1%)-supplementec V8 agar [17] at 27°C for 2 days. Paper discs of 8 mm i.d. were used.
[c] See Table 8.2.
[d] *A. xylinum* cells (1 mg/ml in dry weight) were incubated with [^{14}C]glucose (0.2 µCi/ml) and the indicated inhibitors in 1 ml in total of 50 mM phosphate buffer (pH 6.0) at 27°C for 1 hour. The cells were boiled in alkaline conditions as described [16]. Percent inhibition of formation of radioactive alkali-insoluble materials is shown.
[e] The cell free preparation (0.01 mg protein/ml) from *A. xylinum* was incubated at 27°C with UDP-[^{14}C]glucose (1 µCi/ml), the indicated inhibitors and GTP (0.25 µmol/ml) in 0.2 ml in total of 50 mM Tris-HCl (pH 8.5), as described [1]. Percent inhibition of formation of radioactive alkali insoluble materials is shown.

Figure 8.3 / Herbicidal effect of phthoxazolin against velvetleaf (*Abutilon theophrasti*). Several seeds were sown on a pot and grown at room temperature under light-dark control. A phthoxazolin solution was sprayed at 25 g/ over the plants one week after germination. The picture was taken two weeks after the treatment.

phthoxazolin induced round-shaped morphology of *P. parasitica* at the mycelial tips, which resembled the spheroplast-like morphology of *P. oryzae* induced by the chitin synthesis inhibitors polyoxin and nikkomycin. Phthoxazolin (100 µg/ml) inhibited the incorporation of [^{14}C]glucose into the cellulose fraction of resting cells of *A. xylinum* (Table 8.4). It also inhibited the GTP- and polyethylene glycol-stimulated cellulose synthesis from UDP-[^{14}C]glucose by a cell-free extract from *A. xylinum*, which was prepared according to the method of Benziman *et al.* [1, 14] (Table 8.4). A known cellulose synthesis inhibitor, coumarin, inhibited cell-free synthesis of cellulose to the same degree, but did so to a lesser extent in resting cell system. Because the cellulose-synthesizing system of *A. xylinum* resembles that of plants [16], the herbicidal effect of phthoxazolin is assumed to be due to inhibition of cellulose synthesis. All of these findings, together with the highly selective antimicrobial activity, suggest that phthoxazolin is a specific inhibitor of cellulose synthesis in bacterial, fungal and plant systems.

DISCUSSION

Many chemically synthesized herbicides have been practically used. However, some of them impair human health and environmental problems result.

Because cellulose is ubiquitous among plant and algae cell walls, but is not contained in human and animal cells, cellulose biosynthesis would provide a promising target site for safe and non-selective herbicides. However, inhibitors of cellulose biosynthesis of microbial origin have not been documented.

As described, we established a new method based on a selective antimicrobial activity against *Phytophthora parasitica* and discovered two new herbicides, phthoramycin and phthoxazolin using the screening method. Thus, our newly established method has proven useful for screening of cellulose synthesis inhibitiors with herbicidal activity.

From this data, phthoxazolin was shown to be a specific inhibitor of cellulose biosynthesis although phthoramycin is not a specific one. Phthoxazolin is expected to be useful as a new type of herbicide, plant growth regulator, and as a biochemical reagent for the studies on cellulose biosynthesis.

REFERENCES

1. Aloni, Y., Delmer, D. P., and Benziman, M. 1982. Achievement of high rates *in vitro* synthesis of 1,4-ß-glucan: Activation by cooperative interaction of the *Acetobacer xylinum* enzyme system with GTP, polyethylene glycol, and a protein factor. *Proc. Natl. Acad. Sci.* U.S.A. 79: 6448–6452.

2. Bartnicki-Garcia, S. 1968. Cell wall chemistry, morphogenesis and taxonomy of fungi. *Annu. Rev. Microbiol.* 23: 87–108.

3. Ko, W. H. 1978. Heterothallic *Phytophthora*: Evidence for hormonal regulation of sexual reproduction. *J. Gen. Microbiol.* 107: 15–18.

4. Masuma, R., Tanaka, Y., Tanaka, H. and Ōmura, S. 1986. Production of nanaomycin and other antibiotics by phosphate-depressed fermentation using phosphate-trapping agents. *J. Antibiot.* 39: 1557-1564.

5. Mori, T., Takahashi, K., Kashiwabara, M., and D. Uemura. 1985. Structure of oxazolomycin, a novel ß-lactone antibiotic. *Tetrahedron Lett.* 26: 1073–1076.

6. Nakagawa, A., Miura, S., Imai, H., Imamura, N., and Ōmura, S. 1989. Structure and biosynthesis of a new antifungal antibiotic, phthoramycin. *J. Antibiot.* 42: 1324–1327.

7. Ōmura, S., Murata, M., Hanaki, H., Hinotozawa, K., Ōiwa, R., and Tanaka, H. 1984. Phosalacine, a new herbicidal antibiotic containing phosphinothricin. Fermentation, isolation, biological activity and mechanism of action. *J. Antibiot.* 37: 829–835.

8. Ōmura, S., Hinotozawa, K., Imamura, N., and Murata, M. 1984. The structure of phosalacine, a new herbicidal antibiotic containing phosphinothricin. *J. Antibiot.* 37: 939–940.

9. Ōmura, S., Murata, M., Imamura, N., Iwai, Y., and Tanaka, H. 1984. Oxetin, a new antimetabolite from an actinomycete. Fermentation, isolation, structure and biological activity. *J. Antibiot.* 37: 1324–1332.

10. Ōmura, S., Murata, M, Kimura, K., Matsukura, S., Nishihara, T., and Tanaka, H. 1985. Screening for new antifolates of microbial origin and a new antifolate AM-8402. *J. Antibiot.* 38: 1016–1024.

11. Ōmura, S., Tomoda, H., Xu, Q. M., Takahashi, Y., and Iwai, Y. 1986. Triacsins, new inhibitors of acyl-CoA synthetase produced by *Streptomyces* sp. *J. Antibiot.* 39: 1211–1218.

12. Ōmura, S., Tanaka, Y., Hisatome, K., Miura, S., Takahashi, Y., Nakagawa, A., Imai, H., and Woodruff, H. B. 1988. Phthoramycin, a new antibiotic active against a plant pathogen, *Phytophthora* sp. *J. Antibiot.* 41: 1910–1912.

13. Ōmura, S., Tanaka, Y., Kanaya, I., Shinose, M., and Takahashi, Y. 1990. Phthoxazolin, a specific inhibitor of cellulose biosynthesis, produced by a strain of *Streptomyces* sp. *J. Antibiot.* 43: 717–719.
14. Ross, P., Weinhouse, H., Aloni, Y., Michael, D., Weinberger-Ohana, P., Mayer, R., Braun, S., de Vroom, E., van der Marel, G. A., van Boom, J. H., and Benziman, M. 1987. Regulation of cellulose synthesis in *Acetobacter xylinum* by cyclic diguanylic acid. *Nature* 325: 279–281.
15. Sakurai, T., Kihara, T., and Isono, K. 1983. Structure of cytovaricin-acetonitrile (1:1), $C_{47}H_{80}O_{16} \cdot C_2H_3N$. *Acta Cryst.* C39: 295-297.
16. Swissa, M., Aloni, Y., Weinhouse, H., and Benziman, M. 1980. Intermediary steps in *Acetobacter xylinum* cellulose synthesis: Studies with whole cells and cell-free preparation of the wild-type and a cellulose-less mutant. *J. Bacteriol.* 143: 1142–1150.
17. Tomoda, H. and Ōmura, S. 1990. New strategy for discovery of enzyme inhibitors: Screening with intact mammalian cells or intact microorganisms having special functions. *J. Antibiot.* 43: 1207–1222.

9

Discovery and Identification of a Novel Fermentation-Derived Insecticide

Herbert A. Kirst,* Karl H. Michel, Eddie H. Chio, Raymond C. Yao, Walter M. Nakatsukasa, LaVerne D. Boeck, John L. Occolowitz, Jonathan W. Paschal, and Jack B. Deeter
Lilly Research Laboratories, Eli Lilly and Company, Indianapolis, Indiana 46285

Gary D. Thompson
Discovery Research, DowElanco, Greenfield, Indiana 46140

INTRODUCTION

The isolation of soil microorganisms and screening of their culture broths has been a proven source of novel compounds that possess potentially useful biological activities. This methodology has been established over several decades to be the most productive means for the discovery of new antimicrobial agents (23). Although its application to non-anti-infective targets has been more recent, the methodology of screening fermentation broths is now being vigorously pursued using a wide variety of assays and targets outside the areas of infectious diseases (7, 8, 27, 31).

Insecticides have previously been sought from among the many secondary metabolites produced by microbial fermentations (8, 11, 25). In addition, the fermentation-derived anthelmintic agent, avermectin, and certain of its semisynthetic derivatives have exhibited useful insecticidal activities (3, 10, 21, 24). Due to economic and environmental pressures resulting from the continual development of resistance in insects and the desire for less toxic agents, new insecticidal agents with improved properties are still being sought (28).

ASSAY

One of the indicator assays that has previously been developed to screen compounds for potential insecticidal activity is larvicidal activity against instar larvae of the mosquito, *Aedes aegypti* (4, 12, 32). Other workers have employed a similar assay to either screen culture broths or to study the larvicidal activity of fermentation-derived compounds such as valinomycin, tunicamycin, polymyxin B, and gramicidin S (22, 26). Consequently, screening of culture broths using a mosquito larvicidal assay has been well documented as a viable approach to the discovery of new fermentation products with insecticidal activities.

During our screening of fermentation broths for mosquito larvicidal activity, many previously identified compounds were detected. In addition, a library of previously isolated, fermentation-derived compounds were screened. Among the known compounds exhibiting mosquito larvicidal activity were macrolide compounds such as avermectin, milbemycin, rutamycin, and turimycin H; polyethers and ionophores such as nigericin, lonomycin, macrotetrolide, and A23187; aminoglycoside antibiotics such as gentamicin and neomycin; and members of other structural classes of fermentation products such as antimycin A, netropsin, phyllomycin, piericidin, streptimidone, and tubercidin.

PRODUCING ORGANISM

In the course of screening, an active culture broth containing an unidentified compound was detected and assigned the number A83543. The producing microorganism responsible for the activity had been isolated from a soil sample collected in the Virgin Islands. The organism was subsequently characterized and determined to be a new species within the genus *Saccharopolyspora*. Based upon its taxonomic features, the name *Saccharopolyspora spinosa* has been assigned to it (20).

Although species of *Saccharopolyspora* were not previously well represented among organisms which produce important fermentation products, this situation has been changing. The producer of erythromycin, the most important macrolide antibiotic, has been reclassified from *Streptomyces erythreus* to *Saccharopolyspora erythraea* (19). An unusual bicyclic derivative of erythromycin has been recently obtained from the *Saccharopolyspora* species L53-18 (33). Various aminoglycosides as well as an unusual macrolide, nargenicin, have been produced by strains of *S. hirsuta* and *S. hirsuta* subsp. *kobensis* (2, 9, 13, 17). Other publications have reported discoveries of additional species of *Saccharopolyspora* and fermentation products isolated from their culture broths. These include texazone from *S. taberi* (18), menaquinone derivatives from *S. rectivirgula* [ex. *Faenia*] (6, 18), YL-0358 from *S. griseopurpurea* (14), and coumamidines from a species designated as AB 1167L-65 (16).

ISOLATION OF ACTIVE COMPONENTS

Isolation of the active components from the culture broth of A83543 was performed by a combination of extractive and chromatographic procedures. The crude fermentation broth was filtered, the biomass was thoroughly washed with water, and the filtrates were discarded. The biomass was then extracted with methanol and the resultant mixture was filtered. The filtrate was collected, concentrated under reduced pressure, and extracted three times with diethyl ether. The combined ether extracts were concentrated and separated by reversed-phase HPLC on a Rainin "Autoprep" instrument utilizing a Lobar RP-8 column (size B, E. M. Industries, Inc.), eluting with a methanol-acetonitrile-water mixture. This procedure was employed for twelve preparative cycles through the Autoprep system. The partially purified A83543 factors were further purified by reversed-phase HPLC on the Rainin "Autoprep" using a Dynamax C-18 preparatory chromatography column (21.4 mm × 25 cm). This column was similarly eluted with a methanol-acetonitrile-water mixture to yield pure components of the A83543 complex.

The A83543 factors were characterized by analytical HPLC on a Nova C18 column (Waters, 4.5 × 100 mm). The eluent was methanol-acetonitrile-water, using a gradient from 35:35:30 to 45:45:10 plus 0.05% ammonium acetate, with a run time of fifteen minutes. Detection was effected by a UV monitor set at 250 nm. Under these conditions, elution of the individual factors occurred in the following order, with relative HPLC retention times (in minutes) noted in parentheses: pseudoaglycone (2.55), A83543C (2.62), A83543B (4.22), A83543H (5.29), A83543F (shoulder on back of A83543H peak), A83543G (6.49), A83543E (6.57), A83543A (8.97), and A83543D (11.62).

Factor A was the major component of the A83543 complex, occurring to the extent of approximately 80–90%, depending upon the particular fermentation conditions. Factor D was the second most abundant factor, constituting approximately 10–15% of the complex. The remaining factors were relatively minor constituents of the complex.

STRUCTURAL ELUCIDATION

High resolution mass spectrometric analysis of A83543A established its molecular weight as 731 and its empirical formula as $C_{41}H_{65}NO_{10}$. Similar analyses of the other factors showed that all were close analogues of A83543A. Factor D differed from factor A by an additional methylene unit, with a molecular weight of 745 and an empirical formula of $C_{42}H_{67}NO_{10}$. Factor G was isomeric with factor A. Factors B, E, F, H, and J all yielded identical results and differed from factor A by the absence of a methylene unit, with molecular weights of 717 and empirical formulas of $C_{40}H_{63}NO_{10}$. Factor C differed from these latter factors by the absence of another methylene unit, with a molecular weight of 703 and an empirical formula of $C_{39}H_{61}NO_{10}$. The pseudoaglycone of A83543A was derived from the loss of an aminosaccharide moiety, with a molecular weight of 590 and an empirical formula of $C_{33}H_{50}O_{9}$.

By the application of MS/MS techniques, factors A-F, H and J were determined to contain the aminosugar forosamine, a well known component of the macrolide antibiotic spiramycin. This aminosugar was absent in the factor designated as the pseudoaglycone. In contrast, A83543G was determined to contain a different but isomeric aminosaccharide which was proposed to be ossamine, an aminosugar previously isolated from the fermentation product ossamycin (30). Compounds containing aminosugars with these empirical compositions suggested structures within the macrolide class; however, literature searches through various databases of natural products did not reveal any previously known molecules possessing these empirical formulas and properties.

A83543A exhibited a UV chromophore in ethanol solution with a λ_{max} of 243 nm (ϵ of 9000); no shifts were observed upon change of pH to either acidic or basic conditions. Titration studies in 66% aqueous dimethylformamide solution indicated the presence of a single basic amino group with a pK_a value of approximately 7.8. A83543A was poorly soluble in water at pH values above 7; although it was soluble in dilute aqueous acids, it was susceptible to hydrolysis of forosamine as the pH was lowered. It was readily soluble in organic solvents such as methanol, ethanol, chloroform, dichloromethane, ethyl acetate, acetonitrile, and acetone. The other factors exhibited similar properties, indicating that their structures were closely related to that of A83543A. The combined results from these physical chemical characterization studies indicated that the A83543 factors were novel members of the macrolide family of compounds.

NMR experiments including ^1H homonuclear decoupling, ^{13}C DEPT, 2D one bond heteronuclear correlation, and 2D long range heteronuclear correlation (FULCOUP) were combined with mass spectral fragmentation data to provide the overall structure of A83543A (see figure 9.1). The relative stereochemistry at carbon atoms 3, 4, 7, 11, and 12 was determined from coupling constants and difference NOE experiments.

A83543A crystallized from ethanol water solution, yielding a white crystalline product. A single crystal x-ray diffraction study confirmed the gross structure which had been established and determined the relative stereochemistry at carbon

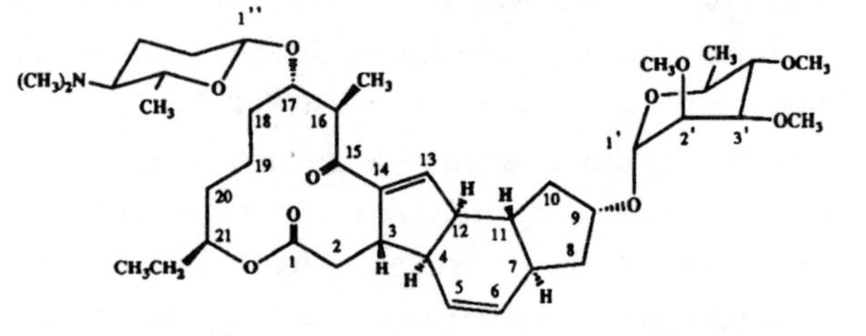

Figure 9.1 / Structure of A83543A depicting absolute stereochemistry and numbering

atoms 9, 16, 17, and 21. The structure of A83543A is depicted in figure 9.1.

The structure of A83543A contained a twelve-membered macrocyclic lactone to which was fused a 5,6,5-*cis-anti-trans*-tricyclic ring system (octahydro-*as*-indacene skeleton). Embedded within this tetracyclic ring system were an α,β-unsaturated ketone and an isolated double bond. Also present were two hydroxyl groups, which were glycosylated respectively with an aminosugar (forosamine) and a neutral sugar (2,3,4-tri-O-methylrhamnose).

The absolute stereochemistry of A83543A was established by acidic hydrolysis and subsequent isolation of forosamine, which was simultaneously compared with a sample that had been isolated by similar means from spiramycin (Sigma Chemical Co.). The two samples of forosamine were identical in all respects, including sign and degree of optical rotation, thereby showing that forosamine in A83543A was D-(+)-forosamine, the same as in spiramycin (1). Combined with the relative stereochemistry determined from x-ray analysis, the absolute stereochemistry was thereby established as that depicted in figure 9.1. The numbering system used in figure 9.1 for the tetracyclic ring system follows the established conventions which are used for macrolide antibiotics. The neutral sugar was given the "prime" designation since it is attached to the hydroxyl group on the lower numbered carbon atom, while the aminosugar was given the "double prime" designation.

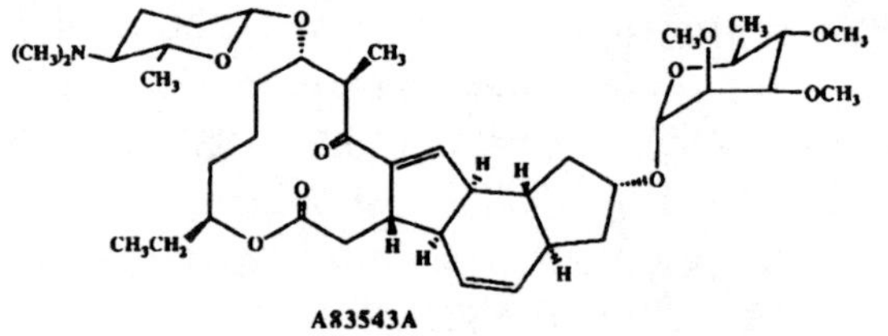

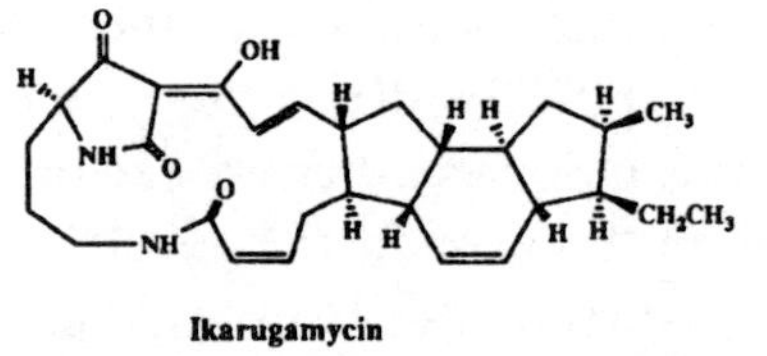

Figure 9.2 / Comparison of structures of A83543A, ikarugamycin, and capsimycin

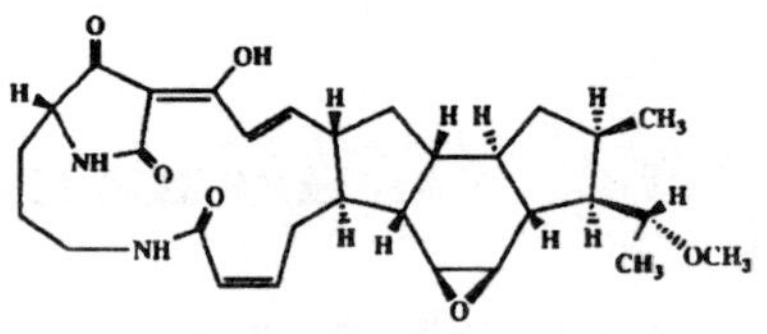

Literature searches failed to uncover similar molecules and indicated that A83543A was structurally unique. The compounds most closely related in structure were ikarugamycin and capsimycin, which are depicted in figure 9.2. Although ikarugamycin also contains a 5,6,5-*cis-anti-trans*-tricyclic ring system, it is fused to a macrocyclic lactam that additionally contains a tetramic acid moiety embedded within it (15). Furthermore, the absolute stereochemistry of the tricyclic rings of A83543A and ikarugamycin are opposite to each other, and the two compounds differ in their substituent pattern at several places; in addition, ikarugamycin does not possess any saccharides on its tetracyclic framework (15). Capsimycin is another previously-known compound which contains a 5,6,5-tricyclic ring system and is proposed to be closely related in structure to ikarugamycin (29).

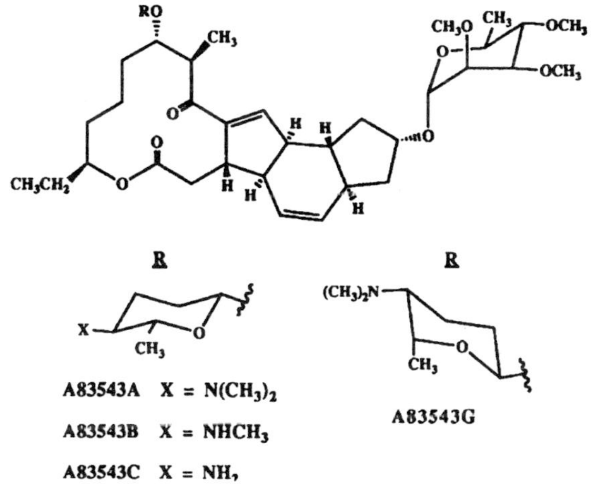

Figure 9.3 / Structures of A83543 factors A, B, C, and G

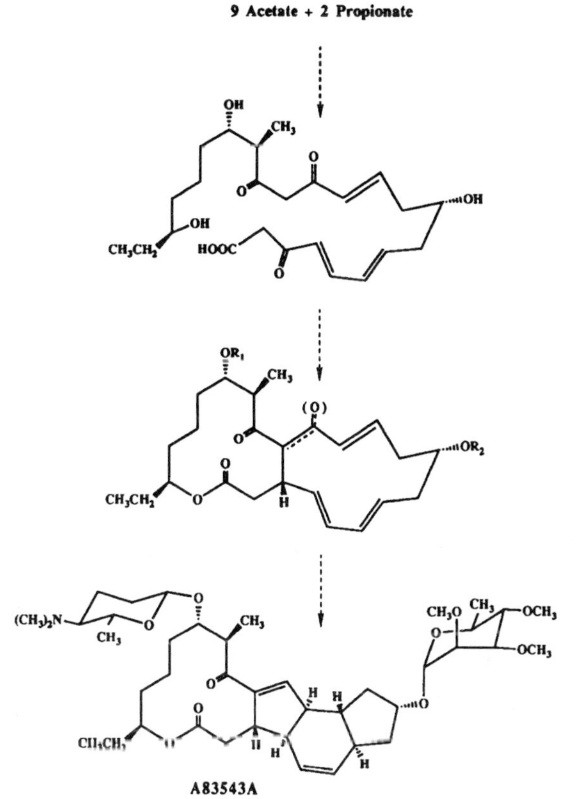

MINOR FACTORS

On the basis of mass spectrometric and NMR spectroscopic studies, A83543B and A83543C were determined to be the N-demethyl and N,N-didemethyl derivatives of A83543A, respectively (see figure 9.3). As previously indicated, A83543G possessed ossamine rather than forosamine as the aminosugar. The structure depicted in figure 9.3 best represents the NMR data for this factor.

The macrolide structure of A83543A suggested that its biosynthesis followed the usual polyketide pathways involving condensation and subsequent modification of nine acetate units and two propionate units. This mode of assembly would produce a substituted unsaturated hydroxyacid of 23 carbon atoms in length such as that illustrated in figure 9.4. The tetracyclic ring system could then be produced by the combination of three ring-forming steps: lactonization, an intramolecular Claisen condensation, and an intramolecular Diels-Alder reaction. Intramolecular cyclizations have been proposed for both ikarugamycin and another unusual macrocyclic lactone, nargenicin (5, 15). The exact order in which these three ring-forming steps might occur in A83543 has not

Figure 9.4 / One potential biosynthetic pathway for A83543A

yet been determined. One possible sequence in which lactonization occurs first and the intramolecular Diels-Alder reaction occurs last is depicted in figure 9.4.

The origin of factors D, E, and F, whose structures were deduced from their mass and NMR spectra (see figure 9.5), were readily explained by this polyketide biosynthetic scheme. A83543D contains an additional methyl group at C-6, which would arise if the polyketide synthase substituted propionate for acetate during this step in the biosynthetic cycle. Similarly, A83543E possessed a methyl group rather than an ethyl group at C-21, which would result if the biosynthetic starter unit were acetate rather than propionate. A83543F lacked the methyl group found in A83543A at C-16; this compound would be produced if acetate rather than propionate were incorporated at this

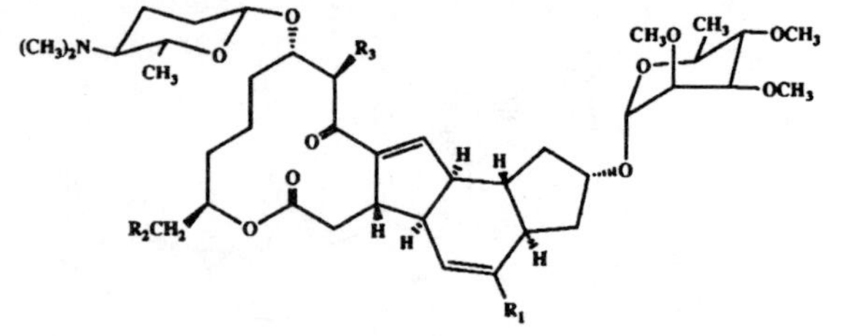

	R₁	R₂	R₃
A83543A	H	CH₃	CH₃
A83543D	CH₃	CH₃	CH₃
A83543E	H	H	CH₃
A83543F	H	CH₃	H

Figure 9.5 / Structures of A83543 factors A, D, E, and F

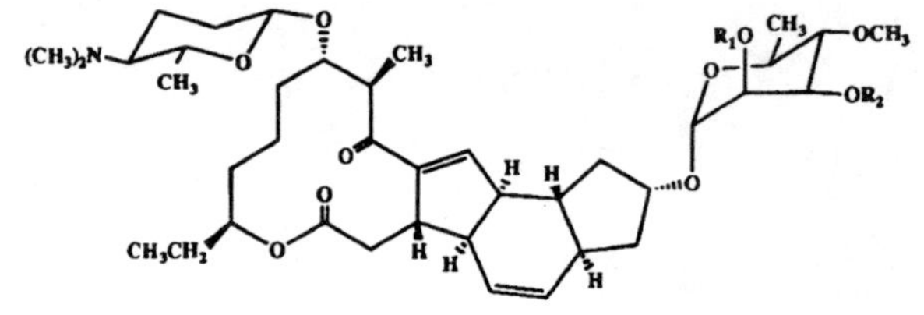

	R₁	R₂
A83543A	CH₃	CH₃
A83543H	H	CH₃
A83543J	CH₃	H

Figure 9.6 / Structures of A83543 factors A, H, and J

step in the polyketide assembly. The structure of factor H was determined to be the 2'-O-demethyl derivative of A83543A (see figure 9.6); the absence of the 2'-O-methyl resonance in its NMR spectrum was especially significant. A83543J was isomeric with A83543H and was established as the 3'-O-demethyl derivative of A83543A (figure 9.6) by similar analysis of NMR spectra. Consequently, different factors have been isolated from the A83543 complex which differ from factor A in their degree of either N-, C-, or O-methylation.

ACTIVITY

Several of the individual factors exhibited good activity *in vitro* against fourth instar mosquito larvae after 24 hr incubation at a concentration of 0.312 ppm. Factors A, B, C, D, and E caused 60%, 60%, 60%, 30%, and 80% mortality, respectively, in this assay; factors F and G were not active at this concentration. Against first instar mosquito larvae, A83543A possessed an MIC value of 0.016 μg/ml. The demonstration of good mosquito larvicidal activity indicates that the individual components responsible for the activity detected in culture A83543 have been successfully isolated and identified.

SUMMARY

Screening of culture broths for mosquito larvicidal activity has led to the discovery of A83543, a family of related factors produced by fermentation of a novel microbial species, *Saccharopolyspora spinosa*. The factors responsible for the biological activity were isolated from culture broths by extraction and reversed-phase HPLC chromatography. Structures of the individual factors were elucidated by a combination of mass spectrometric, NMR and UV spectroscopic, and x-ray crystallographic studies. The active compounds possess a structurally unique tetracyclic ring system consisting of a 5,6,5-*cis-anti-trans*-tricyclic moiety fused to a 12-membered macrolide. Substituents on the tetracyclic ring system include an aminosugar and a neutral sugar. Nine individual factors which differ predominantly in their degree of N-, O-, or C-methylation have been isolated.

The discovery of this new family provides one more example of success in finding novel compounds with biological activity from screening of culture broths of random soil microorganisms. Although this technology has long been employed, it continues to be a useful tool in discovery research, especially since it is proving highly applicable in therapeutic areas other than those associated with infectious diseases.

ACKNOWLEDGEMENTS

The authors gratefully acknowledge the critical assistance given by many of our colleagues and associates within the Lilly Research Laboratories and DowElanco Discovery Research. We especially thank J. W. Martin, D. M. Berry, P. J. Baker, J. S. Mynderse, O. W. Godfrey, T. E. Eaton, V. M. Daupert, L. W. Crandall, F. P. Mertz, D. K. Baisden, D. W. Norton, T. K. Elzey, N. D. Jones, and L. C. Creemer for their excellent technical support and assistance.

REFERENCES

1. Albano, E. L. and Horton, D. 1969. A synthesis of 2,3,4,6-tetradeoxy-4-(dimethylamino)-D-*erythro*-hexose (forosamine) and its D-*threo* epimer. *Carbohyd. Res.* 11: 485–495.
2. Awata, M., Satoi, S., Muto, N., Hayashi, M., Sagai, H., and Sakakibara, H. 1983. Saccharocin, a new aminoglycoside antibiotic. *J. Antibiotics* 36: 651–655.
3. Babu, J. R. 1988. Avermectin: biological and pesticidal activities. In: Biologically Active Natural Products (Cutler, H. G., ed.), pp. 91–108, American Chemical Society, Washington, D.C.
4. Brown, A. W. A. and Pal, R. 1971. Insecticide Resistance in Arthropods. 2nd edition. World Health Organization, Geneva, Switzerland.
5. Cane, D. E. and Yang, C.-C. 1984. Biosynthetic origin of the carbon skeleton and oxygen atoms of nargenicin A_1. *J. Amer. Chem. Soc.* 106: 784–787.
6. Collins, M. D., Kroppenstedt, R. M., Tamaoka, J., Komagata, K., and Kinoshita, T. 1988. Structures of the tetrahydrogenated menaquinones from *Actinomadura angiospora, Faenia rectivirgula,* and *Saccharothrix australiensis. Curr. Microbiol* 17: 275–280.
7. Demain, A. L., Somkuti, G. A., Hunter-Cevera, J. C., and Rossmoore, H. W., eds. 1989. Novel Microbial Products for Medicine and Agriculture. Elsevier Science Publishers, Amsterdam.
8. Deshpande, B. S., Ambedkar, S. S., and Shewale, J. G. 1988. Biologically active secondary metabolites from *Streptomyces. Enzyme Microb. Technol.* 10: 455–473.
9. Deushi, T., Iwasaki, A., Kamiya, K., Kunieda, T., Mizoguchi, T., Nakayama, M., Itoh, H., Mori, T., and Oda, T. 1979. A new broad-spectrum aminoglycoside antibiotic complex, sporaricin. *J. Antibiotics* 32: 173–179.
10. Dybas, R. A. 1989. Abamectin use in crop protection. In: Ivermectin and Abamectin (Campbell, W. C., ed.), pp. 287–310, Springer-Verlag, New York.
11. Heisey, R. M., Mishra, S. K., Putnam, A. R., Miller, J. R., Whitenack, C. J., Keller, J. E., and Huang, J. 1988. Production of herbicidal and insecticidal metabolites by soil microorganisms. In: Biologically Active Natural Products (Cutler, H. G., ed.), pp. 65–78, American Chemical Society, Washington, D.C.
12. Henrick, C. A., Staal, G. B., and Siddall, J. B. 1973. Alkyl 3,7,11-trimethyl-2,4-dodecadienoates, a new class of potent insect growth regulators with juvenile hormone activity. *J. Agric. Food Chem.* 21: 354–359.
13. Ikeda, Y., Kondo, S., Kanai, F., Sawa, T., Hamada, M., Takeuchi, T., and Umezawa, H. 1985. A new destomycin-family antibiotic produced by *Saccharopolyspora hirsuta. J. Antibiotics* 38: 436–438.
14. Imai, Y., Suzuki, K., Miyazaki, S., Nagai, K., Tsunoda, N., Morioka, M., Fujita, S., Furuya, T., and Numazaki, Y. 1988. Antitumor antibiotic YL-0358M-A and its manufacture with *Saccharopolyspora*. Jpn. Kokai 63–44,581. *Chem. Abstr.* 109: 127327m.
15. Ito, S. and Hirata, Y. 1977. The structure of ikarugamycin, an acyltetramic acid antibiotic possessing a unique *as*-hydrindacene skeleton. *Bull. Chem. Soc. Japan* 50: 1813–1820.
16. Jackson, M., Karwowski, J. P., Theriault, R. J., Kohl, W. L., Humphrey, P. E., Sunga, G. N., Swanson, S. J., and Villarreal, R. M. 1989. Coumamidines, new broad spectrum antibiotics of the cinodine type. *J. Antibiotics* 42: 527–532.
17. Kamiya, K., Deushi, T., Iwasaki, A., Watanabe, I., Itoh, H., and Mori, T. 1983. A new aminoglycoside antibiotic, KA-5685. *J. Antibiotics* 36: 738–741.
18. Korn-Wendisch, F., Kempf, A., Grund, E., Kroppenstedt, R. M., and Kutzner, H. J. 1989. Transfer of *Faenia rectivirgula* Kurup and Agre 1983 to

the genus *Saccharopolyspora* Lacey and Goodfellow 1975, elevation of *Saccharopolyspora hirsuta* subsp. *taberi* Labeda 1987 to species level, and amended description of the genus *Saccharopolyspora*. *Int. J. Syst. Bacteriol. 39:* 430–441.
19. Labeda, D. P. 1987. Transfer of the type strain of *Streptomyces erythraeus* (Waksman 1923) Waksman and Henrici 1948 to the genus *Saccharopolyspora* Lacey and Goodfellow 1975 as *Saccharopolyspora erythraea* sp. nov., and designation of a neotype strain for *Streptomyces erythraeus*. *Int. J. Syst. Bacteriol. 37:* 19–22.
20. Mertz, F. P. and Yao, R. C. 1990. *Saccharopolyspora spinosa* sp. nov. isolated from soil collected in a sugar mill rum still. *Int. J. Syst. Bacteriol. 40:* 34–39.
21. Miller, T. W., and Gullo, V. P. 1989. Avermectins and Related Compounds. In: Natural Products Isolation. Journal of Chromatography Library, Vol. 43, (Wagman, G. H. and Cooper, R., eds.), pp. 347–376, Elsevier Science Publishers, Amsterdam.
22. Mishra, S. K., Keller, J. E., Miller, J. R., Heisey, R. M., Nair, M. G., and Putnam, A. R. 1987. Insecticidal and nematicidal properties of microbial metabolites. *J. Indust. Microbiol. 2:* 267–276.
23. Mitscher, L. A., Drake, S., Gollapudi, S. R., and Okwute, S. K. 1987. A modern look at folkloric use of anti-infective agents. *J. Nat. Prod. 50:* 1025–1040.
24. Mrozik, H., Eskola, P., Linn, B. O., Lusi, A., Shih, T. L., Tischler, M., Waksmunski, F. S., Wyvratt, M. J., Hilton, N. J., Anderson, T. E., Babu, J. R., Dybas, R. A., Preiser, F. A., and Fisher, M. H. 1989. Discovery of novel avermectins with unprecedented insecticidal activity. *Experientia 45:* 315–316.
25. Neumann, R. and Peter, H. H. 1987. Insecticidal organophosphates: nature made them first. *Experientia 43:* 1235–1237.
26. Nickerson, K. W. and Schnell, D. J. 1983. Toxicity of cyclic peptide antibiotics to larvae of *Aedes aegypti*. *J. Invertebr. Pathol. 42:* 407–409.
27. Omura, S. 1988. Search for bioactive compounds from microorganisms-strategies and methods. In: Biology of Actinomycetes, pp. 26–31, Japan Scientific Societies Press, Tokyo.
28. Pickett, J. A. 1988. Chemical pest control-the new philosophy. *Chem. Brit.* 1988: 137–142.
29. Seto, H., Yonehara, H., Aizawa, S., Akutsu, H., Clardy, J., Arnold, E., Tanabe, M., and Urano, S. 1979. Structural studies of capsimycin and biosynthetic studies of ikarugamycin. *Chem. Abstr. 92:* 211459u.
30. Stevens, C. L., Gutowski, G. E., Bryant, C. P., Glinski, R. P., Edwards, O. E., and Sharma, G. M. 1969. The isolation and synthesis of ossamine, the aminosugar fragment from the fungal metabolite ossamycin. *Tetrahedron Lett. 1969:* 1181–1184.
31. Umezawa, H. 1987. Studies on antibiotics and enzyme inhibitors. *Rev. Infect. Dis. 9:* 147–164.
32. World Health Organization. 1963. W.H.O. Technical Report Serial: Number 265.
33. Yaginuma, S., Morishita, A., Muto, N., Ishizawa, K., Hayashi, M., and Saito, T. 1989. Antibiotic L53–18A and process for preparation thereof. European patent application 379–395 (assigned to Toyo Jozo KK, July 25, 1990).

10

Novel Avermectins by Mutational Biosynthesis

K. S. Holdom[1]; D. Beck[2]; C. J. Dutton[1]; S. P. Gibson[1]; A. C. Goudie[1]; E. W. Hafner[2]; B. W. Holley[2]; S. E. Lee[2]; H. A. I. McArthur[2]; M. S. Pacey[1]; J. C. Ruddock[1]; R. G. Wax[2]; W. C. Wernau[2]; J. D. Bu'Lock[3] and M. K. Richards[3]

Pfizer Central Research, [1]Sandwich, England, and [2]Groton, Connecticut, USA; [3]Department of Chemistry, University of Manchester, England

A mutant of Streptomyces avermitilis, *designated I-3 (ATCC 53567) has been isolated which lacks functional branched chain 2-oxo acid dehydrogenase activity and is unable to degrade the branched-chain amino acids isoleucine and valine into their corresponding methylbutyric and isobutyric acids. These fatty acids are essential precursors for avermectin biosynthesis, and the mutant is therefore unable to synthesise avermectins unless these acids are supplied to the fermentation. When alternative organic acids or biosynthetic precursors are added to the fermentation, novel avermectins are produced, in which the added substrate is incorporated at the C-25 position. In an initial series of experiments over 800 substrates were tested and 44 were incorporated to give novel C-25 substituted avermectins. Structure-incorporation relationships (SIR) are discussed. The new avermectins retained activity against the nematode* Caenorhabditis elegans.

INTRODUCTION

The production of novel antibiotics by the technique of precursor directed biosynthesis was first demonstrated by feeding phenoxyacetic acid to a penicillin producing organism, leading to the formation of penicillin V. Many new antibiotics produced by this technique have been reported [10], but the presence of the parent compound often interferes with the detection of any new analogues, and preference of the biosynthetic enzymes for the natural precursor can reduce incorporation of the unnatural substrate.

In 1969, Shier and his colleagues [13] overcame these problems by coupling the technique of directed biosynthesis with a mutant of the neomycin producing organism, *Streptomyces fradiae*, that was unable to produce neomycin unless supplemented with the essential precursor, 2-deoxystreptamine (2DS). By feeding analogues of 2DS to the mutant, new antibiotics called hybrimycins were formed. The terms "mutational biosynthesis" [8] or "mutasynthesis" [11] have been used to describe this elegant technique. Since then a number of new antibiotics have been produced covering a range of structural classes [10], though usually by feeding substrates which are analogues of intermediates in the later stages of biosynthesis. None of these processes has been commercialised, probably reflecting either the synthetic inaccessibility of the precursors or poor incorporation. We have applied the technique of mutasynthesis to the avermectins where we perceived significant opportunities for the preparation of new antiparasitic agents.

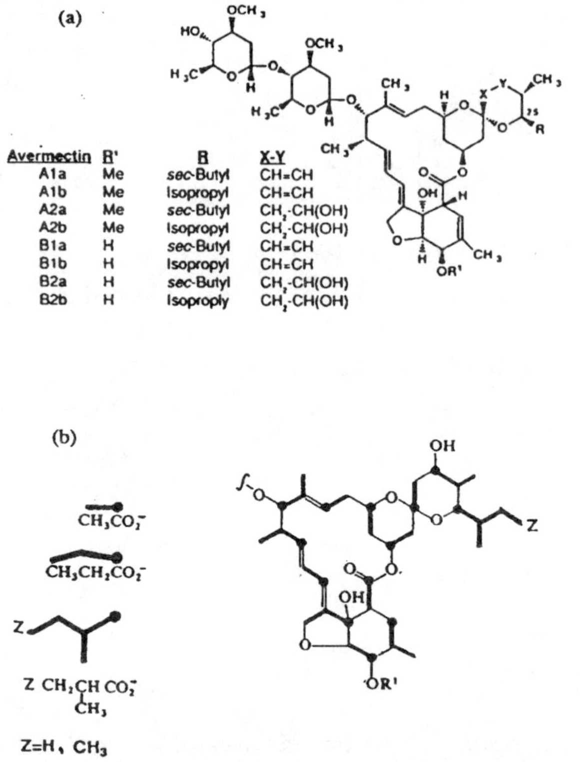

Avermectin	R'	R	X-Y
A1a	Me	*sec*-Butyl	CH=CH
A1b	Me	Isopropyl	CH=CH
A2a	Me	*sec*-Butyl	CH_2-CH(OH)
A2b	Me	Isopropyl	CH_2-CH(OH)
B1a	H	*sec*-Butyl	CH=CH
B1b	H	Isopropyl	CH=CH
B2a	H	*sec*-Butyl	CH_2-CH(OH)
B2b	H	Isopropoly	CH_2-CH(OH)

Figure 10.1 / (a) Avermectin structure and (b) biosynthetic origin of the ring carbon atoms.

NEW AVERMECTINS BY MUTATIONAL BIOSYNTHESIS

The avermectins (Fig. 10.1a) are potent, broad spectrum antiparasitic agents, discovered by Merck in 1979 [7a,b] and produced by the unique microorganism, *S. avermitilis*. They are used exclusively in animal health care for control of internal and external parasites, and have been evaluated in humans for treatment of onchocerciasis.

Biosynthetic studies [4,12] have shown that the macrolide ring is formed by the condensation of one branched chain fatty acid (isobutyric or 2 methyl butyric), seven acetate and five propionate units *via* a traditional polyketide pathway (Fig. 10.1b). Ring O-methylation, and glycosylation are late stages in the biosynthesis. The C-25 side chain is formed from an α-branched acyl-CoA unit derived from the branched chain 2-oxo acids of leucine and valine, which presumably initiates polyketide biosynthesis and produces the "a" and "b" series respectively. The biosynthesis of the avermectins can be represented schematically as Fig. 10.2a. Modification of the C-25 side-chain appeared to provide the most promise for generating novel avermectins through a mutasynthetic approach for two reasons. Firstly, avermectins with C-25 side chains derived from both 2-methylbutyric and isobutyric acid occur naturally, indicating at least some lack of enzyme specificity within the polyketide synthetase complex. Secondly, genetic control of the biosynthesis of these branched-chain fatty acid precursors is unlikely to be associated with the polyketide complex; consequently, mutational

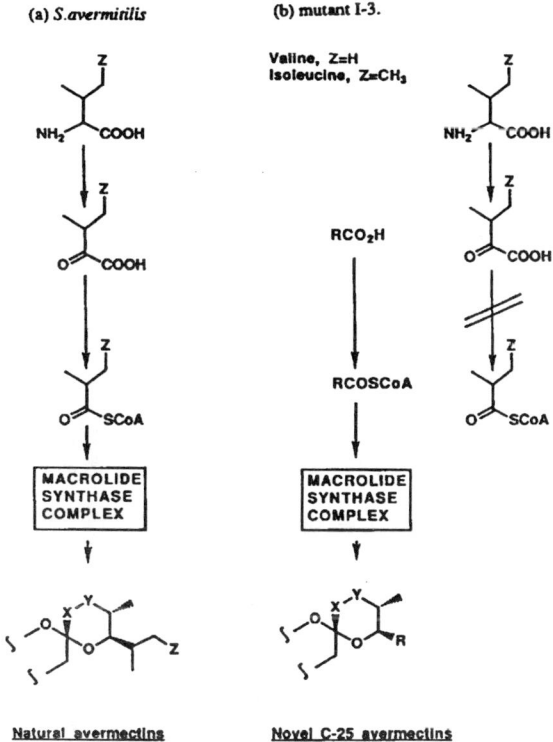

Figure 10.2 / Origin of the C-25 group in (a) *S. avermitilis* and (b) mutant I-3.

blocks in their biosynthesis should not interfere with macrolide ring formation. It would, therefore, be anticipated that a mutant of *S. avermitilis* blocked in the biosynthesis of the branched-chain fatty acyl-CoA starter units would be incapable of producing avermectins, unless the fatty acid precursors were supplied from the medium. Activation of exogenously-supplied fatty acids to form the required Coenzyme A intermediate should occur through standard transacylation [9]. It was further envisaged that alternative organic acids could substitute for the natural precursors and lead to the production of novel avermectins (Fig. 10.2b).

We therefore undertook to search for such a mutant of *S. avermitilis*. Examination of the biosynthesis and degradation of isoleucine/valine (Fig. 10.3) shows that a mutant deficient in the decarboxylation step of the branched-chain 2-oxo acid dehydrogenase activity would be unable to form the acyl CoA starter units but would not be impaired in the biosynthesis of isoleucine/valine which are required for growth. However, further examination of this pathway also indicates that such a decarboxylase negative mutant might be unable to metabolise pyruvate or α-oxoglutarate because of a shared enzyme system, as occurs in *B. subtilis* [6]. Furthermore, branched chain fatty acids are important components of membrane lipids in Gram positive organisms, and such mutations in *B. subtilis* are lethal [17].

RESULTS

An intensive screening programme was carried out to obtain a decarboxylase negative mutant of *S. avermitilis*. A high throughput screening procedure was developed using a 96-well microtitre format (Fig. 10.4), to detect $^{14}CO_2$ release from labelled substrate [16].

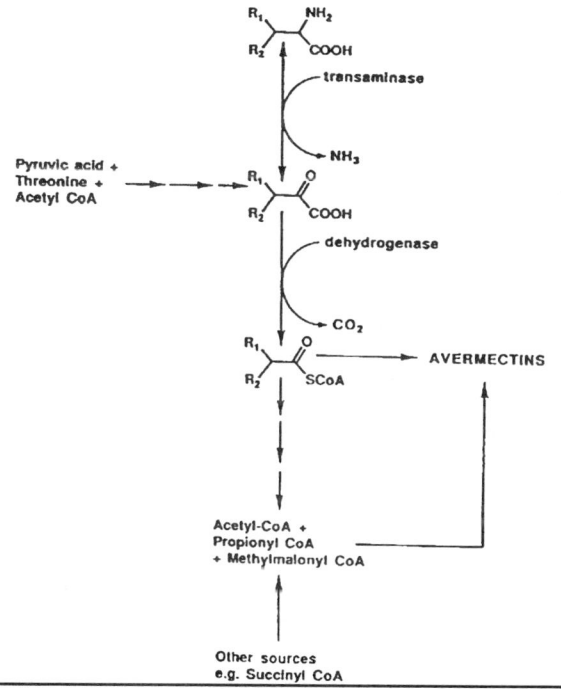

Figure 10.3 / Branched chain amino acid degradation and avermectin biosynthesis.

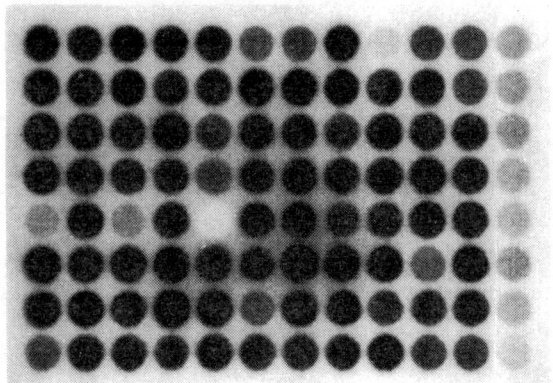

Figure 10.4 / Identification of 2-oxo acid dehydrogenase negative mutant I-3. Mutants were generated by treatment of spores with N-methyl-N'-nitro-N-nitrosoguanidine (NGT) at 1 mg/ml for 60 mins at pH 9.0. Mutants were permeabilised with 5% toluene and incubated with [^{14}C1]-2-oxoisocaproic acid, 2.5 µCi/ml, 10µCi/µmole. Released $^{14}CO_2$ was trapped in filter paper soaked in $Ba(OH)_2$, and visualized by radio autography. Light areas represent putative mutants which do not release $^{14}CO_2$.

The putative mutant, designated I-3, was subsequently confirmed by more detailed enzyme studies (Fig. 10.5) to be decarboxylase negative for the 2-oxo acids derived from isoleucine, leucine and valine. By analogy with the nomenclature used for similar mutants of *Pseudomonas putida* [15] the mutation in I-3 is designated bkd-11. Decarboxylase activity was retained with pyruvate (Fig. 10.5) and α-oxoglutarate (results not shown), indicating that the mutation had not compromised the dehydrogenase complexes for these two oxo acids. As expected the decarboxylase block prevents the mutant from growing on the three branched chain amino acids as sole carbon source, but did not inhibit their biosynthesis and the mutant is not auxotrophic for these amino acids. Utilisation of other carbon sources was not impaired, indicating that the lack of branched chain fatty acids was not an absolute requirement for growth. Subsequent

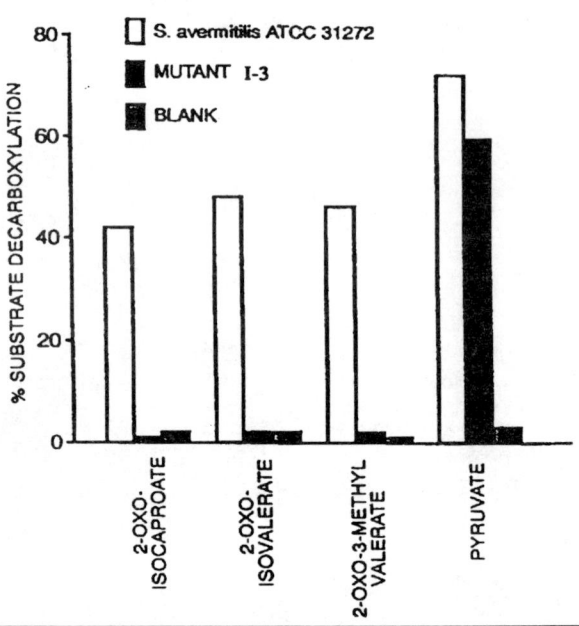

Figure 10.5 / 2-oxo acid dehydrogenase activity in *S. avermitilis* and mutant I-3. Toluene permeabilised cells were incubated in Tris-HCl buffer pH 7.5 containing: thiamine pyrophosphate 0.4 mM; coenzyme A 0.11 mM, nicotinamide adenine dinucleotide 0.6 mM; dithiothreitol 2.6 mM; $MgCl_2$ 4.1mM; and ^{14}C labelled substrate, 8200 dpm, 10µCi per µmole, and incubated at 30° for 2 h. Released $^{14}CO_2$ was trapped by paper containing Hyamine Hydroxide (IM solution of methyl benzethonium hydroxide in methanol) and radioactivity measured in Aquasol (Beckman).

studies showed that the unsaturated fatty acid content in I-3 was significantly increased over the parent *S. avermitilis* (manuscript in preparation).

As predicted, the mutant I-3 was incapable of producing avermectins when grown in submerged fermentation (Fig. 10.6a). The appearance of oligomycin, a known coproduced metabolite of *S. avermitilis*[3], and unidentified polyenes, shows that the mutant is still able to make other polyketide products. When the mutant was supplemented with the natural substrates isobutyric acid and 2-methylbutyric acid, the natural `b' and `a'

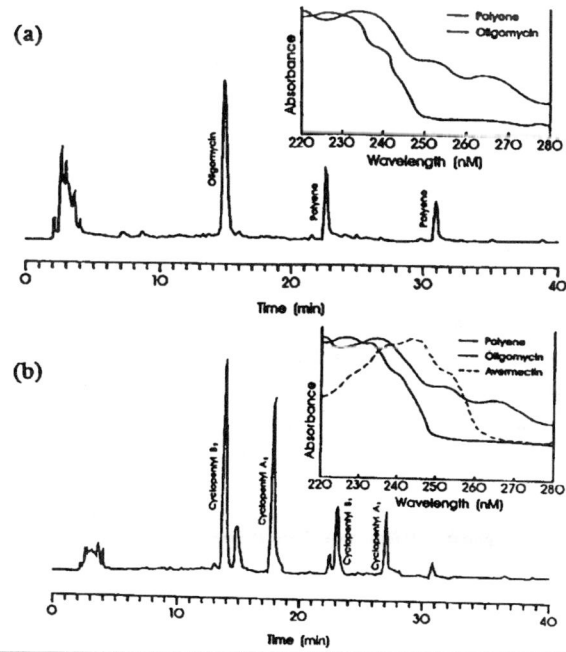

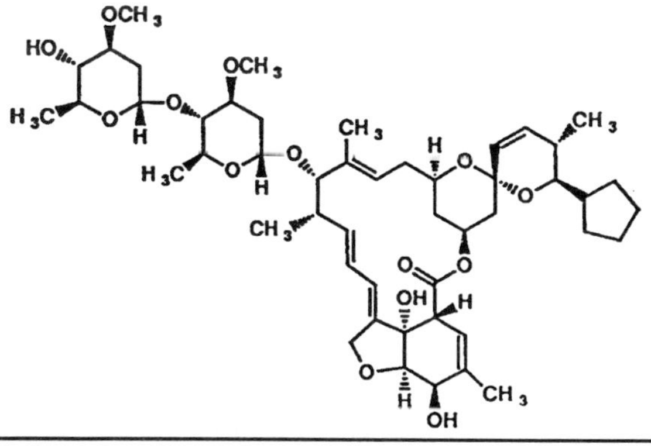

Figure 10.7 / Structure of C-25 cyclopentyl avermectin B1.

Figure 10.6 / HPLC analysis of acetone/dichloromethane extract of a fermentation of mutant I-3 (a) unsupplemented and (b) supplemented with cyclopentane carboxylic acid 0.2 g/l. Fermentation medium: Starch 8%; Pharmamedia 0.7%; $CaCO_3$ 0.7%; K_2HPO_4 0.1%; glutamic acid 0.06%; $FeSO_4 7H_2O$ 0.001%; $ZnSO_4$ 0.00001%; $MnSO_4$ 0.001%. Chromatography Conditions: Hewlett Packard 1090A LC; Beckman Ultrasphere C-18, 5 micron column, 4.6 mm x 25cm; Linear gradient methanol: water 80:20 to 95:5 over 40 mins. Flow rate 0.85 ml/min. Insert: UV chromophore for the avermectin, oligomycin, and polyene structural classes.

avermectins were produced respectively. Further, the addition of an alternative precursor to the fermentation, for example cyclopentane carboxylic acid, induced production of novel avermectins (Fig. 10.6b) as shown by their characteristic UV chromophore. Extensive structure elucidation studies confirmed that the added cyclopentane carboxylic acid had been incorporated into the C-25 position (Fig. 10.7). We were also able to show that all of the four possible members of this novel series were produced, representing the C-25 cyclopentyl avermectin A1, A2, B1 and B2 structures.

We subsequently used a spontaneous, morphologically stable mutant of I-3 (ATCC 35368) to extensively investigate the incorporation of other organic acids into the avermectin structure. Compounds were added at a concentration of 0.2 g/l to flask fermentations as the free acid, ester or sodium salt, up to 96 hours after inoculation. New compounds were identified by HPLC analysis of acetone/methylene chloride extracts of mycelium and confirmed as avermectins by their characteristic UV chromophore, which was determined on-line by diode array UV detection. Structure identity of each new compound was determined on pure compound, prepared by larger scale fermentation using NMR and particularly mass spectroscopy where fragmentation patterns[1] preserve the C-25 functionality (Fig. 10.8). Of over 800 substrates tested in our first series of experiments, 44 were incorporated at the C-25 position to produce novel avermectins [2]. Some of these substrates were oxidative precursors of the corresponding acid, and produced the same novel avermectin as the free acids themselves. Subsequent to our work, the incorporation of alternative branched-chain fatty acids by the parent strain of *S. avermitilis* to produce new C-25 substituted avermectins has been reported [5].

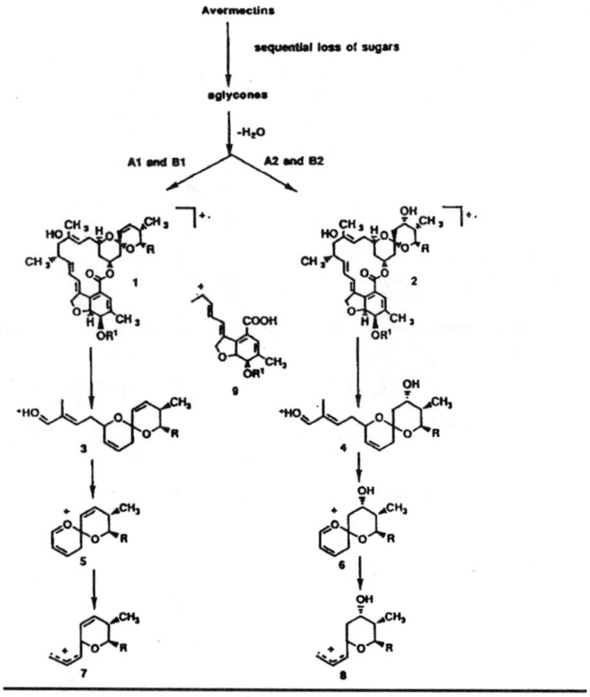

Figure 10.8 / Avermectin mass spectral fragmentation.

DISCUSSION
STRUCTURE-INCORPORATION RELATIONSHIPS

The range of substrates accepted (Table 10.1) suggests some structure-incorporation relationships (SIR). There seems to be a maximum size of about 8 carbons and carbon-carbon double and triple bonds are accepted. Compounds containing S or O are accepted as substrates if these atoms are present as ethers or contained in aromatic rings. Alkyl or alkoxy substituents are permissable but polar substituents such as hydroxy or amino were not accepted. The mutant is also capable of converting alcohols, esters or amides to the corresponding acid prior to incorporation into an avermectin. The results show that the enzyme which initiates avermectin biosynthesis and the polyketide synthase complex can accept a wide range of acyl CoA starter units. All the novel C-25 avermectins retained activity against the free living nematode *C. elegans*[14], at a concentration of 0.1µg/ml indicating the extensive modification possible at this position without loss of biological activity.

CONCLUSION

The mutasynthesis approach has shown particular utility for these complex macrolides in the modification of peripheral substituents, rather than the macrolide ring itself. The new avermectins produced are commercially inaccessible via semisynthesis and this important technique has provided a new route to investigate the structure-activity relationships of a biologically important class of compounds.

ACKNOWLEDGEMENTS

The authors wish to thank their colleagues in the Animal Health Biology and Natural Products groups and the Analytical Chemistry Department for their invaluable assistance in this work.

REFERENCES

1. Albers-Schonberg A., Arison B. H., Chabala J. C., Douglas A. W., Eskola P., Fisher M. H., Lusi A., Mrozik H., Smith J. L. and Tolman R. L. 1981. Avermectin structure determination. *J. Amer. Chem. Soc.* 103: 4216–4221.
2. Bu'Lock J. D., Goudie A. C., Holdom K. S. and Gibson S. P. 1986. New antiparasitic avermectin and milbemycin derivatives. EP-214-731-A.
3. Bu'Lock J. D., Morris G. A. and Richards M. K. 1986. Biosynthetic origins of the large macrolide, oligomycin A. *Tetrahed. Lett.* 27: 2917–2920.
4. Cane D. E., Liang T. C., Kaplan L., Mallin M. K., Schulman M. D., Hensens, O. D., Douglas, A. W. and Albers-Schongerg, G. 1983. Biosynthetic origin of the carbon skeleton and oxygen atoms of the avermectins. *J. Amer. Chem. Soc.* 105: 4110–4112.

Table 10.1 / Examples of carboxylic acids producing new C-25 substituted avermectins and their HPLC retention times.

Carboxylic acid or precursor	HPLC retention time (minutes)			
	B2	A2	B1	A1
cyclopropane carboxylic acid	8.3	10.6	15.0	19.3
cyclobutane carboxylic acid	11.7	14.3	19.9	23.4
cyclopentane carboxylic acid	14.2	17.4	23.0	26.7
cyclohexane carboxylic acid	16.9	20.3	26.0	29.6
cycloheptane carboxylic acid	18.6	22.6	29.0	33.3
2-thienoic acid	8.5	11.3	15.9	19.9
thietane-3-carboxylic acid	8.3	10.6	14.9	17.8
4-tetrahydropyran carboxylic acid	8.1	10.4	14.5	16.3
1-cyclohexene carboxylic acid	14.7	17.8	24.1	27.7
2,3-dimethylbutyric acid	14.9	17.1	24.0	27.4
2-ethylbutyric acid	15.2	18.1	24.6	28.3
methylthiolactic acid	9.1	11.4	15.1	18.4
2-methyl-4-methoxybutyric acid	10.0	12.3	16.5	20.0
2-methylpent-4-enoic acid	12.9	15.8	20.9	24.8
tiglic acid	11.1	13.6	19.2	22.8
2-methylpent-4-ynoic acid	8.9	11.5	16.2	19.5
cyclobutane methanol	11.7	14.3	19.9	23.4
aminomethyl cyclobutane	11.7	14.3	19.9	23.4
ethyl cyclobutane carboxylate	11.7	14.3	19.9	23.4
cyclohexane carboxamide	16.9	20.3	26.0	29.6
cyclohexane propionic acid	16.9	20.3	26.0	29.6

5. Chen T. S., Arison B. H., Gullo V. P. and Inamine E. S. 1989. Further studies on the biosynthesis of the avermectins. *J. Ind. Microbiol.* 4: 231–238.

6. Lowe P. N., Hodgson J. A. and Perham R. N. 1983. Dual role of a single multienzyme complex in the oxidative decarboxylation of pyruvate and branched chain 2-oxo acids in *Bacillus subtilis*. *Biochem J.* 215: 133–140.

7a. Miller T. W., Chaiet L., Cole D. J., Cole L. J., Flor J. E., Goegelman R. T., Gullo V. P., Joshua H., Kempt A. J., Krellwitz W. R., Monaghan R. L., Ormond R. E., Wilson K. E., Albers-Schonberg G. and Putter I. 1979. Avermectins, a new family of potent anthelmintic agents: isolation and chromatographic properties. *Antimicrob. Agents Chemother.* 15: 368–371.

7b. Miller T. W. and Gulo V. P. 1989. Natural Products Isolation (Eds. Wagman G. and Cooper R.), Chapter 8, Elsevier New York.

8. Nagaoka K., and Demain A. L. 1975. Mutational biosynthesis of a new antibiotic, streptomutin A, by an idiotroph of *Streptomyces griseus*. *J. Antibiotics* 28: 627–635.

9. Nunn W. D. 1986. A molecular view of fatty acid catabolism in *Escherichia coli*. *Microbiol. Rev.* 50: 179–192.

10. Okami Y. and Hotta K. 1988. Search and discovery of new antibiotics In: Actinomycetes in Biotechnology (Goodfellow M., S. T. Williams and M. Mordarski, eds.) pp. 33–67, Academic Press.

11. Reinhart K. L. Jr. 1977. Mutasynthesis of new antibiotics. *Pure and Applied Chem. 49:* 1361–1384.
12. Schulman M. D., Valentino D. and Hensens O. 1986. Biosynthesis of the avermectins by *S. avermitilis. J. Antibiotics 39:* 541–549.
13. Shier W. T., Reinhart K. L. Jr., and Gottlieb D. 1969. Preparation of four new antibiotics from a mutant of *Streptomyces fradiae. Proc. Natl. Acad. Sciences, USA 63:* 198–204.
14. Simpkin K. G., and Coles G. L. 1981. The use of *Caenorhabditis elegans* for anthelmintic screening. *J. Chem. Technol. Biotechnol. 31:* 66–69.
15. Sykes P. J., Burns G., Menard J., Hatter K. and Sokatch J. R. 1987. Molecular cloning of genes encoding branched-chain keto acid dehydrogenase of *Pseudomonas putida.* J. Bacteriol. *169:* 1619–1625.
16. Tabor H., Tabor C. W. and Hafner E. W. 1976. Convenient method for detecting $^{14}CO_2$ in multiple samples: application to rapid screening for mutants. *J. Bacteriol. 128:* 485–486.
17. Willecke K., and Pardee A. B. 1971. Fatty acid requiring mutant of *Bacillus subtilis* defective in branched-chain **α**-keto acid dehydrogenase. *J. Biol. Chem. 246:* 5264–5272.

Biological Modification of Cyclosporins by the Multifunctional Enzyme Cyclosporin Synthetase

Alfons Lawen and Horst Kleinkauf
From the Institute of Biochemistry and Molecular Biology TU Berlin, Franklinstr. 29, D-1000 Berlin, FRG.

Cyclosporins were separated on silica gel HPTLC plates. Solvents were water-saturated EtOAc (I), EtOAc-MeOH-H$_2$O, 100:5:5 (in volume) (II) or diisopropylether-CHCl$_3$-MeOH, 6:3:1 (in volume) (III). The running span was 2 × 10 cm (I) or 10 cm (II and III). HPLC was performed as described in (17). MeAhdo, N-methyl-(+)-2-amino-3-hydroxy-4,4-dimethyloctanoic acid; AllylGly, allylglycine; MeCyclohexylAla, N-methyl-L-ß-cyclohexyl-alanine.

INTRODUCTION

The nonribosomal synthesis of peptides ranging from three to 20 residues is generally catalyzed by multienzymes (12). A recent survey summarizes about 300 known basic types arranged in 90 groups (31). Biosynthesis involves the activation of amino (imino)- or hydroxy acid carboxyls as adenylates, their stabilization as activated intermediates on the enzyme surface, and their covalent step-by-step polymerization. By this covalent mechanism no free intermediates are formed, and only the final peptide is released by cyclization or hydrolysis. Modified amino acids can have for example a D-configuration, are N-methylated, or carry hydroxyl functions. In examples, studied so far, epimerization may occur before activation by a racemase or epimerase (cyclosporin (2), virido-grisein (28)) or in the activated state of the amino acid (gramicidin S (30), δ-(L-α-aminoadipate)-L-cysteinyl-D-valine (ACV) (29)). N-methylation is generally observed exclusively in the activated amino acid state (enniatin, beauvericin, actinomycin, and cyclosporin) (3). Hydroxylation has been studied in one case (viridogrisein), and occurs on the amino acid level before activation (23).

Acylation of peptides is catalyzed by defined enzymes from adenylates (actinomycin (10, 11), enterobactin (24)) or CoA-esters (polymyxin (14, 16), alamethicin (22)) directly to the respective first activated amino acid preceding peptide elongation.

To obtain peptide analogs, the relevant pathway including direct precursors should be known (13). The organization of the peptide templates into multienzyme systems may then permit the synthesis of modified structures by the introduction of the respective precursor analogs. The use of such *in vitro* systems avoids the possible metabolization of such analogs leading to complex peptide mixtures to be separated. Instead defined products are usually obtained. Multienzyme systems can be detected by their unusual sizes, often being excluded by conventional polyacrylamide gel systems. Recent sequence data obtained from ACV-synthetases from several fungi (5, 19, 26) and tyrocidine—as well as gramicidin S synthetases from *Bacillus brevis* (15, 20, 21, 32) indicate extensive similarities between these eukaryotic and prokaryotic proteins. In addition similarities have been noted to enterochelin synthetase components entE and entF, to plant and fungal acyl CoA-synthetases, and to firefly luciferase (26). All of these enzymes form acyl adenylates, and may transfer these to either CoA or pantetheine. This pantetheine is a cofactor detected in peptide synthetases, probably involved in the transport of intermediates between adjacent active sites.

Apparently a carboxyl activating region in such an enzyme system requires a space of about 1000 amino acid residues. Enzymes just contributing acyl adenylates are of smaller size. Thus the sizes of such enzymes appear to have been considerably underestimated in the past, due to the unavailability of appropriate marker proteins. For example the tripeptide forming ACV synthetases have sizes of about 425 kDa instead of 250 kDa previously estimated by SDS-gel electrophoresis.

We herein describe the performance of the largest isolated multienzyme polypeptide to date, forming the cycloundecapeptide cyclosporin.

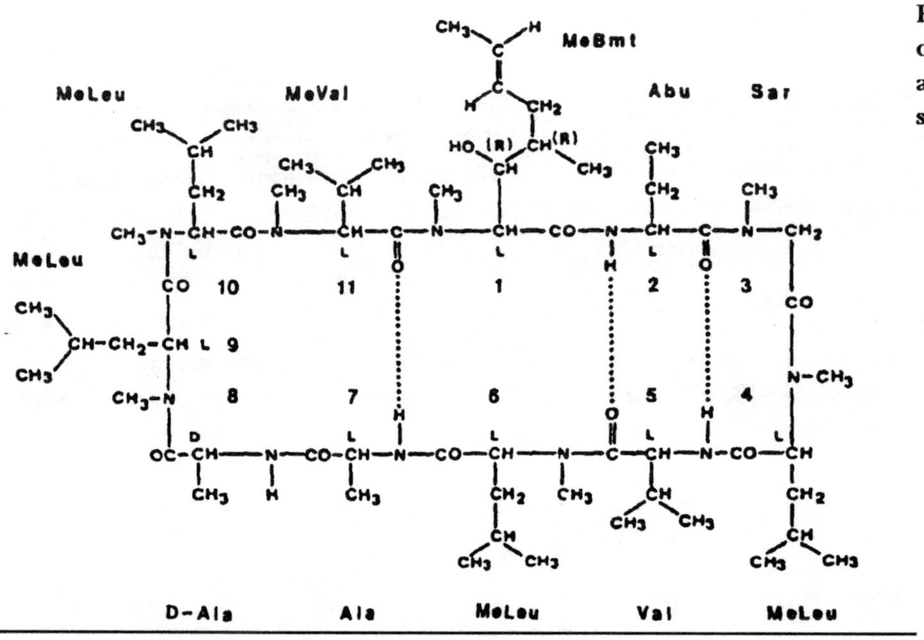

Figure 11.1 / Structure of CyA. Abu, 2-aminobutyric acid; Sar, sarcosine.

CYCLOSPORIN

Cyclosporin A (CyA; Fig. 11.1) is a cyclic undecapeptide which exerts antifungal, antiparasitic, antiinflammatory and immunosuppressive activities (4). It is commonly used in human transplantation surgery and in the treatment of autoimmune diseases (9, 25).

CyA is the main component, to date, from the twenty-five naturally occuring cyclosporins which have been described to be produced by the fungus *Beauveria nivea* (previously designated as *Tolypocladium inflatum*). All cyclosporins have the basic structure shown in Fig. 11.1; substitutions of amino acids have been observed in positions 1, 2, 4, 5, 7, and 11, unmethylated peptide bonds have been reported for positions 1, 4, 6, 9, 10, and 11 (28). By feeding amino acid precursors to the fungus, several cyclosporins with substitutions in positions 1, 2, and 8 have been obtained (27).

In 1984, Wenger described the first total chemical synthesis of cyclosporin (33), which led to the synthesis of several hundred analogues of CyA (7, 8). Nevertheless CyA possesses highest immunosuppressive activity of all cyclosporins.

CYCLOSORIN SYNTHETASE

Cyclosporin A has been shown to be synthesized by a nonribosomal mechanism (2, 34). The enzyme system responsible for the cyclosporin synthesis has been characterized to be one single multienzyme polypeptide (18) with a molecular mass of at least 650 kDa (Fig. 11.2). Like many other peptide and depsipeptide synthetases cyclosporin synthetase contains 4'-phosphopantetheine as the prosthetic group.

This enzyme activates all constituent amino acids of CyA in their unmethylated form as adenylates (measured by amino acid dependent ATP/pyrophosphate exchange) and binds these covalently as thioesters. At this stage, *N*-methylation of the amino acids 1, 3, 4, 5, 9, 10, and 11 takes place, followed by peptide bond formation and cyclization. Thus, at least 40 reaction steps are carried out by one polypeptide chain (18).

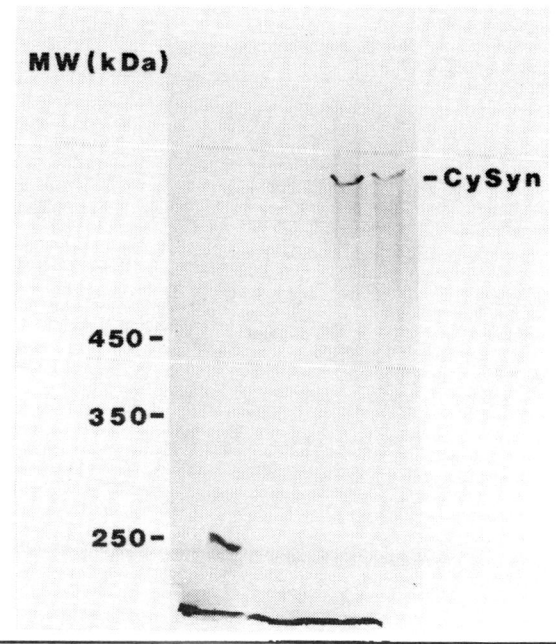

Figure 11.2 / Molecular mass estimation of cyclosporin synthetase in 3% polyacrylamide gel. Extrapolation of molecular weights of the calibration proteins enniatin synthetase (250 kDa), linear gramicidin synthetase (350 kDa) and tyrocidin synthetase III (450 kDa) results in a molecular weight of about 650 kDa for cyclosporin synthetase (CySyn).

S-adenosyl-L-methionine acts as a methyl donor in cyclosporin biosynthesis. The methyltransferase activiti(es) of cyclosporin synthetase is similar to the previously described enniatin synthetase (1), an integral part of the enzyme as demonstrated by using a photolabeling method for methyltransferases (18). The *in vitro* synthesis of an unmethylated CyA is not possible. Nevertheless with cyclosporin synthetase isolated from a blocked mutant of *B. nivea* which synthesizes the diketopiperazine *cyclo*-(D-Ala-MeLeu) instead of CyA, an unmethylated product could be obtained (Fig. 11.3).

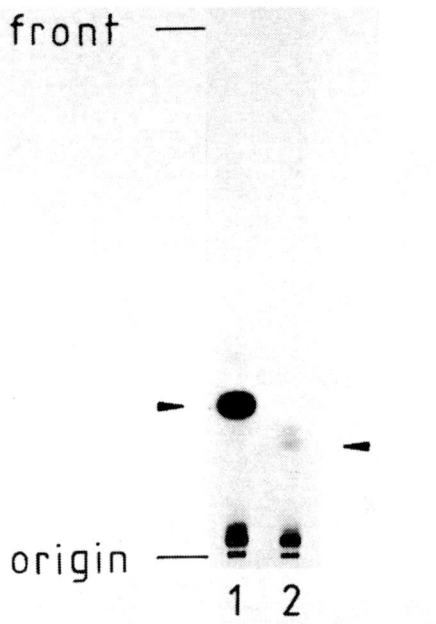

Figure 11.3 / Enzymatic synthesis of cyclo-(D-Ala-Leu). The enzyme fractions were incubated for 2 hours at 25°C with [^{14}C]-Leu, D-Ala and with (1) or without (2) S-adenosyl-L-methionine. The left arrow indicates the position of cyclo-(D-Ala-MeLeu), whereas the right arrow shows the position of cyclo-(D-Ala-Leu). (Dittmann, J. and Lawen, A., unpublished).

These experiments were carried out by incubating at suboptimal temperature for seven days partially purified enzyme preparations from the high-producer strain of *B. nivea* 7939/45 together with all the necessary amino acids in their unmethylated form (including one D-amino acid), ATP, MgCl$_2$ and S-adenosyl-L-methionine. Suboptimal temperature was chosen in order to stabilize the enzyme and to minimize S-adenosyl-L-methionine degradation. The cyclosporins were extracted and purified chromatographically. Table 11.1 summarizes chromatographic data of some of the cyclosporins identified as products of *in vitro* synthesis experiments.

As is shown exemplary for [D-Abu^8]CyA in Fig. 11.4 this method is helpful to synthesize new cyclosporins not obtainable *in vivo*. Experiments to obtain [D-Abu^8]CyA by feeding D-aminobutyric acid to the fungus resulted in stimulation of CyA production, probably due to the action of a race-

As in the case of enniatin synthetase the unmethylated product is synthesized more slowly than the methylated one (6). On account of the synthesis of the unmethylated product, cyclosporin synthetase can be considered as a peptide synthetase joined together with a methyltransferase(s) unit(s).

IN VITRO SYNTHESIS OF NEW CYCLOSPORINS

With a partially purified cyclosporin synthetase preparation CyA was obtained together with homologues which either occur naturally or had been obtained by precursor feeding experiments (2). Recently it became possible to synthesize *in vitro* a number of new cyclosporins which were not obtainable *in vivo* (17).

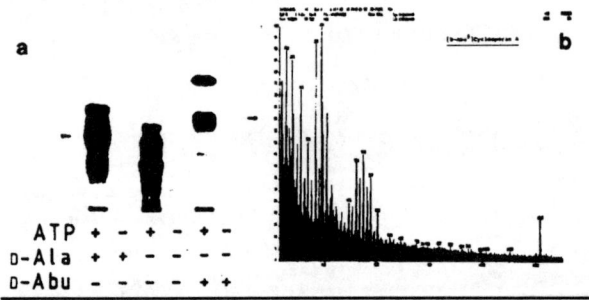

Figure 11.4 / *In vitro* synthesis of [D-Abu^8]CyA. (a) When all constitutive amino acids of CyA are incubated together with ATP, Mg^{2+}, S-adenosyl-L-[^{14}C-methyl]-methionine and the enzyme fraction, the main reaction product is CyA (left arrow). Is D-alanine (D-Ala) exchanged by D-Abu, the new [D-Abu^8]CyA is synthesized (right arrow). The TLC separation of EtOAc-extractable reaction products is shown as autoradiogram. (b) FAB-mass spectrum of [D-Abu^8]CyA, showing the correct molecular ion peak (1216 (M+H)+). (Lawen, A. and Traber, R., unpublished).

Table 11.1 / Chromatogaphic data of enzymatically formed cyclosporins

Cyclosporin	TLC R_f-value solvent I	solvent II	solvent III	HPLC α-value
CyA	0,37	0,44	0,42	10,00
CyO = [MeLeu^1,Nva$_2$]CyA	0,65	0,59	0,68	13,95
Dihydro-CyA	0,40	0,47	0,49	12,18
[MeAhdo^1]CyA	0,43	0,48	0,55	13,74
[MeCyclohexylala^1]CyA	0,48	0,54	0,57	19,79
[Allylgly^2]CyA	0,53	0,55	0,48	10,14
CyQ = [Val^4]CyA	0,19	0,28	0,20	4,84
CyM = [Nva^{2,5}]CyA	0,53	0,57	0,58	13,61
[Nva^{2,5},MeNva^{11}]CyA	0,56	0,61	0,58	14,34
[Nva^5,MeNva^{11}]CyA	0,38	0,46	0,45	10,20
[aIle^5,aMeIle^{11}]CyA	0,55	0,57	0,53	15,77
[aIle^{5,11}]CyA^1	0,30	0,42	0,41	10,67
CyU = [Leu^6]CyA	0,45	0,48	0,52	8,92
CyV = [Abu^7]CyA	0,48	0,53	0,49	11,97
[D-Abu^8]CyA	0,43	0,47	0,51	13,09
[ß-chloro-D-Ala^8]CyA	0,49	0,54	0,47	11,73

Table 11.2 / Immunosuppressive activity of new cyclosporins

Cyclosporin	Biosynthesis	Activity
Cyclosporin A (Sandimmun^R)	natural	+++
Cyclosporin A	enzymatic	+++
[Nva^{2,5},MeNva^{11}]CyA	enzymatic	++
[Nva^5,MeNva^{11}]CyA	enzymatic	++(+)
[aIle^5,aMeIle^{11}]CyA	enzymatic	++
[aIle^{5,11}]CyA	enzymatic	+
[D-Abu^8]CyA	enzymatic	++

+++ = strong immunosuppressive activity
++ = moderate activity
+ = weak activity
(from (17))

mase; whereas the new cyclosporin could be obtained in sufficient amounts enzymatically. Since for the new cyclosporins no references exist, it was necessary to produce sufficient quantity for structural proof. Preliminary structural proof was done by FAB mass spectrometry and the result is shown for [D-Abu^8]CyA in Fig. 11.4b.

A second criterion to confirm a cyclosporin structure is the immunosuppressive activity of this class of cyclopeptides. As can be seen in Table 11.2 all new cyclosporins so far synthesized *in vitro* exert immunosuppressive activity in *in vitro* assays.

REFERENCES

1. Billich, A., and Zocher, R. (1987) Enzymatic synthesis of cyclosporin A. *Biochemistry 26*, 8417–8423.
2. Billich, A., and Zocher, R. (1987) *N*-methyltransferase function of the multifunctional enzyme enniatin synthetase. *J. Biol. Chem. 262*, 17258–17259.
3. Billich, A., and Zocher, R. (1990) Formation of *N*-methylated peptide bonds in peptides and peptidols. *Biochemistry of Peptide Antibiotics* (H. Kleinkauf and H. v. Döhren, eds.) pp. 57–79, de Gruyter Berlin 1989.
4. Borel, J. F. (1986) Editorial: Ciclosporin and its future. *Prog. Allergy 38*, 9–18.
5. Diez, B., Gutierrez, S., Barredo, J. L., van Solingen, P., van der Voort, L., and Martin, J. F. (1990) The cluster of penicillin biosynthetic genes: Identification and characterization of the *pcb*AB gene encoding the α-aminoadipyl-cysteinyl-valine synthetase and linkage to the *pcb*C and *pen*DE genes. *J. Biol. Chem. 265*, 16358–16365.
6. Dittmann, J., Lawen, A., Zocher, R., and Kleinkauf, H. (1990) Isolation and partial characterization of cyclosporin synthetase from a cyclosporin non-producing mutant of *Beauveria nivea*. *Biol. Chem. Hoppe-Seyler 371*, 829–834.
7. Durette, P. L., Boger, J., Dumont, F. Firestone, R., Frankshun, R. A., Koprak, S. L., Lin, C. S., Melino, M. R., Pessolano, A. A., Pisano, J., Schmidt, J. A., Sigal, N. H., Staruch, M. J. and Witzel, B. E. (1988) A study of the correlation between cyclophilin binding and *in vitro* immunosuppressive activity of cyclosporine A and analogues. *Transplant. Proc. 20, Suppl. 2*, 51–57.
8. Fliri, H. G. and Wenger, R. M. (1990) Cyclosporine: Synthetic studies, structure-activity relationships, biosynthesis and mode of action. *Biochemistry of Peptide Antibiotics* (Kleinkauf, H. and von Döhren, H.; eds.) Walter de Gruyter, Berlin, New York, pp. 245–287.
9. Kahan, B. D. (ed.) (1984) *Cyclosporin: Biological Activity and Clinical Applications*, Grune & Straton Inc., Orlando.
10. Keller, U. (1987) Actinomycin synthetases. Multifunctional enzymes responsible for the synthesis of the peptide chains of actinomycin. *J. Biol. Chem. 262*, 5852-5856.
11. Keller, U., Kleinkauf, H., and Zocher, R. (1984) 4-Methyl-3-hydroxy-anthranilic acid activating enzyme from actinomycin-producing *Streptomyces chrysomallus*. *Biochemistry 23*, 1479–1484.
12. Kleinkauf, H., and von Döhren, H. (1987) Biosynthesis of peptide antibiotics. *Annu. Rev. Microbiol. 41*, 259–289.
13. Kleinkauf, H., and von Döhren, H. (1990) Bioactive peptide analogs: *In vivo* and *in vitro* production. *Progr. Drug Res. 34*, 287–317.
14. Komura, S., and Kurahashi. (1985) Biosynthesis of polymyxin E by a cell-free enzyme system. IV. Acylation of enzyme-bound L-2,4-diaminobutyric acid. *J. Biochem. (Tokyo) 97*, 1409–1417.
15. Krätzschmar, J., Krause, M., and Marahiel, M. A. (1989) Gramicidin S biosynthesis operon containing the structural genes *grs*A and *grs*B has an open reading frame encoding a protein homologous to fatty acid thioesterases. *J. Bacteriol. 171*, 5422–5429.
16. Kurahashi, K., Komura, S., Akashi, K., and Nishio, C. (1982) Biosynthesis of antibiotic peptides polymyxin E and gramicidin A. In *Peptide Antibiotics —Biosynthesis and Functions* (Kleinkauf, H., and von Döhren, eds.), pp. 275–288, de Gruyter Berlin, 1982.
17. Lawen, A., Traber, R., Geyl, D., Zocher, R., and Kleinkauf, H. (1989) Cell-free biosynthesis of new cyclosporins. *J. Antibiot. 42*, 1283–1289.
18. Lawen, A., and Zocher, R. (1990) Cyclosporin synthetase. The most complex peptide synthesizing multienzyme polypeptide so far described. *J. Biol. Chem. 265*, 11355–11360.
19. MacCabe, van Liempt, H., Palissa, H., Unkless, S. E., Riach, M., von Döhren, H., and Kinghorn, J. E. (1991) Molecular characterization of the *Aspergillus nidulans acv*A gene encoding δ-(L-α-Aminoadipyl)-L-cysteinyl-D-valine synthetase, the first enzyme of the penicillin biosynthesis. *J. Biol. Chem. 265*, in press.
20. Mittenhuber, G. (1988) Antibiotikasynthetase gene aus *Bacillus brevis*: Organisation der Gene der Tyrocidinsynthetasen 1 und 2. *Thesis*, TU Berlin.

21. Mittenhuber, G., Weckermann, R., and Marahiel, M. A. (1989) Gene cluster containing the genes for tyrocidine synthetases 1 and 2 from *Bacillus brevis*: Evidence for an operon. *J. Bacteriol.* 171, 4881.
22. Mohr, H. and Kleinkauf, H. (1978) Alamethicin biosynthesis: Acetylation of the amino terminus and attachment of phenylalaninol. *Biochim. Biophys. Acta* 526, 375–386.
23. Okumura, Y. (1990) Directed biosynthesis of neoviridogriseins. *Biochemistry of Peptide Antibiotics* (H. Kleinkauf and H. von Döhren, eds.), pp. 365–378, de Gruyter Berlin 1990.
24. Rusnak, F., Faraci, N. S. and Walsh, C. T. (1989) Subcloning, expression, and purification of the enterobactin biosynthetic enzyme 2,3-dihydroxybenzoate-AMP-ligase: Demonstration of enzyme-bound (2,3-dihydroxybenzoyl) adenylate product. *Biochemistry* 28, (17) 6827–35.
25. Schindler, R. (ed.) (1985) *Ciclosporin in Autoimmune Diseases*, Springer Verlag, Berlin 6827–6835.
26. Smith, D. J., Earl, A. J., Turner, G. (1990) The multifunctional peptide synthetase performing the first step of penicillin biosynthesis in *Penicillium chrysogenum* is a 421 073 dalton protein similar to *Bacillus brevis* peptide antibiotic synthetases. *EMBO J.* 9, 2743–2750.
27. Traber, R., Hofmann, H., and Kobel, H. (1989) Cyclosporins—New analogues by precursor directed biosynthesis. *J. Antibiot.* 42, 591–597.
28. Traber, R, Hofmann, H., Loosli, H.-R., Ponelle, M, and von Wartburg, A. (1987) Neue Cyclosporine aus *Tolypocladium inflatum*. Die Cyclosporine K-Z. *Helv. Chim. Acta*, 70, 13–36.
29. van Liempt, H., von Döhren, H., and Kleinkauf, H. (1989) δ-(L-α-Aminoadipyl)-L-cysteinyl-D-valine synthetase from *Aspergillus nidulans*. *J. Biol. Chem.* 264, 3680–3684.
30. Vater, J. (1990) Gramicidin S synthetase. *Biochemistry of Peptide Antibiotics* (H. Kleinkauf and H. v. Döhren, eds.) pp. 33–55, de Gruyter Berlin 1990.
31. von Döhren, H. (1990) Compilation of peptide structures—A biosynthetic approach. *Biochemistry of Peptide Antibiotics* (H. Kleinkauf and H. von Döhren, eds.), pp. 411–507, de Gruyter Berlin 1990.
32. Weckermann, R., Fürbass, R., and Marahiel, M. A. (1988) Complete nucleotide sequence of the *tyc*A gene encoding the tyrocidine synthetase 1 from *Bacillus brevis*. *Nucl. Acids Res.* 16, 11841.
33. Wenger, R. M. (1984) Synthesis of cyclosporine. Total synthesis of `cyclosporin A' and `cyclosporin H,' two fungal metabolites isolated from the species *Tolypocladium inflatum* GAMS. *Helv. Chim. Acta* 67, 502–525.
34. Zocher, R., Nihara, T., Paul, E., Madry, N., Peeters, H., Kleinkauf, H., and Keller, U. (1986) Biosynthesis of cyclosporin A: Partial purification and characterization of a multifunctional enzyme from *Tolypocladium inflatum*. *Biochemistry* 25, 550–553.

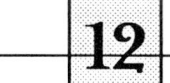

Semduramicin: Design and Preparation of a New Anticoccidial Ionophore by Semi-Synthesis and Mutasynthesis

E. A. Glazer, W. P. Cullen, G. M. Frame, A. C. Goudie*, D. A. Koss, J. A. Olson, A. P. Ricketts,
E. J. Tynan, N. D. Walshe*, W. C. Wernau and T. K. Schaaf
Central Research Division, Pfizer Inc, Groton, CT 06340

A novel polyether ionophore, UK-58,852, exhibiting potent anticoccidial activity but suboptimal toleration in chickens was discovered as part of an effort to identify fermentation-derived products for the treatment of coccidiosis in poultry. We sought to improve the use animal toleration of UK-58,852 with a series of single step chemical transformations designed to introduce structural and physicochemical features consistent with reduced systemic exposure to drug, while retaining anticoccidial potency. This work culminated in the discovery of semduramicin, an ionophore displaying a biological profile warranting further anticoccidial evaluation under field conditions. While initially derived semi-synthetically, we sought a process, based on biosynthetic considerations, for the production of semduramicin by direct fermentation. Successful completion of our mutation program marked the first reported example of an ionophore improved by synthesis and subsequently prepared by direct fermentation.

---

*Central Research Division, Pfizer Inc., Sandwich, U.K.

INTRODUCTION

Avian coccidiosis, an enteric protozoal infection caused by pathogenic species of *Eimeria*, is a ubiquitous problem in the high intensity rearing systems characteristic of the poultry industry [6]. This has resulted in a dependence on anticoccidial feed additives affording prophylactic disease control. Fermentation-derived polyether carboxylic acid ionophores, such as salinomycin and monensin, have been the dominant class of anticoccidial feed additives for nearly two decades. They have achieved this position because they provide excellent disease control and are relatively refractory to resistance development [7, 10, 11].

The anticoccidial activity of the ionophores is thought to be related to their ability to disrupt cationic cross-membrane gradients [10], a mechanism which is not specific to the parasite and can lead to toxicity in the chicken host. Thus, there are few ionophores among more than 120 known members of this class which are sufficiently well-tolerated to be commercially useful. Our objective was to find a novel polyether ionophore which offers improvements such as increased potency or enhanced therapeutic index and growth performance relative to the major anticoccidial drugs in this series, salinomycin and monensin.

A fermentation screening program which employed the Gram-positive indicator organism *Bacillus stearothermophilus* was initiated to specifically highlight new polyether ionophores. This involved an initial examination of broths for the presence of a metabolite which produced a double zone of inhibition characteristic of ionophores. Secondary testing using a *B. stearothermophilus* ionophore sensitive/resistant pair was then conducted to further support the presence of a polyether ionophore.[*] Given the importance of oral activity for an anticoccidial agent, candidate broths were then screened for activity in feed in a chicken model [2] prior to chemical resolution.

[*]Unpublished methodology developed at Central Research Division, Pfizer Inc., Nagoya, Japan.

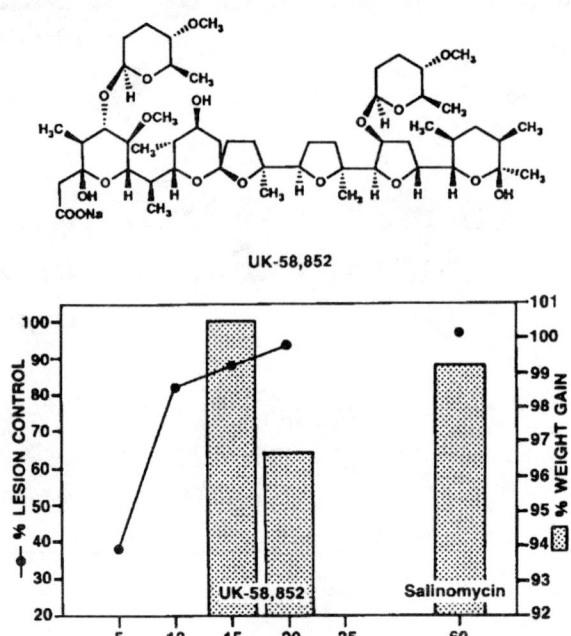

Figure 12.1 / Relationship of anticoccidal (*E. tenella*) and toleration (weight gain in uninfected chicks) for UK-58,852 and salinomycin.

While this work led to the discovery of several novel compounds [3, 4, 5], few stood out as having excellent potency and broad spectrum. One of these, UK-58,852 [13], is effective in controlling the five major pathogenic species of *Eimeria* in our laboratory models when given continuously at 20 ppm in feed, which is one-third the use level of salinomycin. Unfortunately, while UK-58,852 at 20 ppm provides optimal lesion control, it causes weight suppression at this concentration after administration for only three weeks in feed (Figure 12.1). Although this performance deficit appears relatively modest, it becomes more severe over the 6–7 week broiler growth cycle (data not shown) and renders UK-58,852 unsuitable for possible commercial use. Rather than abandon this attractive but flawed natural product, we chose to attempt to improve its suboptimal therapeutic index through synthetic modification. Our approach, based on rational drug design and execut-

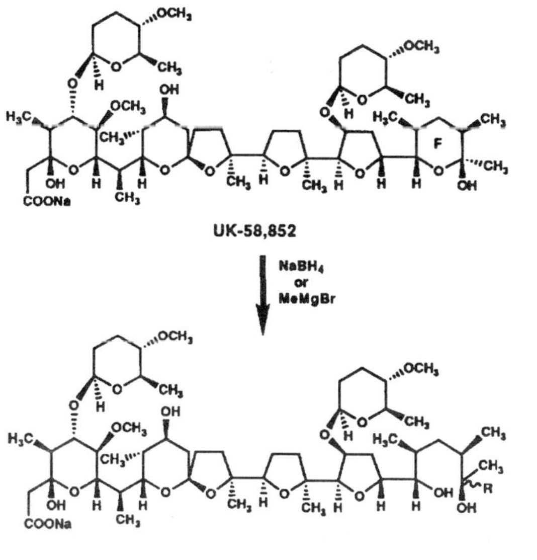

Figure 12.2 / Properties of UK-58,852 F-ring analogs

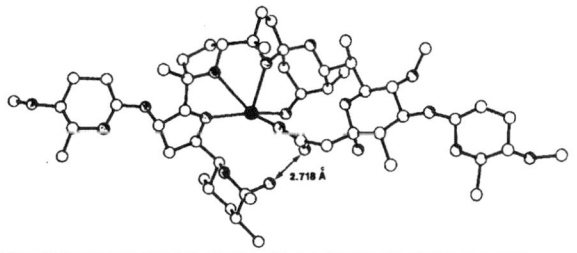

Figure 12.3 / Three dimensional representation of UK-58,852 (silver salt)

ed by synthetic and mutasynthetic methodologies, is reported herein.

RESULTS

Synthetic modification of UK-58,852 targeted a series of novel, polar analogs. Based on literature reports [18, 19] that major structural changes are inconsistent with retention of biological activity, we chose to emphasize relatively simple, single step transformations. Initial efforts involved selective opening of the F-ring hemiketal of UK-58,852 with sodium borohydride [19] or Grignard reagents such as methyl magnesium bromide [8]. As shown in Figure 12.2, analogs 1 and 2 are substantially more polar than UK-58,852 (reversed phase HPLC). Another approach to polar analogs of UK-58,852 involved the hydrolytic cleavage of the A and E-ring deoxy sugars to unmask additional hydroxyl functionality (Fig. 12.3).

Acid mediated hydrolysis of UK-58,852 was conducted under conditions which resulted in the formation of two polar ionophores (Figure 12.4). Thus, treatment of UK-58,852 with p-toluenesulfonic acid (1.1 equivalents) in acetonitrile/water (95/5) for 4 hours afforded monoglycone 3 as the predominant hydrolysis product. The structure of monoglycone 3 was unambiguously assigned by X-ray crystallography (Figure 12.5). A more extended reaction time (24 hours) gave comparable amounts of both 3 and the corresponding aglycone 4.

As shown in Figures 12.2 and 12.4, synthetic modification of UK-58,852 provided a series of ionophores exhibiting a wide range of polarities and anticoccidial activities. However, the monoglycone 3, now designated semduramicin [16, 17], stood out as having the most promising anticoccidial profile (Figure 12.6).

As described above, semduramicin can be synthesized by acid hydrolysis of UK-58,852. While this process is efficient for laboratory scale preparations, we sought a method which would be amenable to large scale production of this new ionophore for more extensive biological testing. We therefore investigated the possibility of obtaining semduramicin directly from fermentation. Although the biosynthetic pathway for semduramicin has not been established, we hypothesized that glycoside formation may occur last, as in the biosynthesis of erythromycin [9, 14]. Accordingly, we undertook a mutation program [1, 15] with the

UK-58,852

3 (SEMDURAMICIN)

4

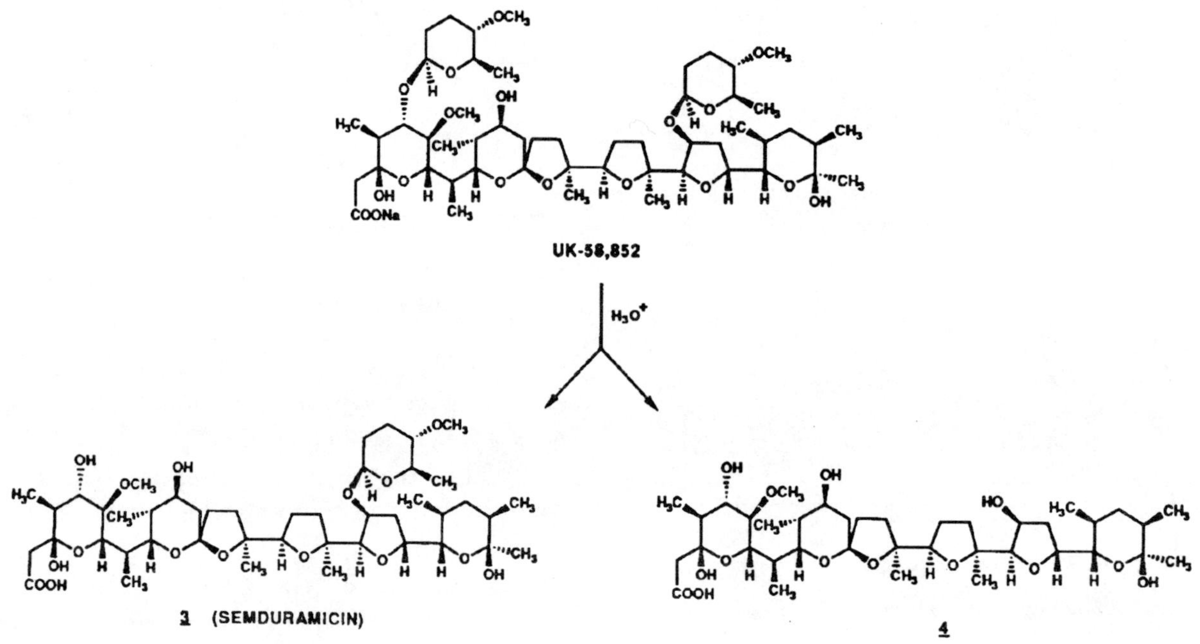

Compound	MEC[a] (ppm)	RT[b]
UK-58,852	20	10.4
3	⩽25	6.9
4	>50	4.3

[a]minimum effective concentration in feed in chickens infected with *E. tenella*
[b]reversed phase HPLC retention time in minutes (90:10 CH_3CN/H_2O; Partisil C8 10 μ column)

Figure 12.4 / Products resulting from acid hydrolysis of UK-58,852

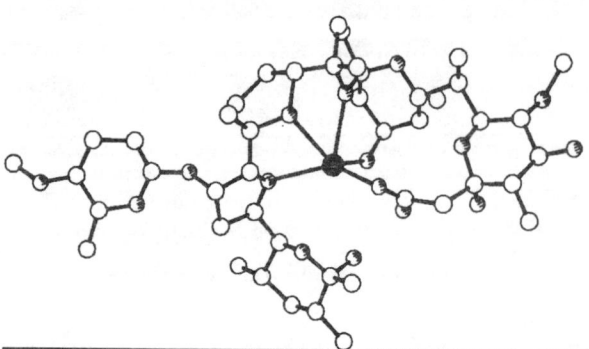

Figure 12.5 / Three dimensional representation of semduramicin (silver salt)

UK-58,852 producing organism (*Actinomadura roseorufa,* ATCC39697) designed to generate a mutant blocked in the putative glycosyl transferase 2 (Figure 12.7). The desired mutant was obtained in two steps. An interim, partially blocked mutant producing both mono and diglycone (~9:1 ratio) was selected following treatment of the initial UK-58,852 producing culture with the mutagenic agent N-methyl-N′-nitro-N-nitrosoguanidine. Subsequent mutation of this interim culture afforded the direct semduramicin producing organism, free of co-produced UK-58,852. The large scale fer-

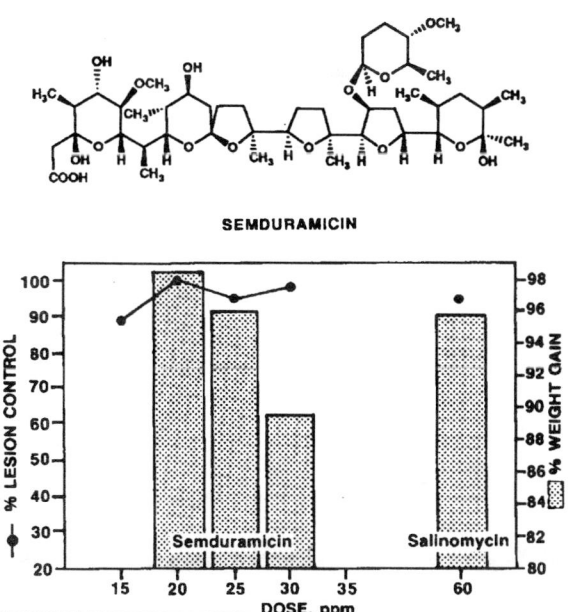

Figure 12.6 / Relationship of anticoccidial efficacy (*E. tenella*) and toleration (weight gain in uninfected chicks) for semduramicin and salinomycin.

mentation of semduramicin was readily accomplished using this mutant organism as the inoculum. Recovery of the natural product involved extraction of whole broth with methylisobutyl ketone, followed by silica gel chromatography and recrystallization. Semduramicin produced by fermentation was identical in all respects to that obtained semi-synthetically from UK-58,852.

DISCUSSION

We reasoned that improved toleration and an attendant expanded therapeutic ratio might be achieved by reducing the exposure of the chicken to the ionophore by increasing the drug clearance rate. We sought to accomplish this objective by targeting analogs of UK-58,852 with enhanced polarity. The ring opened analogs *1* and *2* are more polar than UK-58,852, but are also much less potent than the parent compound against cecal coccidiosis (*Eimeria tenella*). The X-ray structure

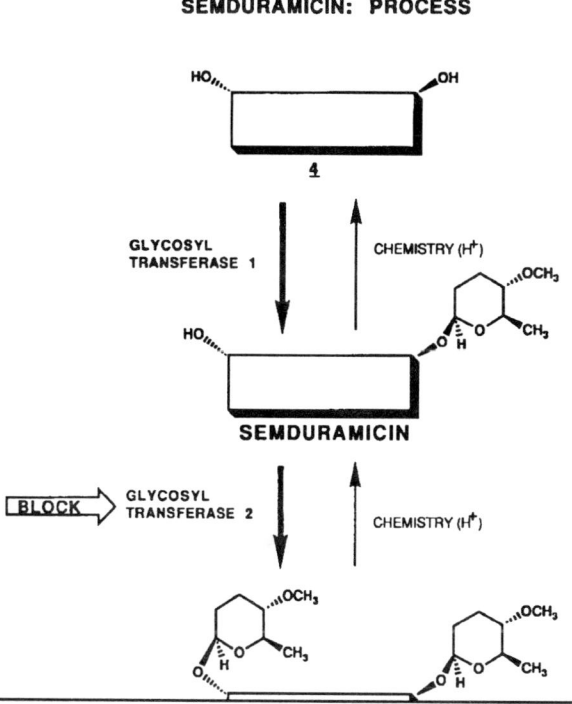

Figure 12.7 / Hypothetical biosynthetic glycosylation of aglycone 4.

of UK-58,852 (Figure 12.3) shows that the oxygen of the F-ring hydroxyl is within 2.718 Å of the carboxyl oxygen and can therefore participate in the formation of a strong hydrogen bond. This interaction could contribute to the conformational stability of the ionophore-cation complex. We reasoned that the reduction in potency seen with F-ring analogs might reflect a disruption in this strong hydrogen bond and, thus, in the ion binding properties of the ionophore. Subsequent transformations, therefore, targeted the deoxy sugars located at the periphery of the molecule, distal to the ion-binding cavity.

The most polar hydrolysis product, aglycone *4*, lacked significant anticoccidial activity either *in vivo* (MEC >50 ppm) or surprisingly, *in vitro*. Whether the high polarity of *4* results in an inability to penetrate lypophilic membranes with a consequential loss of intrinsic anticoccidial activity is unknown. However, semduramicin (*3*) is intermediate in polarity, and afforded broad spectrum disease control at ≤25 ppm, comparable in potency to the parent ionophore in our primary *in vivo* models. Further evaluation of semduramicin was conducted in secondary screens designed to establish whether evaluation under use conditions was warranted [12]. These tests showed that semduramicin at ≤25 ppm is both highly efficacious and very well tolerated, with an anticoccidial profile (Figure 12.6) at least comparable to that of salinomycin, the current industry standard. The improved toleration of semduramicin is consistent with metabolism studies* which show a substantially increased drug clearance rate relative to that of UK-58,852.

CONCLUSION

Having discovered a potent, broad spectrum anticoccidial ionophore (UK-58,852) by fermentation, we were disappointed to find that it lacked toleration in chickens comparable to that of key commercial agents. Rather than abandon this natural product, we chose to intervene synthetically to correct the flaw in toleration. Reasoning that a broadened therapeutic index might result from an enhanced drug clearance rate, we targeted UK-58,852 analogs with increased polarity. This work culminated in the discovery of semduramicin, a novel polyether ionophore with an anticoccidial profile in our model studies at least comparable to salinomycin, yet more than twice as potent. While initially derived by semisynthesis, a successful mutation program afforded a process for the direct production of semduramicin by fermentation. This is the first example of an ionophore improved by synthesis and subsequently produced by direct fermentation.

---

*Lynch, M. J. Central Research Division, Pfizer Inc., Groton, CT, personal communication.

ACKNOWLEDGEMENT

We would like to recognize the contribution of Pfizer Central Research, Nagoya, Japan, to the discovery of UK-58,852 and particularly Junsuke Tone and Hiroshi Maeda.

REFERENCES

1. Baltz, B. H. 1986. Chapter 14. Mutagenesis in *Streptomyces* spp. In: Manual of Industrial Microbiology and Biotechnology (Demain, A. L. and N. A. Solomon., Eds.) pp. 184–190, American Society for Microbiology, Washington, D.C.
2. Chappel, L. R., H. L. Howes and J. E. Lynch. 1974. The Site of Action of a Broad Spectrum Aryltriazine Anticoccidial, CP-25,415. *J. Parasitol.* **60**: 415–420.
3. Cullen, W. P., J. Bordner, L. H. Huang, P. M. Moshier, J. R. Oscarson, L. A. Presseau, R. S. Ware, E. B. Whipple, Y. Kojima, H. Maeda, S. Nishiyama, J. Tone, K. Tsukuda, K. S. Holdom and J. C. Ruddock. 1990. CP-60,993, A New Dianemycin-like Ionophore Produced by *Streptomyces hygroscopicus* ATCC 39305: Fermentation, Isolation and Characterization. *J. Indus. Microbiol.* **5**: 365–374 and references cited therein.
4. Dirlam, J. P., A. M. Belton, J. Bordner, W. P. Cullen, L. H. Huang, Y. Kojima, H. Maeda, H. Nishida, S. Nishiyama, J. R. Oscarson, A. P. Ricketts, T. Sakakibara, J. Tone and K. Tsukuda. 1990. CP-84,657, A Potent Polyether Anticoccidial Related to Portmicin and Produced by *Actinomadura* sp. *J. Antibiot.* **43**: 668–679.
5. Dirlam, J. P., L. Presseau-Linabury and D. Koss. 1990. The Structure of CP-80,219, A New Polyether Antibiotic Related to Dianemycin. *J. Antibiot.* **43**: 727–730.
6. Fernando, M. A. 1982. Chapter 7. Pathology and Physiology. In: The Biology of the Coccidia (P. L. Long, Ed.) pp. 287–327, University Park Press, Maryland.
7. Grafe, U., R. Schlegel and M. Bergholz. 1984. *Polyether Antibiotics. Pharmazie* **39**: 661–670.
8. Hammen, P., S. Grabley, G. Seibert and I. Winkler. 1988. New Nigericin Derivatives. European Patent Application 336–248-A.

9. Masamune, S., S. B. Gordon and J. W. Corcoran. 1977. Macrolides. Recent Progress in Chemistry and Biochemistry. *Angew. Chem. Int. Ed. Engl.* **16**: 585–607.
10. McDougald, L. R. 1982. Chapter 9. Chemotherapy of Coccidiosis. In: The Biology of the Coccidia (P. L. Long, Ed.) pp. 373–427, University Park Press, Maryland.
11. McDougald, L. R. and W. M. Reid. 1988. Anticoccidial Drug Market: 1987. *Broiler Industry* **40**: 100–108.
12. Ricketts, A. P., L. R. Chappel, E. A. Glazer, T. T. Migaki and J. A. Olson. SIM Conference, October 14–17, 1990, Sarasota, Florida.
13. Ruddock, J. C., L. R. Chappel, W. P. Cullen, K. S. Holdom, H. Maeda, J. Tone, E. Whipple and D. J. Williams. UK-58,852, a Novel Ionophore. Fermentation, Isolation and Biological Properties. SIM Conference, October 14–17, 1990, Sarasota, Florida.
14. Tsou Hwei-ru, R. S. Rajan, R. Fiala, P. C. Mowery, M. W. Bullock, D. B. Borders, J. C. James, J. H. Martin and G. O. Morton 1984. Biosynthesis of the Antibiotic Maduramicin. *J. Antibiot.* **37**: 1651–1663.
15. Tynan, E. J., T. H. Nelson, R. A. Davies and W. C. Wernau. The Direct Production of Semduramicin by a Mutant of *Actinomadura roseorufa* Huang sp. nov. SIM Conference, October 14–17, 1990, Sarasota, Florida.
16. USAN Council. 1989. List No. 310, New Names. Clinical Pharmacology and Therapeutics. **46**: 483.
17. USAN Council. 1990. List No. 313, New Names. Clinical Pharmacology and Therapeutics. **47**: 91.
18. Wells, J. L., J. Bordner, P. Bowles and J. W. McFarland. 1988. Novel Degradation Products from the Treatment of Salinomycin and Narasin with Formic Acid. *J. Med. Chem.* **31**: 274–276.
19. Westley, J. W. 1983. Chapter 2. Chemical Transformations of Polyether Antibiotics. In: Polyether Antibiotics, Vol. 2 (Westley, J. W., Ed.) pp. 51–86, Marcel Dekker, New York.

13

Stereospecific Microbiological Chiral Reduction of Ketone Groups by a Novel *Aspergillus* Species

Ellen S. Baron and Michael Homann
Microbiological Development Department, Schering-Plough Corp., Union, N.J. 07083, USA.

Fermentation and enzyme technologies provide efficient and economic alternatives to classical chemical processes for the synthesis of pharmaceutical compounds. Of particular interest are applications concerning redox reactions involving chiral compounds such as dilevalol, a vasodilating agent which blocks β-adrenergic receptors. The unique pharmacological properties of this cardiac drug are dependent on the formation of the R R diastereomeric alcohol resulting from the stereospecific reduction of a functionalized ketone group. More than 50 microbial cultures were screened for their ability to catalyze this bioconversion including the identification and subsequent utilization of a unique soil isolate of Aspergillus niveus (ATCC 20922). This microorganism was capable of efficiently catalyzing this stereospecific bioconversion without the concomitant degradation of the desired product. Through improvements in compound production rates and reaction yields an industrial process for the stereospecific synthesis of dilevalol using microbial fermentation was established. This microbial process obviates the need for the recycling of chemical racemates enabling improved economics and reduced environmental hazards. Additional improvements should be forthcoming from continued research including molecular cloning currently under investigation.

INTRODUCTION

The drive to produce stereoisomerically pure drugs is based on the recognition that there are pharmacological differences between drug stereoisomers. Although it is known from numerous examples that optically pure isomers of pharmaceuticals have very different activities, still more than 80% of all chiral pharmaceuticals are used in the racemic form, and at least 30% of all pharmaceuticals have one or more chiral centers [8]. This circumstance has been attributed to the fact that the difference in activity was realized at a comparatively late stage of the development of the drug, or to the fact that effective methods for the separation of stereoisomers were not available.

The consciousness of the importance of stereochemistry was surfaced with thalidomide where only the R isomer has the favorable sedative effect while the S isomer causes prenatal deformities. Numerous examples appear in the literature which document biochemical differences among stereoisomers [8].

A recent review by the PMA Ad Hoc Committee on Racemic Mixtures [5] reveals that in general the focus of chiral pharmaceuticals should be on the single active stereoisomer to avoid the possible negative side effects resulting from racemic mixtures. The stereoisomeric composition of drug substances has become a key issue in the development, approval and clinical use of pharmaceuticals. The role of stereochemistry in drug development is rapidly evolving based on the accumulation of scientific knowledge and the development of new technologies which has enabled the synthesis, separation, and analysis of individual enantiomers of a particular drug substance.

Contemporary microbial transformations have become one of the major sources utilized to conduct chiral non-racemic organic synthesis. For example, *Saccharomyces cerevisiae* is utilized as a virtual chemical reagent owing to the diverse capability of catalyzing asymmetric synthesis of optically active compounds [11]. Similarly, the NADPH dependent oxidoreductase of *Thermoanaerobium brockii* has been utilized for the enantioselective reduction of a broad spectrum of aldehydes and ketones as well as for the oxidation of primary and secondary alcohols [13]. The wide application of microbial fermentations to conduct stereospecific reactions for complex organic syntheses is primarily due to the diverse capabilities the microorganisms possess to regenerate expensive cofactors such as NADPH as well as provide a stable environment for novel enzymatic catalysis. Reported in this paper is a novel strain improvement program we devised to develop a dilevalol resistant mutant of soil isolate *Aspergillus niveus* (104C) which would be commercially capable of producing dilevalol.

PROCESSES FOR GENERATION OF CHIRAL COMPOUNDS—ADVANTAGES AND DISADVANTAGES

Currently, much of the strategy regarding chirality in the chemical industry is still based on classical chemistry, specifically being dependent on resolution technology. The obvious weaknesses of classical resolution lie in the often cumbersome recycling processes needed to make use of the total molecular input. Occasionally recycling is impossible leading to cost and waste disposal problems. Nevertheless, using classical resolution the industry probably exceeds a billion dollars a year and interest in chiral therapeutics has been projected to be in the order of $2 billion by the end of the century [3]. Examples include the manufacture of side chains for many classes of antibiotics, several nervous system compounds, anti-inflammatories, prostaglandins, etc.

Alternatively, a wide assortment of industrially relevant reactions (Figure 13.1) can be accomplished using either microbial or enzymatic processes [10].

Enzymes provide unique features which are sometimes not achievable using standard chemical catalysts. Enzymatic methods can be applied usefully in cases where either substrates or prod-

- Oxidation/reduction with chirality
- Hydroxylation with positional specificity
- Hydrolysis/formation of amide, ester bonds with chiralty
- Addition to double bonds with chirality
- Selective phosphorylation
- Production/separation of cis/trans isomers
- Trans esterification
- Oxygenation

Figure 13.1 / Relevant reactions for microbial or encymatic processes

ucts of reactions are chemically labile. Enzyme-catalyzed reactions offer stereospecificity, regiospecificity, and are effective at low temperatures (0–110C) and are active over a broad range of pH values (pH 2–12). Enzymes are non-toxic and readily degradable (non-polluting), and they can be produced in large quantities.

Enzymes can also often substitute for reactions requiring strong acids and bases. This translates into a money savings obviating the need for the use of specially resistant materials in reaction vessels. It also means that the presence of large amounts of salts (which would otherwise have to be removed after the reaction) can be eliminated. This can offer significant cost savings and lead to a minimization of environmental pollutants.

Since enzymes are proteins they have their limitations. Enzymes are sensitive to elevated temperatures, extremes in pH, exposure to harsh chemicals, and sometimes, sensitivity to organic solvents. It is not possible to use enzymes under conditions which denature proteins or in the presence of chemicals which specifically inhibit a particular enzyme activity.

Enzymes also have their drawbacks due to limited stability, poor productivity and conversion yields, narrow specificity (one substrate per enzyme) [2] and dependence on co-factor regeneration. Finally, the cost of enzymes may preclude their application in process stages that function well with inexpensive chemicals. Consequently, selective bioconversions using whole cells can offer several advantages over isolated enzymes in that they are generally economical, the enzymes are stable in their natural environment, cofactor recycling is done automatically and there is a broad substrate specificity. Practical problems, peculiar to scale-up, have been greatly reduced with the introduction of immobilization technology.

Immobilized or entrapped non-viable cells for *in situ* enzymatic transformations have been shown to provide attractive industrial process alternatives [16].

Some disadvantages of using whole cell systems still have to be overcome. These include the fact that the reactions often involve the use of substrates at low molar concentrations (frequently in the range of millimolar levels), and reaction yields and compound production rates are often low. Additionally, whole cell systems sometimes exhibit low enantioselectivity or stereospecificity owing to the operation of more than one enzyme as well as low yields due to catabolism of the product. Despite these disadvantages, and as evidenced by the large numbers of publications which have appeared in recent years, the clearly perceived usefulness of these systems, and the advantages of using them, outweigh the disadvantages.

Outlined in Figure 13.2 are some of the advantages/disadvantages of the process technologies previously discussed. Overall, biotransformations are being applied with increasing frequency toward the synthesis of chiral compounds because they are more efficient and cost effective than chemical or physical methods which also tend to be environmentally burdening.

DILEVAVOL PRODUCTION BY FERMENTATION USING *ASPERGILLUS NIVEUS*

Dilevalol is (R,R)-(-)-2-hydroxy-5-{1-hydroxy-2-[(1-methyl-3-phenylpropyl)amino]ethyl} benzamide hydrochloride, one of four optical isomers of

	Chemical Synthesis	Fermentation	Enzymatic Technology
Cost	Low—High	Low—Moderate	Low—High
Yield	Low—High	Low—High	Moderate—High
Time	Rapid	Slow	Rapid
Selectivity	Low	High	High
Versatility	Narrow	Broad	Narrow
Toxic By-Products	Solvents, heavy metals, acids, bases, etc.	Potential toxins process-related contaminants	Few
Environmental	Waste problem (cost & handling	Minimal impact	Minimal impact
Conditions*	Severe	Mild	Mild

*Temperature, pH, pressure, aqueous vs. non-aqueous, flammable

Figure 13.2 / Advantages and disadvantages of various process technologies.

labetalol. This compound exhibits both vasodilating and beta-blocking activities. It has the structural formula:

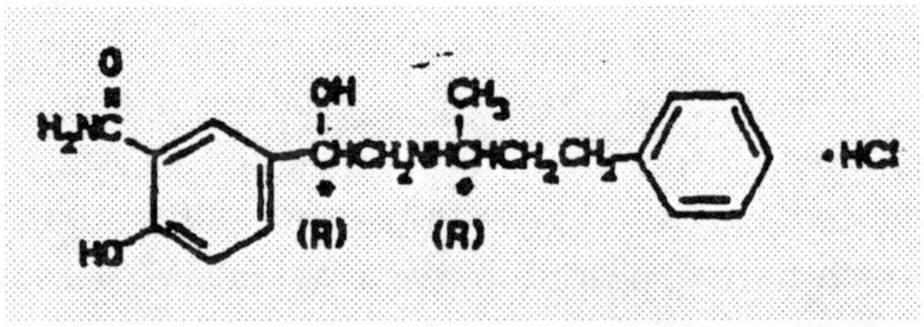

The current chemical process for the synthesis of dilevalol involves the reduction of R-aminoketone resulting in a 50:50 racemic mixture containing the desired and undesired stereoisomeric configurations of dilevalol. Separation of the single desired configuration (R-R) from the undesired stereoisomer (S-R) is a labor and cost intensive operation. Our goal was to modify this production process by developing an efficient economical microbial bioprocess for the stereospecific reduction of R-aminoketone to dilevalol.

DEVELOPMENT OF A FERMENTATION FOR THE BIOCONVERSION OF R-AMINOKETONE TO DILEVALOL

Initial studies were devoted to identifying a microorganism capable of producing the R-R diastereomeric alcohol from the reduction of R-aminoketone hydrochloride. More than 50 cultures were screened for this activity including a soil isolate of *Aspergillus niveus* which was found to be capable of stereospecifically reducing R-aminoketone without further catabolism of the product. Summarized below are some of the key developments achieved. Details of the strain selection and enzyme characterization of the bioconversion reaction will be published elsewhere [6, 7].

MATERIALS AND METHODS

- *Media:* Composition of all media are given in Table 13.1 and are described in Homann et al. [7].
- *Microbial Bioconversion:* Conidia (5×10^6) were inoculated into seed medium (100 mls/300 ml flask) and allowed to germinate and grow vegetatively for 48 hours at 34°C

Table 13.1 / Dilevalol Production Media (g/l)

	DLV-1	DLV-2	DLV-3	DLV-4	DLV-5
Potassium Nitrate	—	—	—	—	10.0
Yeast Extract	5.0	2.0	10.0	10.0	—
Glucose	9.0	10.0	18.0	18.0	18.0
Glycerol	—	—	—	5.0	5.0
KH_2PO_4	2.5	2.5	2.5	2.5	2.5
l-Lysine	—	—	1.0	1.0	1.0
$MgSO_4 \cdot 7H_2O$	—	—	0.25	0.25	0.25
$ZnSO_4 \cdot 7H_2O$	—	.001	0.05	0.05	0.05
$FeSO_4$	—	.001	—	—	—
$CuSO_4$	—	.002	—	—	—
$MnSO_4$	—	.00012	—	—	—
$NaMoO_4$	—	.00005	—	—	—
Post-sterile pH:					
Flask	5.1	5.2	5.1	5.1	5.1
Fermenter	—	—	5.3	5.3	—

Dilevalol Seed Media (g/l)

Dilevalol Sporulation Media (g/l)

	SIM-6	SIM-15	VEZINA
Soy Flour	35.0	18.0	—
Potato Dextrin	50.0	50.0	—
WPD-650			
Cerelose	5.0	5.0	27.5
Calcium Carbonate	5.0	5.0	—
$CoCl_2 \cdot 6H_2O$	0.002	0.002	—
NaCl	—	—	250
Corn Steep	—	—	5
Molasses	—	—	50
Post-sterile pH	7.1	6.9	6.0

with agitation (250 rpm, 2 inch stroke). A 3–5% inoculum (vol/vol) was transferred from the seed cultures into production media (100 mls/300 ml flask) and grown at 34°C for approximately 24 hours with agitation (300 rpm) followed by addition of R-aminoketone hydrochloride (3–16 g/l R-aminoketone base). R-aminoketone hydrochloride was bioconverted to dilevalol for a period of 48–72 hours, at 34°C with agitation (300 rpm). Fermentation samples were extracted with acidified methanol and analyzed by HPLC as described by Homann et al [7]. The stereospecific R-R configuration of dilevalol was confirmed by Mr. M. Ruggeri (Schering-Plough Research, Chemical Development, Union).

- *Mutagenesis:* Aspergillus niveus strain 104C was used as the initial parent strain. Germinating conidia grown briefly in DLV-1 medium were chemically mutagenized using the alkylating agent NTG ((N-methyl-N-nitro-)N-nitrosoguanidine) in phosphate buffer (0.1M) as described by Homann et al. [7].
- **In Vitro** *Enzyme Assays:* Dilevalol synthesis was measured in the presence of 100 mM MOPS buffer pH 7.5, 2 mM NADPH, 4 mM RAK-HCl and 0.1–1.0 mg/ml sample protein at 30°C for a period of 60 min as described by Homann et al [6]. Protein concentration was determined by the method of Bradford [1].
- *RAK Reductase Purification:* Whole cells were disrupted and membranous material removed by differential centrifugation, followed by purification using anion exchange and dye-ligand chromatography as described by Homann et al [6].
- *Analysis of R-aminoketone-reducing Protein:* Affinity purified enzyme was concentrated and subjected to isoelectric focusing (pH 4.2–4.6). The location of protein bands with reductase activity was identified *in situ* by a coupled colorimetric assay whereby menadione (5mM) serves as an electron donor in the conversion of nitroblue tetrazolium (0.5 mg/ml) to formazan which turns the corresponding protein band purple [12, 14]. To confirm this fraction was also capable of reducing R-aminoketone to dilevalol, a parallel region of the IEF gel (non-stained) was removed and assayed for activity *in vitro*. The gel slice containing the active enzyme was subjected to SDS-polyacrylamide electrophoresis [9] and subsequently soaked in KCl to visualize protein precipitates rather than staining. The opaque protein bands were removed and electro-eluted using 3-cyclohexylamino-1-propane-sulfonic acid buffer without methanol. The protein solution was concentrated approximately ten-fold in a Centricon filtration unit and submitted for N-terminal sequence analysis.
- *Isolation of the A.* niveus *trpC Gene:* Recombinant plasmids possessing *A. niveus* chromosomal DNA fragments were transformed into *Escherichia coli* strain MC1066 according to Hanahan and Meselson [4]. Those plasmids containing the *A. niveus trpC* gene were identified by complementation of the *trpC* mutation of MC1066 by selecting for growth on minimal medium without tryptophan.
- *Transformation of A.* nidulans: The *A. niveus trpC* gene was transformed into *A. nidulans* (*trpC*) according to the method of Yelton et al [17]. Transformants were selected for growth on minimal medium without tryptophan.
- *Determination of A.* niveus *Genome Size:* To establish the genome size of *A. niveus,* an analysis based on the method of van Gorcom et al. [15] was employed. Known amounts of *A. niveus* chromosomal and plasmid DNA (possessing the cloned *trpC* gene of *A. niveus*) were dot-blotted onto nitrocellulose filters and probed with a P^{32}-labeled fragment of the *trpC* region. The activity present in the spots was quantitated in a scintillation counter and the approximate genome size calculated according to the formula illustrated in Figure 13.3.

RESULTS

ISOLATION OF *A. NIVEUS* DILEVALOL RESISTANT STRAINS

Soil isolate *Aspergillus niveus* strain DL104C was capable of stereospecifically bioconverting 1 g/l R-aminoketone (base) to dilevalol with 80% efficiency. At R-aminoketone (RAK) levels of 3 g/l, cell viability and growth decreased, as well as the efficiency of bioconversion (25%). Cell viability and growth were also reduced by dilevalol. In order to establish a viable industrial bioprocess increased levels of conversion of RAK to dilevalol had to be achieved. Since RAK was found to be chemically unstable in semi-solid agar plate medi-

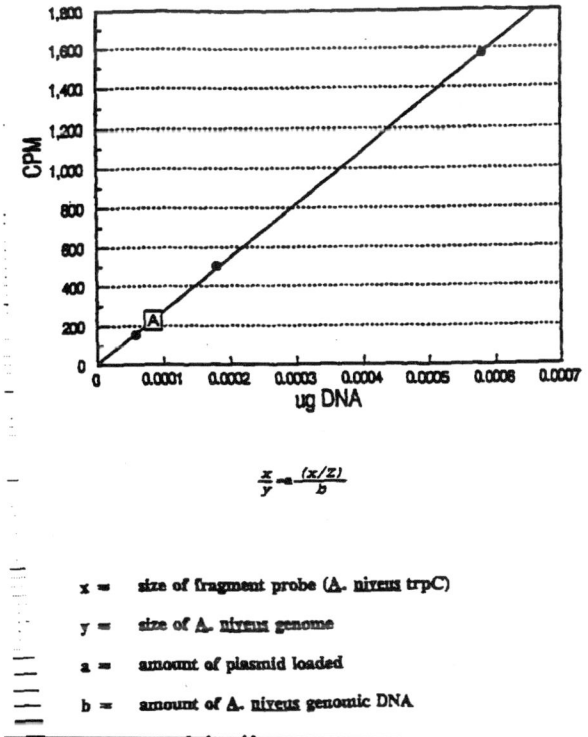

Figure 13.3 / Determination of *Aspergillus niveus* genome size.

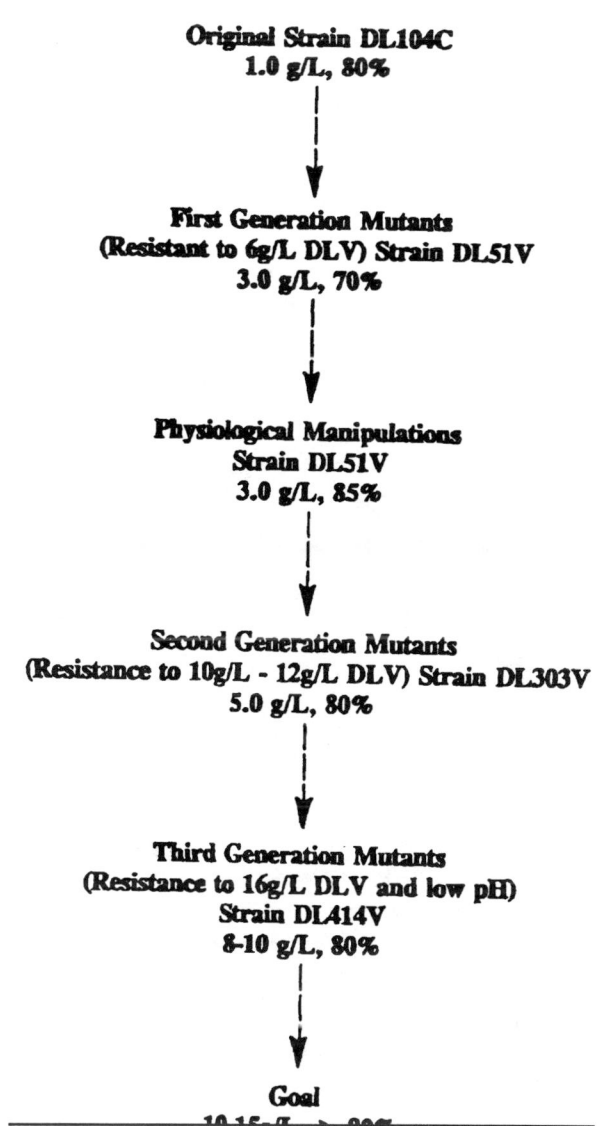

Figure 13.4 / *A. Niveus* dilevatol resistant strains.

um, we elected to select for mutant colonies resistant to increasing levels of dilevalol.

Following mutagenesis with the alkylating agent (N-methyl-N-nitro) N nitrosoguanidine, three generations of dilevalol resistant mutants were isolated, each being selected at progressively elevated levels of dilevalol (Figure 13.4). The final selection was also conducted in a low pH medium at the limit of dilevalol solubility. This selective environment was considered advantageous since a low pH fermentation would minimize the chemical decomposition of R-aminoketone (see Figure 13.5). The resulting mutant, strain DL414V was capable of bioconverting 10g/L R-aminoketone to dilevalol with ≥80% efficiency. The bioconversion capability of this mutant was stable following several successive sporulation and serial propagations under non-selective conditions (without dilevalol) in flask fermentations indicating that this mutant was a viable candidate for fermentation scale-up.

FERMENTATION SCALE-UP

Considerable research was conducted at the 10L scale to answer practical questions aimed at achieving a reproducible and scaleable process. Details of the process optimization and scale-up will be published elsewhere. Fermentation temperature was optimized to 34°C. The chemical decomposi-

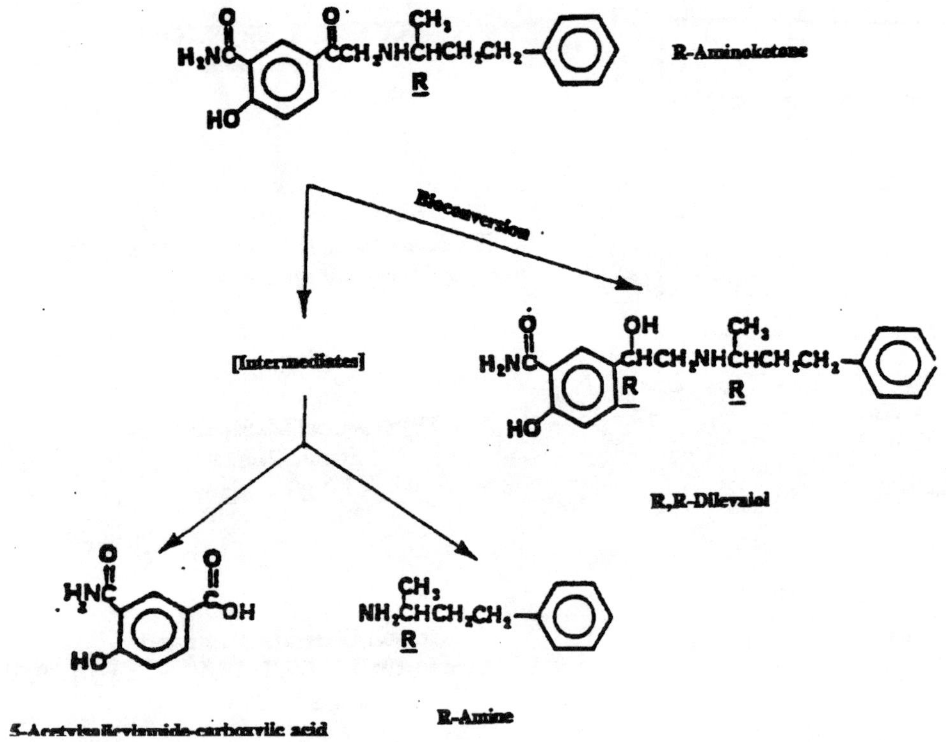

Figure 13.5 / Chemical decomposition of R-aminoketone.

tion of R-aminoketone was shown to decrease with decrease in pH. As can be seen in the schematic of the dilevalol bioconversion fermentation profile (Figure 13.6) the pH was optimized to allow spontaneous change, initially reaching a minimum of 4.7 and a gradual increase to around 5.5 where the pH was maintained in the fermentor. Carbon dioxide evolution rate (CO_2) peaks at log 14 and then starts to form a shoulder which coincided with the rise in pH at log 19. CO_2 then decreases asymptotically for the remainder of the fermentation. As evident from the figure, dissolved oxygen forms a pattern which is nearly complementary to the CO_2 evolution rate. Several salient findings were made. First, the timing of R-aminoketone addition was very critical in getting satisfactory bioconversion. R-aminoketone addition time was coupled to an observation of a 0.5 unit rise in pH just prior to RAK addition with a concomitant plateau in CO_2 evolution. Second, there was a problem in scale-up of oxygen transfer due to the high viscosity of the mycelial broth (mixing and mass transfer were

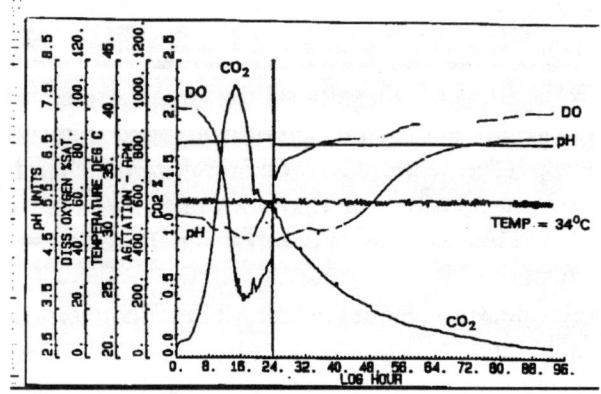

Figure 13.6 / Fermentation profile of a typical batch.

poor). Initial aeration and agitation rates were based on providing equal impeller tip speed at various scales of operation. This was found not to provide sufficient oxygen transfer, therefore the agitation rate was empirically raised until a dissolved oxygen level of 30% could be maintained. Finally, there were no scale-up problems encountered in seed train propagation.

ISOLATION OF THE RAK REDUCTASE GENE FROM *A. NIVEUS*

The strategy to clone the reductase gene is outlined in Figure 13.7.

The *A. niveus trpC* gene was isolated by complementation of an *E. coli*(*trpC-*) auxotroph transformed with a plasmid library constructed from *A. niveus* chromosomal DNA (4–10 kb) and plasmid pBR322. Regretfully the reductase gene could not be isolated in a similar manner by shot-gun cloning in *E. coli* since RAK was cyto-toxic to *E. coli* cells. Alternatively, the related filamentous fungus *Aspergillus nidulans* was evaluated as a possible host for gene cloning and was shown to be resistant to RAK toxicity as well as incapable of producing dilevalol.

The *A. niveus trpC* gene efficiently transformed an auxotrophic strain of *A. nidulans* (trpC-) demonstrating that *A. niveus* genes can be expressed in an alternate fungal host. However, attempts to isolate the RAK reductase gene by shot-gun cloning in *A. nidulans*, using the *A. niveus trpC* gene, as a selectable marker were abandoned when the apparent genome size of *A. niveus* was determined (Figure 13.3). Using the *A. niveus trpC* gene as a probe, the chromosomal genome size of *A. niveus* was calculated to be 4.6×10^8 bp which is approximately 10-fold larger than the genome size reported for *A. nidulans* (1×10^7 bp). The immense size of the *A. niveus* genome prohibited efficient shot-gun cloning of the RAK reductase gene and necessitated random screening of thousands of *trpC+* transformants required to isolate a single reductase positive transformant.

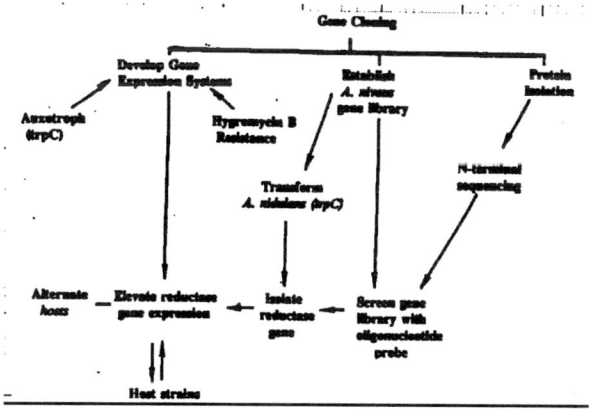

Figure 13.7 / Schematic of gene cloning strategies employed in the cloning of the RAK reductase in *Aspergillus niveus*.

Consequently, a more specific cloning strategy was devised whereby the purification of the RAK reductase enzyme would enable the generation of an oligonucleotide probe for gene cloning.

Partially pure enzyme was subjected to isoelectric focusing. The RAK reductase protein was identified *in situ* and the protein was cut out of the IEF gel and subjected to native, and subsequently, to SDS-PAGE electrophoresis. The purified *A. niveus* keto-reductase migrated on SDS-PAGE with an apparent subunit molecular weight of 40,000. The isolated protein band from the SDS-PAGE gel was electro-eluted and subjected to amino acid analysis (Figure 13.8). Amino-terminal sequence determination and amino acid analysis were performed at the Schering Microchemistry facility. Attempts at sequencing the amino terminus of the purified protein revealed that the amino terminus was blocked.

Comparison of the published amino acid composition of various NADPH-dependent reductases compared with the R-aminoketone reductase and calculation of the mole percentage of each amino acid is presented in Figure 13.9. The RAK reductase appears to have some similarity to human carbonyl reductase on the basis of the absence of cysteine and tryptophan, and the relatively high percentage of glycine. The RAK

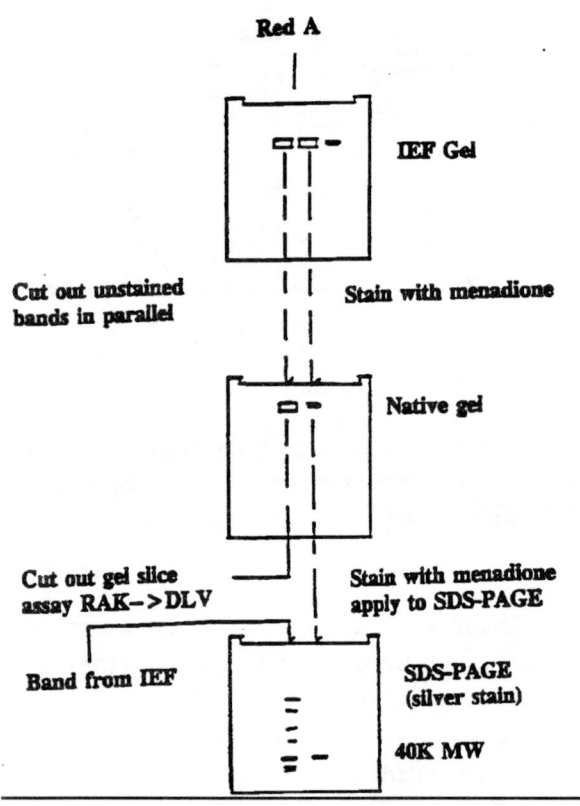

Figure 13.8 / Direct correlation between a 40,000 dalton protein and DLV reductase activity.

reductase has considerably more alanine and less glutamine and glutamic acid and asparagine and aspartic acid. Of the reductases which have to date been cloned, there is no obvious similarity to the RAK reductase at the amino acid (and presumably, the DNA) level.

An aliquot of the isolated enzyme protein band (the putative RAK reductase) was electroeluted from an SDS-PAGE gel and subjected to trypsin cleavage. Following HPLC chromatography approximately 12 peaks were observed, of which three were of sufficient separation and intensity to allow for their isolation and sequencing.

Two oligonucleotide probes were synthesized (Figure 13.10) from two peptides which yielded the least degenerate codon sequences. Both oligonucleotide mixes hybridized exclusively to *A. niveus* chromosomal DNA. Probe A hybridized with EcoR1 fragments of 5.7 and 4.3kb and HindIII fragments of 3.6 and 2.6kb. Probe B hybridized to EcoR1 fragments of 9.3 and 8.6kb and to HindIII bands of 8.6 and 8.1kb with the smallest band, in both cases, being the most intense. Duplicate filters of over 120,000 transductants were prepared and probed separately with oligonucleotide mix A and B under conditions of high stringency (hybridization, 6X SSC, 42°C; wash, 6X SSC, 50°C). Six hybridizing clones on duplicate filters were visible. Although no overlapping clones were evident upon restriction digestion of isolated plasmid DNA, a number of fragments in common were present. Southern blot hybridization of these DNA samples along with select other cosmids was performed (Figure 13.11). One cosmid (18-1) hybridized very strongly to an 8.6kb EcoR1 and an 8.1kb HindIII fragment, while faint bands could be seen in cosmids 18-5 and 18-6.

DISCUSSION

Early research endeavors attempted to isolate the RAK reductase gene by subtractive hybridization utilizing cDNA derived from mRNA of putative induced cultures (which had been exposed to RAK) and hybridizing them to cDNA derived from non-induced cultures (which were not exposed to RAK); in this manner genes which were constitutive in both non-induced and induced cultures could be separated from inducible genes. However, this strategy was subsequently shown to be inadequate since enzyme analysis of cell free extracts from cultures grown in the absence and presence of RAK both possessed RAK reductase activity, revealing that enzyme activity as well as gene expression were constitutive. Consequently, development of appropriate expression vectors and selectable markers which could be used in *A. niveus* to conduct successful gene cloning was investigated.

Figure 13.9 / Comparison of the amino acid composition of NADPH-dependent reductases.

MOLE %

	DLV	BAR[1]	HAR[2]	RAR[3]	HAD[4]	HCR[5]	CCR[6]	BPS[7]	FLC[8]
A Ala	11.7	6.3	6.2	6.7	9.1	7.4	6.8	6.1	4.9
N Asn	}5.1	4.4	4.9	3.9	3.8	}8.6	}10.2	4.5	5.3
D Asp		7.3	4.9	6.7	5.0			5.4	7.6
C Cys	0	2.2	2.3	2.8	2.2	0	1.5	2.2	1.8
Q Gln	}7.2	3.8	4.3	4.9	4.1	}11.9	}10.6	1.9	2.7
E Glu		6.6	7.5	6.0	6.9			8.0	5.8
G Gly	10.9	4.7	5.2	3.9	5.3	8.6	10.6	4.8	4.0
H His	2.9	2.8	2.9	2.5	2.8	2.6	1.5	3.2	3.1
I Ile	4.9	6.3	5.9	5.7	4.4	4.8	4.2	5.1	3.1
K Lys	2.3	8.9	8.2	8.1	6.9	5.6	7.6	9.6	7.6
L Leu	10.2	9.8	10.5	9.5	11.0	9.7	9.5	11.5	13.3
M Met	2.8	0.95	1.6	1.4	1.2	1.1	1.5	0.96	1.8
F Phe	3.9	3.8	3.6	3.9	2.8	4.1	3.4	4.5	4.0
P Pro	7.8	6.6	6.9	7.1	6.3	5.2	3.8	6.1	5.8
S Ser	8.5	4.1	5.6	4.6	4.7	6.7	6.1	6.7	7.1
T Thr	6.7	4.1	4.9	4.2	3.8	6.3	5.7	2.9	1.8
W Trp	0	1.9	2.3	1.8	2.2	0	1.5	0.96	1.3
Y Tyr	2.8	4.4	3.9	4.6	4.7	1.1	1.5	3.8	4.4
V Val	6.8	7.6	8.2	8.1	7.5	10.0	9.5	8.3	8.4
R Arg	6.2	3.2	3.6	3.5	3.1	6.3	4.5	3.2	6.2

1 BAR = Bovine lens aldose reductase (Chung and LaMendola, 1989)
2 HAR = Human aldose reductase (Chung and LaMendola, 1989)
3 RAR = Rat lens aldose reductase (Chung and LaMendola, 1989)
4 HAD = Human liver aldehyde reductase (Chung and LaMendola, 1989)
5 HCR = Human carbonyl reductase (Wermuth, 1981)
6 CCR = Chicken carbonyl reductase (Hara, 1982)
7 BPS = Bovine prostaglandin F synthase (Chung and LaMendola, 198)
8 FLC = Frog lens rhocrystallin (Chung and LaMendola, 1989)

At the time of this report, both *A. nidulans* and *A. niveus* have been transformed with 18-1 DNA, containing the putative clone reductase gene, and results are still pending. The ultimate cloning of the bioconverting gene will enable future studies to better elucidate the regulation of this enzyme as well as investigate the engineering of an improved strain for increasing the efficiency of dilevalol synthesis by microbial fermentation.

Early studies employing the transcription and translation inhibitors Actinomycin D and cycloheximide revealed that *de novo* protein synthesis during late exponential growth was essential for the bioconversion activity. Interestingly, bioconversion activity *in vivo* appeared to undergo a brief lag (6–8 hours) between the time of RAK addition and the detection of dilevalol synthesis suggesting that perhaps the presence of substrate

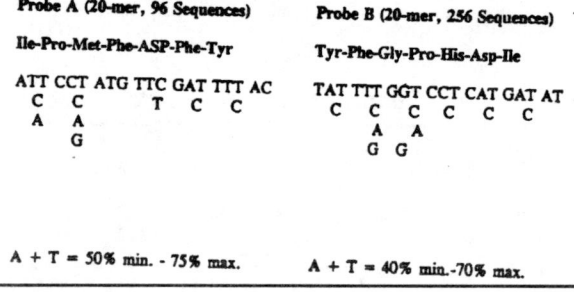

Figure 13.10 / Derivation of oligonucleotide mixes from tryptic fragments of RAK reductase.

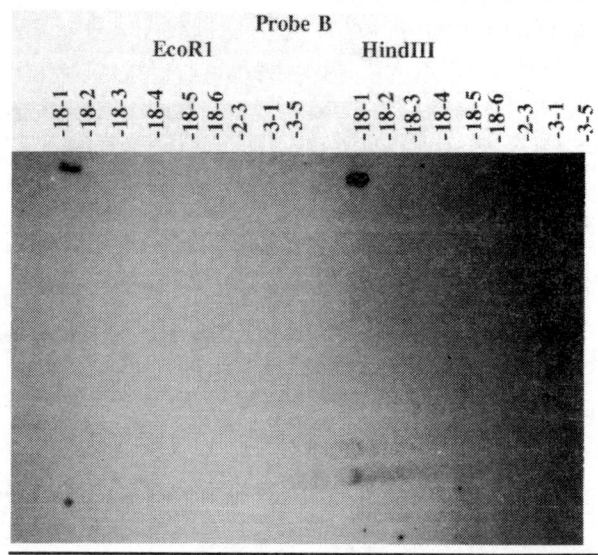

Figure 13.11 / Southern blot analysis of restriction digests of recombinant cosmids. Purified cosmid DNA was digested with Eco RI or HindIII and electrophoresed in a 0.7% TBE agarose gel for 20 hours at 35V. DNA fragments were transferred to nitrocellulose filters according to Southern (1975) and hybridized with the oligonucleotide mix indicated.

was acting as inducer. In addition, resuspension studies involving whole washed cells in various synthetic media indicated that the bioconversion reaction was dependent upon a readily utilized energy source such as glucose. However, several studies indicated that utilization of RAK would not occur until the glucose concentration of the culture was reduced to low levels. Addition of glucose to a culture prior to or directly after RAK loading, resulted in a time dependent inhibition of dilevalol synthesis. When the glucose was exhausted from the medium the amount of dilevalol synthesized increased to a similar level as seen in cultures without glucose spiking [7].

Elucidation of the regulation of the events occurring during the bioconversion was not possible until an *in vitro* enzyme assay was established for measuring the reduction of RAK to dilevalol. *In vitro* enzyme activity was shown to be dependent on the cofactor NADPH, the substrate RAK and active protein. The product of the reaction co-migrated with standard dilevalol as detected by thin-layer chromatography and co-eluted with standard exogenously added dilevalol without peak splitting as detected by HPLC indicating that the product synthesized by the *in vitro* reaction was dilevalol. The requirement for NADPH confirmed our earlier studies which indicated that an energy or reducing source was needed to synthesize dilevalol. The *in vitro* reduction of RAK to dilevalol does not occur if NADH is added as the co-enzyme of the reaction and catalysis does not require the presence of a metal cofactor [6]. Subcellular localization studies revealed that the enzyme responsible for the bioconversion reaction was located in the cytoplasm. The enzyme has been substantially purified and the enzymological properties characterized [6].

The development of the RAK reductase *in vitro* assay enabled us to better elucidate the events occurring during bioconversion of RAK. Cell extracts derived from whole cells grown in medium with and without RAK possessed similar *in vitro* activity indicating that the ability to synthesize dilevalol was not induced by exposure to RAK. The need for protein synthesis became apparent when a time course study was performed comparing the growth phase of the culture and the *in vitro* enzyme activity. We discovered that bioconversion activity was present throughout the growth of the culture. However, an approximate 3-fold increase in enzyme activity occurred during the transition period of growth from late exponential phase into stationary phase where activity was maintained at a maximum. Apparently when cycloheximide and Actinomycin D were added during late exponential phase, before, during or directly after RAK loading, the normal increased progression of enzyme protein synthesis was prohibited. Reducing the synthesis of the bioconversion enzyme during this transitional time period dramatically reduced the cellular enzymatic capability. Only the amount of enzyme synthesized prior to the addition of inhibitors was utilized to synthesize dilevalol which resulted in the low levels of dilevalol which were detected [1].

Interestingly, the transitional increase in bioconversion activity during late exponential growth coincided with the exhaustion of glucose in the medium. Examination of *in vitro* activity of cultures grown in the presence or absence of excess glucose revealed that glucose at elevated levels resulted in a decreased enzyme specific activity indicating that the bioconversion reaction was derepressed in the absence of glucose [1].

CONCLUSION

Stereospecific reduction of R-aminoketone HCl to dilevalol has been obtained using *Aspergillus niveus* cultures containing a novel NADPH dependent ketone reductase. We have obtained extensive physiologic and biochemical knowledge of the organism and have been successful in optimizing fermentation parameters including aeration, medium composition, seed culture propagation, substrate feeding, temperature and pH. Additionally, the enzyme involved in the specific bioconversion has been elucidated. Through the successful implementation of strain development techniques, strains with high bioconversion potential have been generated. Several distinct advantages over the previous commercial chemical resolution process for dilevalol are noteworthy.

Selective bioconversion using *Aspergillus niveus* was economical, the enzyme was stable in its native environment with high activity, cofactor recycling was done automatically, the substrate could be added at high molar concentrations, the reaction yields and compound production rates were demonstrated to be industrially useful, there was very high stereo- and regioselectivity, and, finally, there was no further metabolism of the product.

As evidenced by the now large numbers of publications which have appeared in recent years, the perceived usefulness of whole cell systems, and the advantages of using them clearly outweigh the disadvantages. It is also evident that there is a vast reservoir of untapped, undiscovered and unsynthesized whole cell and/or enzyme variants with properties to suit most bioprocess engineer's dreams. A search in nature coupled with modern methods of genetic manipulation and protein engineering, now and in the future, make the biogeneration of complex chemical entities a viable option and a new and important arena for research and in the production of industrially important material.

Finally, it should be quite heartening to those of us in the field to realize that neither the existent microbes nor enzymes are perfect, thus leaving still many problems unsolved. To the willing scientists and engineers there still remains an abundance of exciting and promising research in the future.

ACKNOWLEDGEMENTS

The authors thank Ms. Lynda Ludwig for her skillful assistance and gratefully acknowledge the following individuals for their scientific and intellectual input: Robert Vail, Steven Goldberg, Ph.D., William Charney, Ph.D. and Nancy McGraw, Microbiological Development; Steve Gruber, Microchemistry; Mario Ruggeri, Chemical Development; Barr Bauer, Ph.D., Computational Medicinal Chemistry; Yair Alroy, Ph.D. and Eugene Schaefer, Ph.D., Biochemical Engineering; and

Vijay Singh, Ph.D. and Dennis Rendeiro, Fermentation Process Development, all in the Research Division, Schering-Plough Corporation.

REFERENCES

1. Bradford, M. M. (1976). A rapid and sensitive method for the quantitation of microgram quantities of protein utilizing the principle of protein-dye bindings. *Anal. Biochem.* 72: 248–254.
2. David G. G. Crout and Markus Christen. Bio-transformations in Organic Synthesis, R. Scheffold (ed.) Modern Synthetic Methods, Vol 5. Springer-Verlag Berlin Heidelberg p. 1–114 (1989).
3. Dinsmoor, R. (1990). Biofinance Conference Emphasizes Emerging Markets and Ways to Draw Japanese Investors. *Genetic Engineering News*, Vol. 10, 4: 1–11.
4. Hanahan, D. and Meselson, M. (1980). A protocol for high density screening in X1776. *Gene* 10: 63–69.
5. Holmes, K. D., *et al.* 1990. Comments on Enatiomerism in the Drug Development Process, *Pharmaceutical Technology*, 46.
6. Homann, M. J., Vail, R. B and Goldberg. S. (1990b). Isolation and Characterization of a NADPH-dependent Oxidoreductase from *A. niveus* for the Synthesis of the Antihypertensive Agent Dilevalol by Stereospecific Reduction (Manuscript in preparation).
7. Homann, M. J., Vail, R. B. and Brian, W., (1990a). Development and Characterization of a Specific Bioconversion for the Synthesis of the Antihypertensive Agent Dilevalol by *Aspergillus niveus* (Manuscript in preparation).
8. Kumkumian, C. S., The Use of Stereochemically Pure Pharmaceuticals in Drug Stereochemistry, Wainer, I. W. and Drayer, D. E. (Eds). Marcel Dekker, Inc. New York p. 299–310 (1988).
9. Moos, M., Nguyen, N. Y., and Liu, T. Y. (1987). Reproducible high-yield sequencing of proteins electrophoretically separated transferred to an inert support. *J. Biol. Chem.* 263: 6005–6008.
10. Polastro, E. E., *et al.* (1989). Enzymes in the Fine Chemical Industry: Dreams and Realities—Advantages and Disadvantages of Enzyme Applications. Arthur D. Little International, Inc. *Bio/Technology* 7, 12: 1238–41.
11. Seebach, D., Sutter, M. A., Weber R. H. and Zuegar, M. F. (1984). Yeast Reduction of Ethyl Acetoacetate: (5)-(+)-Ethyl-3-hydroxybutanoate. *Org. Synthesis* 63: 1–9.
12. Simon, K., Chaplin, E. R., and Diamond, I. (1977). *Anal. Biochem.* 79: 571–574.
13. Sonnleitner, B., Giovannini, F. and Fiechter, A. (1985). Stereospecific Reductions of Ketones and Oxo-acid Esters Using Continuously Growing Cultures of *Thermanaerobium brockii. J. Biotech* 3: 33–45.
14. Tarmy, E. M., and Kaplan, N. O. (1968). *J. Biol. Chem.* 243: 2579–2586.
15. Van Gorcom, R. F. M., Pouwels, P. H., Goosen, T., Visser, J., Vander Brock, H. W. J., Hamer, J. E. Timberlake, W. E., and Vander Hondel, C. A. M. J. J. (1985). Expression of an *Escherichia coli* β-galactosidase fusion gene in *Aspergillus nidulans. Gene* 40: 99–106.
16. Vojtisek, V. (1989). Immobilized Microbial Cell Aggregates for Biotransformation of Antibiotics, Amino Acids, Sugars, and other Compounds, *Biotech. and Applied Biochem.* 11: 1–10.
17. Yelton, M. M., Hamer, J. E., and Timberlake, W. (1984). Trnasformation of *Aspergillus nidulans* by using a *trpC* plasmid. *Proc. Natl. Acad. Sci U.S.A.* 81: 1470–1474.

14

Searching for Anthelmintics in the Post-Avermectin Era

Christopher L. Haber[1], David P. Thompson[2], Howard A. Whaley[1], B. Lamar Lee[2], Thomas F. Brodasky[1], Charles K. Marschke[1], Jerry W. Bowman[2], Timothy G. Geary[2], Eileen M. Thomas[2], Richard D. Conklin[2], Veronica H. Wiley[1], Charlotte L. Heckaman[1], Derek L. Rosa[1] and Raymond J. Zielinski[1]

[1]Chemical and Biological Screening; [2]Parasitology Research, Upjohn Laboratories, The Upjohn Company, Kalamazoo, MI 49001

INTRODUCTION

The avermectins are a family of macrocyclic lactones (Figure 14.1) produced by the soil microorganism *Streptomyces avermitilis* (6). A 21-chapter volume has recently been published, rich with details of the discovery, production, chemistry, pharmacology and commercial applications of avermectins (9). *S. avermitilis* was isolated by researchers under the direction of Dr. Satoshi Omura at the Kitasato Institute (Tokyo). The avermectins were discovered during an intense screening effort by researchers at Merck, Sharp & Dohme Research Laboratories for microbial products having anthelmintic activities (6). Avermectins were subsequently found to be insecticidal as well (43). The commercial products ivermectin (22,23-dihydroavermectin B_{1a}) and abamectin (avermectin B_{1a}) have been successfully marketed as anthelmintics for veterinary applications. Their spectrum of target parasites, lack of toxicity and infrequent dosing schedule have made them anthelmintic market leaders.

A family of compounds similar to avermectin was identified by Sankyo researchers using a screen for insecticides. These were the milbemycins (53), and they resemble the avermectins in structure (Figure 14.1) and efficacy. The first milbemycin-producer described was *S. hygroscopicus* subsp. *aureolacrimosus*, and it produced

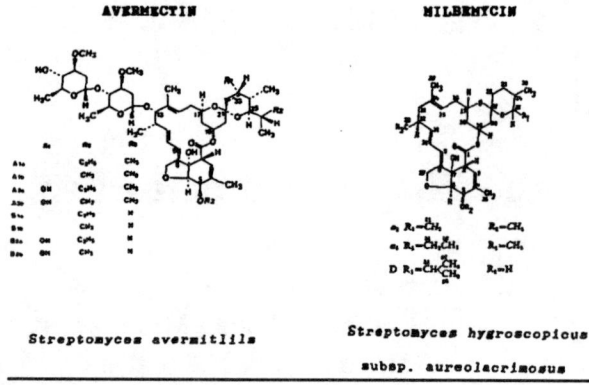

Figure 14.1 / Avermectin and milbemycin structures and producing organisms.

no fewer than 13 different milbemycins (53). Whereas no producers of avermectin other than *S. avermitilis* have been described, milbemycins have been isolated from several other soil organisms. These include *S. cyaneogriseus* subsp. *noncyanogenus* (10), *S. thermoarchaensis* (57) and *Streptomyces* sp. E225 (26).

The mode of action of avermectins has recently been reviewed (46, 55), and is unique among anthelmintics. Although the mechanism is not clearly understood, the compounds appear to act by increasing chloride conductance across muscle membranes in arthropods and nematodes, resulting in alterations in neuromuscular functions. The site of action is the peripheral components of the target organism's neuromuscular system. In mammals, avermectin-sensitive tissue is restricted to the central nervous system where it is generally protected from the compounds by the blood/brain barrier. This provides a very favorable therapeutic ratio in the field.

The avermectins and milbemycins are very good broad-spectrum anthelmintics, and any new entries in this market will have to compete effectively with them in efficacy and spectrum. It is therefore very important in a microbial products screening regimen to be able to identify such compounds early in the process. This allows novel compounds to undergo accelerated development and known (but highly active) entities to be dropped before resources are expended unnecessarily for further characterization. The avermectins were discovered by feeding fermentation products to mice which had been infected with the parasite *Nematospiroides dubius*. The endpoint was clearance of the parasite (7). This *in vivo* screen required a huge commitment of resources, but was ultimately successful. The knowledge gained through the study of avermectins and milbemycins has been applied to the development of *in vitro* anthelmintic screens. This review will describe several such screens for anthelmintics, some general and some very specific for the avermectin/milbemycin family of compounds. The progress of a *Streptomyces hygroscopicus* producer of milbemycins through these screens will be detailed for illustrative purposes.

IN VITRO WHOLE ORGANISM SCREENS

FREE-LIVING NEMATODES

Until recently, *in vitro* tests for activity against target parasites have been lacking. This was due to the difficulty encountered when trying to maintain and/or propagate these parasites outside of their mammalian hosts. In the absence of suitable target parasites for large scale screening, nematocidal activity has been measured using free-living species which are easy to maintain under laboratory conditions.

Foremost among these has been *Caenorhabditis elegans*, a common and extremely well characterized soil nematode. The use of this organism for anthelmintic screening has been described in detail by Simpkin and Coles (12, 51). Briefly, *C. elegans* are inoculated onto an agar medium which contains a lawn of *Escherichia coli*. The nematodes feed on the bacteria, reproduce rapidly and thus provide sufficient numbers for the running of assays. Assays can be carried out by simply adding the sample to be assayed to an agar

Table 14.1 / Inhibition of *Caenorhabditis elegans* motility by ivermectin.

Ivermectin Concentration (nM)	Motility Inhibition[1]		
	Day 1	Day 4	Day 7
57	3.0	3.5	5.0
5.7	1.0	3.0	4.5
0.57	0.0	0.0	0.0

[1]Assay volume = 2.0 ml, inoculated on Day 0 with ca. 30 *C. elegans* at various developmental stages. Scores reflect numbers of motile worms remaining after the indicated time: 0 = >50 worms; 1 = 21–50 worms; 2 = 11–20 worms; 3 = 6–10 worms; 4 = 1–5 worms; 5 = 0 worms. Results are the average of duplicate assays.

plate and observing the motility of the worms relative to control plates. A more quantitative approach involves the continued propagation of the *C. elegans* in a broth medium, using a known number of organisms (mixed stages) as the initial inoculum. The broth medium contains *E. coli* and salts, as well as antibiotics and antifungal agents to inhibit microbial growth. Hours to days after addition of the experimental samples, numbers of motile worms relative to controls are recorded. This method can be automated in a 96-well format, with all sample and media additions carried out by an automatic pipetting workstation, allowing for extremely high throughput of samples.

The avermectins and milbemycins are quite toxic to *C. elegans*, completely arresting movement within minutes at a 5 µM concentration. When the assay is run over several days, activity can be detected at 5.7 nM (Table 14.1). When used as a general fermentation screen, this assay has been effective in detecting producers of avermectins and milbemycins. The organism *Streptomyces cyaneogriseus* ssp. *noncyanogenus* which produces the LL-F28249 family of milbemycins was discovered using this indicator organism (10).

Other soil nematodes, such as *Turbatrix aceti* and *Caenorhabditis briggsae* (56) have been used to screen for anthelmintic agents, although with less success than with *C. elegans*. The advantages to these screens remain the ease of propagating the organisms, their adaptability to high-volume screening, their sensitivity to the modern broad-spectrum anthelmintics, and their low cost. Their major disadvantage is that they lack specificity. As *in vitro* whole organism assays, they pick up many general toxins which are common in microbial fermentations. They also do not employ target organisms, and thus detect compounds which may not be active against parasites, or (worse yet) may fail to detect parasite-specific compounds. These screens are generally run as an initial gate through which active fermentations pass prior to being subjected to more specific (and laborious) tests.

ARTEMIA SALINA

Another *in vitro* whole organism screen which has been applied to avermectin research and discovery employs the common brine shrimp *Artemia salina*. Brine shrimp have been used traditionally as an indicator of cytotoxicity (39, 40), and this use has found application in screening for pesticide residues (21), fungal toxins (24, 42) and antitumor compounds (52). They are also among the easiest organisms to grow in the laboratory, requiring only a saline solution and desiccated eggs (both available in pet supply stores). The characteristic of brine shrimp pertinent to this report is the great sensitivity of these crustaceans to avermectins and milbemycins. Since the discovery of avermectins, crustacean models (with their well-characterized neurological systems) have been used to study the mode of action of this class of compounds (20, 61). Stimulation of GABA binding to brine shrimp membranes by avermectin has also been reported (8). Brine shrimp thus provide an extremely easy means for detecting producers of avermectins and milbemycins. They have, in fact, been used as a preliminary indication of activity during structure/activity studies of modified avermectin molecules (3).

Table 14.2 / Inhibition of *Artemia salina* by ivermectin.

Incubation Time (hours)	LD_{50}[1] (nM)
3	2.6
6	3.0
24	2.3

[1]Assay volume = 1.0 ml inoculated with 6 shrimp at time 0. LD_{50} values based on number of non-motile shrimp per assay at the indicated time. Results are the average of triplicate samples.

When incubated for 6 hours, brine shrimp are sensitive to 3 nM ivermectin (Table 14.2). The relatively short assay time is advantageous for two main reasons: Slower-acting general toxins will not be picked up in this time span, and an assay which requires only 6 hours will allow the harvesting and extraction of fermentations on the day which they are determined to be active. All additions of solutions and samples can be automated, and results can be determined using image analysis systems, minimizing operator time requirements and bias. This assay can routinely detect producers of avermectins and milbemycins. As a screen for general anthelmintics, brine shrimp suffer the same lack of specificity against toxins as *C. elegans*. This is taken one step further by the fact that the shrimp are not nematodes and may not detect compounds which specifically affect nematodes. As with *C. elegans*, this screen is best included as an initial indicator of activity which requires confirmation by more specific assays.

TARGET PARASITES

Propagation and/or maintenance of target mammalian parasites outside of their hosts in numbers and for durations sufficient for screening have traditionally been difficult. Recent improvements in growth media, judicious choice of larval stages and customized environmental conditions have all contributed to the successful use of target parasites for *in vitro* anthelmintic evaluation, if not for high volume screening (for recent reviews, see Refs. 11, 27). The list of these parasites includes *Angiostrongylus cantonensis* (48), *Ascaris suum* (47), *Cooperia punctata* (32), *Haemonchus contortus* (4, 18), *Necator americanus* (31), *Nippostrongylus brasiliensis* (28), *Trichinella spiralis* (29) and *Trichostrongylus colubriformis* (19, 45), and the list continues to grow. The methods by which these parasites are employed vary, as do the endpoints of the assays. Effectiveness of samples in the systems listed is determined by loss of motility or impaired larval development. Other endpoints for *in vitro* parasite assays are leakage of radioactively labeled internal constituents (13), acetylcholinesterase secretion (44) and inhibition of aggregation (30).

Automation of these assays is also possible, with change in motility, for instance, being quantifiable with a micromotility meter (2, 18, 19). Results of a motility study using the adult stage of *H. contortus* are shown in Figure 14.2, and indicate that ivermectin can be detected at concentrations as low as 0.01 µM. The assay can be as brief as 1–3 hours without appreciable loss of sensitivity. The failure of ivermectin to completely inhibit movement at any tested concentration emphasizes the advantages of a validated, automated system for measuring relative motilities between samples and controls.

The ability to screen for activity against target parasites adds an additional level of sophistication to *in vitro* whole organism screening. Compounds which have a mode of action specific to the parasite biology can theoretically be detected. High volume screening using these organisms remains an elusive goal, however. Target parasite screens are, of necessity, more labor intensive. They require animals for production of the parasites, and this process consumes time, space and money. More sophisticated equipment may be required for accurate quantitation of results. Obtaining

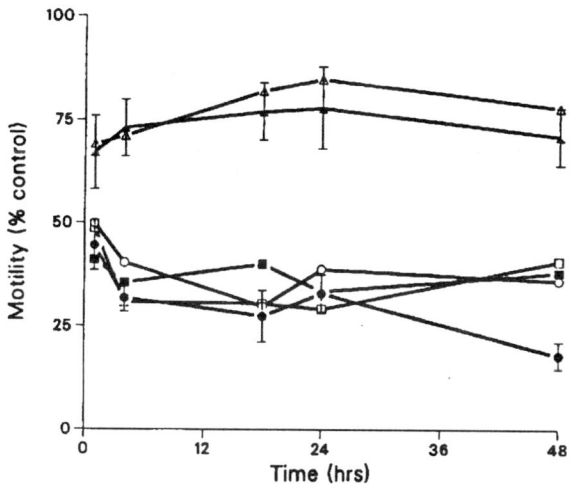

Figure 14.2 / Concentration- and time-dependent effects of ivermectin on adult *Haemonchus contortus*. Five healthy, adult worms were incubated with different ivermectin concentrations in 2.5 ml volumes. Readings were taken at the indicated time points using an automated micromotility recording system (2). Results are the mean ±1 standard error in 8 separate preparations (sample error bars shown). Ivermectin concentrations: ■ = 0.0001 µM; ■ = 0.001 µM; ■ = 0.01 µM; ■ = 0.1 µM; O = 1.0 µM; O = 10.0 µM.

sufficient parasites for continuous, high volume screening is still not feasible. Thus, these screens are better suited for low volume screening, or as secondary screens for fermentations already active in more general screens. Such combination whole organism *in vitro* screening has been reported for determining if new fermentation products are anthelmintic (1). In this case, inhibition of motility of the free-living *T. aceti* or *H. contortus* larvae and prevention of *H. contortus* egg hatch were used. Screening of culture extracts for inhibition of *H. contortus* larval motility has been used successfully in characterizing the milbemycins produced by *Streptomyces* sp. E225 (26).

TARGETED *IN VITRO* SCREENS

In addition to whole organism screens, many other *in vitro* screens have been developed which can detect compounds with particular structural or functional characteristics. Indeed, such regimens have been applied to finding avermectin-like molecules in microbial fermentations. These screens have the advantage of being very specific, and some can be used as primary screens.

IMMUNOLOGICAL SCREENS

The recognition of epitopes by antibodies has given rise to an exciting technology which can recognize minute amounts of a desired compound in a heterogeneous mixture. Such a quality is tailor-made for searching for compounds produced in fermentations which have known structural characteristics. Enzyme-linked immunosorbant assays (ELISAs) have recently been used in screening fermentations for mycotoxins (38), aminoglycoside antibiotics (59) and macrolide antibiotics (60). These assays use antibodies which are prepared against and recognize the compounds of interest and which bind to the compounds under assay conditions. These complexes are detected using additional antibodies to the original immunoglobulins which are conjugated to a reporter enzyme (such as horseradish peroxidase). Following incubation under conditions which allow specific antibody/antigen binding, the bound antibodies are detected by addition of the enzyme substrate. The enzymatic reaction yields a chromophoric product which can be detected spectrophotometrically. ELISAs are sensitive to less than 1 µg/ml, can be adapted to a 96-well format for easy set-up and reading, and can be completed in 2–3 hours.

An alternative to ELISA assays is the radioimmunoassay. Instead of using a reporter enzyme, this assay employs radiolabeled compounds which bind to the prepared antibodies. Compounds in the test samples sharing antigenic features with the labeled compound will compete for binding with the antibodies, resulting in fewer precipitable counts in positive samples. These assays are also very sensitive, and require fewer steps than ELISA assays. They depend, however, on the availability of labeled compounds, and present the problems associated with radioactive reagents.

Immunoassays have been applied to the detection of avermectins. The antibodies prepared to 23-amino-*O*-mycaminosyltylonolide (a 16-membered ring macrolide possessing one amino sugar) were found to have 18% cross-reactivity with a 100 μg/ml solution of avermectin B_1 in an ELISA assay (60). An antiserum to the 4"-hemisuccinate of ivermectin coupled to keyhole limpet hemocyanin has been prepared and characterized for its ability to bind ivermectin (34). The concentration-dependence of competition between [^{3}H]-ivermectin and ivermectin for this antiserum is linear throughout a 2–25 ng/ml ivermectin concentration range. When tested with four avermectin congeners produced by *S. avermitilis*, cross-reactivities in the realm of 50% were obtained (Table 14.3). When a 4 mm diameter agar plug of a 5-day-old *S. avermitilis* culture was allowed to soak in the assay buffer overnight and removed prior to carrying out the assay, competition with [^{3}H]-ivermectin was detected which corresponded to a 31 ng/ml solution of ivermectin. Clearly, this system could be used for screening microbial cultures for the production of avermectins.

RECEPTOR BINDING SCREENS

The binding of receptors to their ligands and subsequent events play an important role in our increasingly molecular level of understanding of chemotherapy. Although the establishment of parasitic infections in mammals may indeed involve receptor/ligand interactions between parasite and host, the identification of these ligands or their specific receptors has not yet

Table 14.3 / Percent cross reaction of anti-ivermectin antiserum with avermectins.

Compound	% Cross Reaction[1]
Avermectin A_1	55
Avermectin A_2	68
Avermectin B_1	49
Avermectin B_2	52
Ivermectin 4" hemisuccinate	114
Tylosin	<0.008

[1]Antiserum prepared to ivermectin 4" hemisuccinate conjugated to keyhole limpet hemocyanin. Cross-reaction percentages are relative to ivermectin and are the average of duplicate samples.

occurred. Drugs which treat such infections successfully, however, have defined several receptors within the parasites which can serve as sites for therapeutic intervention (46). Levamisole and morantel both appear to act at the cholinergic-receptor (33, 35), and piperazine displays properties of a gamma-aminobutyric acid (GABA) agonist (36).

Although the mechanism of avermectin action against nematodes is not clear, it is proposed that avermectin has one or more receptors in neurons or muscle fibers, some of which are associated with GABA-regulated chloride channels (37). Avermectin has a high affinity binding site in *C. elegans* membranes, with a K_d of 0.26 nM and a receptor concentration of 3.53 pmole/mg protein (50). The specific binding site provides an assay for compounds which displace radiolabeled ivermectin. A number of ivermectin derivatives which displace ivermectin in this assay have also been shown to paralyze *C. elegans*, with the correlation coefficient between the assays being 0.975 (50). A competitive inhibitor of ivermectin binding which is structurally quite removed from the avermectins was discovered in a screen of fermentations using *C. elegans* membranes. This compound, cochlioquinone A, is produced by the fungus *Helminthosporium sativum*. It inhibited *C. elegans* motility with an ED_{50} value of 135 µM, and had a K_i for [^{3}H]-ivermectin of 30 µM (49). Although the affinity of cochlioquinone A for ivermectin binding sites is relatively low, the discovery of this compound is a reason to be optimistic that receptor binding screens can be successful in finding new anthelmintic agents.

High affinity binding sites for avermectin have also been found in preparations from rat brain (41) and dog brain (16). Homogenates of rat brain bind [^{3}H]-ivermectin with a K_d of 2.4 nM at 7.4 pmole/mg protein. This homogenate showed concentration-dependent inhibition of ivermectin binding when used to assay an extract of a culture of *S. avermitilis* (Table 14.4). Thus, rat brains can provide sufficient membranous material to screen for inhibitors of ivermectin binding. Such a screen could detect additional compounds which have avermectin-like activity.

Receptor binding assays depend on substantial quantities of tissue as a source of receptors, and a dependable supply of radiolabeled ligand. If these two conditions are met the screens can be automated and carried out in a high volume manner. These screens can be very specific for activities similar to those of the ligand. Indeed, cochlioquinone A was shown to be cross-resistant with ivermectin (49).

ELECTROPHYSIOLOGICAL ASSAYS

The avermectins stimulate chloride conductance across muscle membranes in arthropods (5, 17, 20, 61) and nematodes (25, 36). This ionophoric property can be exploited to develop a screen for microbial products which possess the same mode of action. In arthropods, effects of avermectin are localized at chloride channels, some of these being GABA-associated (61). GABA is the major inhibitory transmitter at axial muscle synapses in arthropods (23) and nematodes (15). Activation of the GABA receptor hyperpolarizes muscle membranes in these organisms, resulting in sustained paralysis. With the exception of *Ascaris* spp., nematodes do not lend themselves easily to microelectrode techniques, but many arthropod neuromuscular systems are well characterized pharmacologically and are ideally suited for these studies.

An assay for avermectin-like activities has been developed using walking leg stretcher muscle fibers from the lined shore crab *Pachygrapsus crassipes* (54). In this electrophysiological assay, changes in membrane resistance and membrane potential are measured in response to the addition of samples, followed by addition of the chloride channel blocker picrotoxinin (Figure 14.3). Avermectin-like compounds give a unique response: a time- and concentration-dependent decrease in membrane resistance which is partially (near 50%) recovered following the addition of picrotoxinin (Figure 14.4). In addition, membrane

Table 14.4 / Effects of extracts from *Streptomyces avermitilis* and *Streptomyces lividans* cultures on [^{3}H]-ivermectin binding to rat brain.

Culture	Area of Agar Culture Extracted $(mm^2)^1$	% [^{3}H]-Ivermectin Binding[2]
S. avermitilis	355	10
	35.5	10
	3.55	35
S. lividans	355	90
	35.5	110
	3.55	80

[1]Cultures grown in 100 mm petri dishes containing 20 ml Bennett's agar (10 g/l dextrose, 2 g/l NZ Amine A, 1 g/l beef extract, 1 g/l yeast extract, 15 g/l agar, pH = 7.0). Values represent dilutions of whole plate extracts which were used in the assays.
[2]Relative to binding of 1 nM ivermectin. Results are averages of five samples.

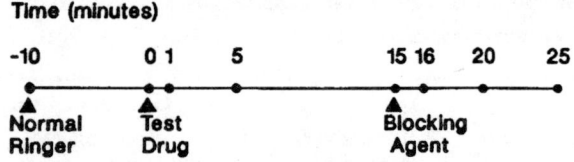

Figure 14.3 / Diagramatic representation of time course for the shore crab muscle resistance assay. Test drug = ivermectin standard solution or microbial culture extract. Blocking agent = 10 µM picrotoxinin. Data recorded for each time point indicated.

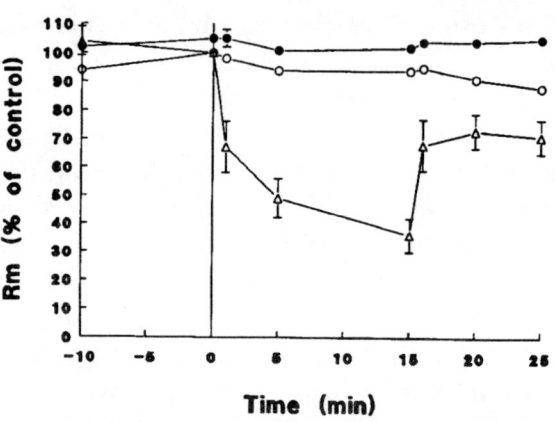

Figure 14.5 / Effects of agar plate extracts of *S. avermitilis* and *S. lividans* on shore crab muscle membrane resistance (22). Cultures were grown on Bennett's agar. Results are the mean ±1 standard error of the percent change in Rm recorded in three separate preparations (sample error bars shown). ● = Uninoculated agar, ○ = *S. lividans*, ▲ = *S. avermitilis*. Reprinted with permission (22).

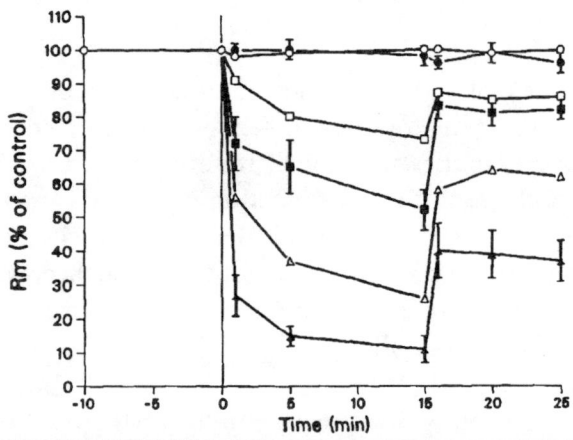

Figure 14.4 / Concentration- and time-dependent effects of ivermectin on shore crab muscle membrane resistance (22). Results are the mean ±1 standard error of the percent change in Rm recorded in three separate preparations (sample error bars shown). ○ = 0.1% DMSO (vehicle), □ = 0.003 µM ivermectin, ■ = 0.01 µM ivermectin, ■ = 0.03 µM ivermectin, ■ = 0.1 µM ivermectin, ■ = 1.0 µM ivermectin. Reprinted with permission (22).

potential recordings show a significant hyperpolarization upon addition of ivermectin. It is this hyperpolarization of the membrane, coupled with the partial recovery of membrane resistance upon the addition of picrotoxinin, which make up the specific "footprint" of avermectin in this assay. The assay can detect avermectin in extracts of *S. avermitilis* cultures, whereas a non-producing *S. lividans* strain gives no response (Figure 14.5). This assay has been put in place as a follow-up to a whole-organism primary screen to detect compounds with avermectin-like activity (22). It is sensitive (to 0.01 µM ivermectin), and can accommodate crude culture extracts. It is labor-intensive and time-consuming, and thus can handle only a limited number of cultures. But its specificity for compounds having an avermectin-like mode of action in crude extracts makes it a powerful tool for early identification of these metabolites.

DISCOVERY OF MILBEMYCIN-PRODUCING *STREPTOMYCES HYGROSCOPICUS* UC 8984

We have employed all the screens (at one time or another) in our search for novel anthelmintic compounds produced by microorganisms. Each has undergone validation tests using a spectrum of known anthelmintics, and the suitability of crude fermentation samples (beers or extracts) for each assay has been determined. What follows is the performance of a single soil isolate in all of these screens, and the identification of the milbemycins which this organism produces.

PRIMARY SCREENS

Streptomyces hygroscopicus UC 8984 was isolated from Michigan soil. A description of the culture is published elsewhere (22). Its activity was initially detected using extracts of agar-grown cultures in the *C. elegans* screen. The extracts from this culture were very active after 4 days of incubation with the nematodes (no motility observed). Subsequent testing of clear fermentation beer with *A. salina* again showed a high level of activity, with complete paralysis detected in less than 3 hours. This activity in two whole organism primary screens provided sufficient incentive to character ize the activity in other, more informative screens.

TARGET PARASITE SCREEN

An extract of UC 8984 was tested against adult *H. contortus* using a motility-sensing *in vitro* assay. In a similar manner to ivermectin (Figure 14.2), this extract inhibited *H. contortus* motility in a time-dependent manner, achieving nearly 90% inhibition in 5 hours. Such activity against a target parasite increased our interest in this isolate, and the organism's activity was tested for properties similar to avermectin in an additional set of assays.

RADIOIMMUNOASSAY

Following the procedures used for an agar culture of *S. avermitilis*, an agar plug of UC 8984 was incubated in assay buffer, removed, and the remaining buffer assayed for [^{3}H]-ivermectin binding competition with anti-ivermectin antiserum. This was repeated several times, consistently yielding no reduction in [^{3}H]-ivermectin binding. This result suggested that compounds produced by UC 8984 were epitopically different from ivermectin and thus unrecognizable by the anti-ivermectin antibodies.

IVERMECTIN BINDING ASSAY

The inhibition of [^{3}H]-ivermectin binding to rat brain homogenates by UC 8984 agar culture extracts was investigated. A dose-dependent competition was seen for binding with the rat brain homogenate, showing [^{3}H]-ivermectin binding decreasing from 62% to 3% of controls with increasing amounts of the culture extracted (see Table 14.4). This indicated a measurable affinity for ivermectin binding sites. K_i values were not determined due to the crudeness of the UC 8984 extract.

ELECTROPHYSIOLOGICAL ASSAY

An extract of agar-grown UC 8984 was tested for its effects on membrane resistance and potential in shore crab (*P. crassipes*) walking leg stretcher muscle fibers. The results (Figure 14.6) showed a rapid decrease in membrane resistance (R_m) which was sustained over time and partially (nearly 50%) reversed following replacement with picrotoxinin after 15 minutes incubation. This is precisely the type of response elicited by extracts from *S. avermitilis* (Figure 14.5) and ivermectin standard solutions (Figure 14.4). Taken along with a 3.7 mV hyperpolarization of the membrane potential (22), this is strongly indicative that UC 8984 metabolites

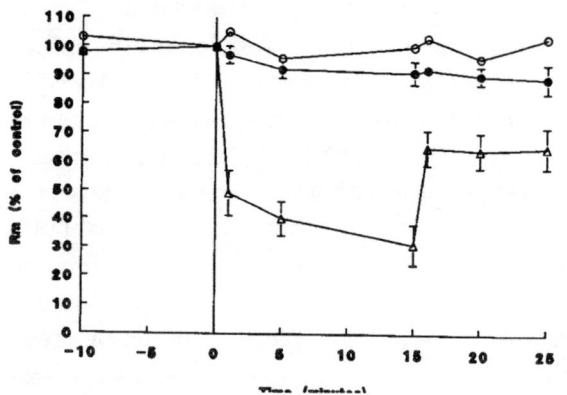

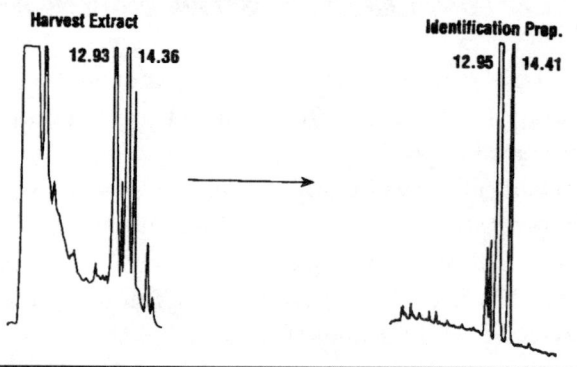

Figure 14.6 / Effects of an agar plate extract of UC 8984 on shore crab muscle membrane resistance (22). Culture was grown on Bennett's agar. Results are the mean ±1 standard error of the percent change in Rm recorded in three separate preparations (sample error bars shown). ○ = 1% DMSO (vehicle), ● = Uninoculated agar, ▲ = UC 8984.

Figure 14.7 / Purification of active components from a 250 liter fermentation of UC 8984 (22). The crude harvest extract was processed to yield the identification prep enriched for the two metabolites indicated in the HPLC elution profile. Reprinted with permission (22).

stimulate chloride conductance in a manner similar to the avermectins. The level of activity observed in this extract is approximately equal to that of 1.0 μM ivermectin (Figure 14.4).

ISOLATION AND IDENTIFICATION OF MILBEMYCINS PRODUCED BY *S. HYGROSCOPICUS* UC 8984

Purification of the active anthelmintic components produced by UC 8984 has been described (22). Briefly, culture extracts were fractionated by thin-layer chromatography. The TLC plates were divided into sections and the sections eluted with methanol. The eluates were assayed against *C. elegans* and the active fractions analyzed by HPLC to determine which metabolites were consistently present. Two such metabolites were thus identified, and these were purified from a 250 liter tank fermentation. An elution profile of these two

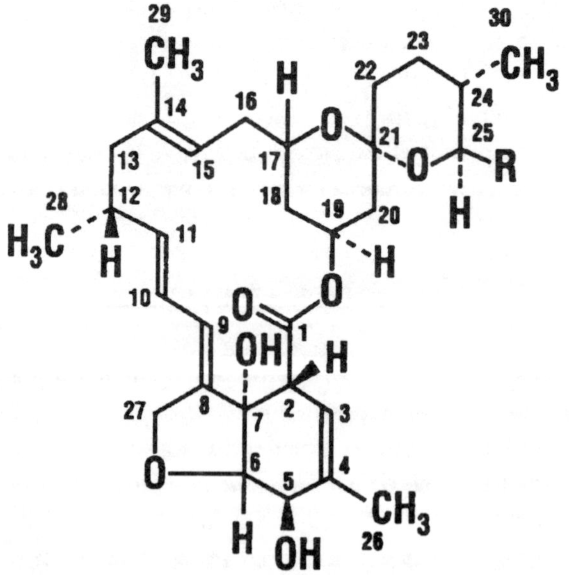

Figure 14.8 Milbemycin structures. R = CH_3: milbemycin α_1, R = CH_2CH_3: milbemycin α_3.

metabolites is shown in Figure 14.7. This preparation was analyzed using ^{13}C NMR, and the metabolites identified as milbemycins α_1 and α_3 (Figure 14.8).

These two milbemycins are members of a large milbemycin family produced by *S. hygroscopicus* subsp. *aureolacrimosus* NRRL 5739 (53). This culture was compared with UC 8984 and found to be similar, but taxonomically distinct (22). It has also been compared to *Streptomyces* strain E225, which produces a related series of milbemycins (26). These two cultures were also found to share some taxonomic similarities. *Streptomyces* strain E225 was identified using a screen which reportedly detected several organisms producing metabolites isolated from *S. hygroscopicus* subsp. *aureolacrimosus* (26). We also routinely discovered producers of the milbemycin series from a variety of sources which are physically similar to UC 8984 (58). A ubiquitous collection of organisms thus exists which share features of physical appearance and milbemycin production.

SUMMARY

The discovery of the avermectins through large scale *in vivo* screening involved dedication of considerable resources and an acceptance of the associated costs. Judicious choices of organisms to be screened also played an important role. Were such an *in vivo* screening program set up today it could take advantage of recent advances in propagating target parasites in rodent hosts. These assays use immunocompromised or nonmedicated jirds which have been infected with the target parasite of interest (14). They are able to detect the presence of avermectins and milbemycins in crude fermentation preparations.

In vitro screening for anthelmintics has become considerably more sophisticated than it was during the search which led to avermectin's discovery. The implementation of target parasite *in vitro* screens allows assessment of activity against commercially important parasites prior to the time-consuming and expensive process of dosing infected animals. Depending on the number of cultures to be screened, these can be used as primary screens, or to characterize leads targeted by less specific whole organism screens.

The avermectins have defined a new mode of action which is highly therapeutic toward helminth infections. They have allowed the development of several different types of *in vitro* screens which can target this action or compounds structurally similar to the avermectins. The examples of immunological, receptor binding and electrophysiological assays show that knowledge gained from the study of avermectins has yielded new screening protocols which can target compounds structurally or functionally similar to avermectins at an early stage. This significant advance permits accelerated development of lead compounds by demonstrating their differences from the avermectin/milbemycin family. It also allows rapid characterization of anthelmintic activity, leading to the dropping of known avermectins or milbemycins. Such was the case for the metabolites produced by *S. hygroscopicus* UC 8984.

The success of the avermectins has shown (again) that animal health products from microbial sources can be highly efficacious, as well as profitable. Most major pharmaceutical companies continue to search for new and better products for this market. The intensity of this search, as well as its ingenuity and sophistication, will certainly continue to increase.

ACKNOWLEDGEMENT

We thank Dr. George A. Conder for critical reading of this manuscript and for helpful suggestions.

REFERENCES

1. Ashton, R. J., Kenig, M. D., Luk, K., Planterose, D. N., Scott-Wood, G. 1990. MM 46115, A new antiviral antibiotic from *Actinomadura pelletieri*: Characteristics of the producing culture, fermentation, isolation, physico-chemical and biological properties. *J. Antibiot.* 43: 1387–1393.
2. Bennett, J. L., Pax, R. A. 1986. Micromotility meter: An instrument designed to evaluate the action of drugs on motility of larvae and adult nematodes. *Parasitology* 93: 341–346.
3. Blizzard, T. A., Ruby, C. L., Mrozik, H., Preiser, F. A., Fisher, M. H. 1989. Brine shrimp (*Artemia salina*) as a convenient bioassay for avermectin analogs. *J. Antibiot.* 42: 1304–1307.
4. Boisvenue, R. J., Brandt, M. C., Galloway, R. B., Hendrix, J. C. 1983. *In vitro* activity of various anthelmintic compounds against *Haemonchus contortus* larvae. *Vet. Parasitol.* 13: 341–347.
5. Bowman, J. W., Lee, B. L., Whaley, H. A., Thompson, D. P. 1991. Effects of dihydroavermectin B_{1a} and analogs on stretcher muscle of the lined shore crab, *Pachygrapsus crassipes. Comp. Biochem. Physiol.* In Press.
6. Burg, R. W., Miller, B. M., Baker, E. E., Birnbaum, J. Curie, S., Hartman, R., Kong, Y-L., Monaghan, R. L., Olson, G., Putter, I., Tunac, J. B., Wallick, H., Stapley, E. O., Oiwa, R., Omura, S. 1979. Avermectins, new family of potent anthelmintic agents: Producing organism and fermentation. *Antimicrob. Agents Chemother.* 15: 361–367.
7. Burg, R. W., Stapley, E. O. Isolation and characterization of the producing organism. See Ref. 9, pp. 24–32.
8. Calcott, P. H., Fatig III, R. O. 1984. Avermectin modulation of GABA binding to membranes of rat brain, brine shrimp and a fungus, *Mucor meihei. J. Antibiot.* 37: 797–801.
9. Campbell, W. C. (ed.). 1989. *Ivermectin and Abamectin.* New York: Springer-Verlag.
10. Carter, G. T., Nietsche, J. A., Hertz, M. R., Williams, D. R., Siegel, M. M., Morton, G. O., James, J. C., Borders, D. B. 1988. LL-L28249 antibiotic complex: A new family of antiparasitic macrocyclic lactones. Isolation, characterization and structures of LL-F28249 α, β, γ, δ. *J. Antibiot.* 41: 519–529.
11. Coles, G. C. 1986. Models of infection for intestinal worms. In *Experimental Models in Antimicrobial Chemotherapy, Vol. 3,* ed. O. Zak, M. A. Sande pp. 333–351, London: Academic.
12. Coles, G. C. 1990. Recent advances in laboratory models for evaluation of helminth chemotherapy. *Brit. Vet. J.* 146: 113–119.
13. Comley, J. C. W., Rees, M. J., O'Dowd, A. B. 1988. Leakage of incorporated radiolabeled adenine—A marker for drug-induced damage of macrofilariae in vitro. *Trop. Med. Parasit.* 39: 221–226.
14. Conder, G. A., Jen, L.-W., Marbury, K. S., Johnson, S. S., Guimond, P. M., Thomas, E. M., Lee, B. L. 1990. A novel anthelmintic model utilizing jirds, *Meriones unguiculatus*, infected with *Haemonchus contortus. J. Parasitol.* 76: 168–170.
15. DelCastillo, J., DeMello, W. C., Morales, T. 1964. Inhibitory action of gamma-aminobutyric acid (GABA) on *Ascaris* muscle. *Experientia* 20: 141–143.
16. Drexler, G., Seighart, W. 1984. Properties of a high affinity binding site for [^{3}H]avermectin B_{1a}. *Eur. J. Pharmacol.* 99: 269–277.
17. Duce, I. R., Scott, R. H. 1985. Actions of dihydroavermectin B_{1a} on insect muscle. *Brit. J. Pharmacol.* 85: 395–401.
18. Folz, S. D., Pax, R. A., Thomas, E. M., Bennett, J. L., Lee, B. L., Conder, G. A. 1987. Detecting *in vitro* anthelmintic effects with a motility meter. *Vet Parasitol.* 24: 241–250.
19. Folz, S. D., Pax, R. A., Thomas, E. M., Bennett, J. L., Lee, B. L., Conder, G. A. 1987. Development and validation of an *in vitro Trichostrongylus colubriformis* motility assay. *Int. J. Parasitol.* 17: 1441–1444.
20. Fritz, L. C., Wang, C. C., Gorio, A. 1979. Avermectin B_{1a} irreversibly blocks postsynaptic potentials at the lobster neuromuscular junction by reducing muscle membrane resistance. *Proc. Nat. Acad. Sci. U.S.A.* 74: 2062–2066.

21. Grosh, D. S. 1967. Poisoning with DDT: Effect on reproductive performance of *Artemia*. *Science,* 155: 592–593.
22. Haber, C. L., Heckaman, C. L., Li, G. P., Thompson, D. P., Whaley, H. A., Wiley, V. H. 1991. Development of a mechanism of action-based screen for anthelmintic microbial metabolites having avermectin-like activity, and the isolation of milbemycin-producing *Streptomyces* strains. *Antimicrob. Agents Chemother.* In Press.
23. Hall, Z. W., Bounds, M. D., Kravitz, E. A. 1970. The metabolism of gamma aminobutyric acid in the lobster nervous system: enzymes in single excitatory and inhibitory axons. *J. Cell Biol.* 46: 290–299.
24. Hartwig, J., Scott, P. M. 1971. Brine shrimp larvae as a screening system for fungal toxins. *Appl. Microbiol.* 21: 1011–1016.
25. Holden-Dye, L., Hewitt, G. M., Wann, K. T., Kvogsgaard-Larsen, P., Walker, R. J. 1988. Studies involving avermectin and the 4-aminobutyric acid (GABA) receptor of *Ascaris* muscle. *Pestic. Sci.* 24: 231–245.
26. Hood, J. D., Banks, R. M., Brewer, M. D., Fish, J. P., Manger, B. R., Poulton, M. E. 1989. A novel series of milbemycin antibiotics from *Streptomyces* strain E225. I. Discovery, fermentation and anthelmintic activity. *J. Antibiot.* 42: 1593–1597.
27. Ibarra, O. F., Jenkins, D. C. 1984. The relevance of *in vitro* anthelmintic screening tests employing the free-living stages of trichostrongylid nematodes. *J. Helminthol.* 58: 107–112.
28. Jenkins, D. C., Armitage, R., Carrington, T. S. 1980. A new primary screening test for anthelmintics utilizing the parasitic stages of *Nippostrongylus brasiliensis, in vitro. Parasitenkd.* 63: 261–269.
29. Jenkins, D. C., Carrington, T. S. 1981. An *in vitro* screening test for compounds active against the parenteral stages of *Trichinella spiralis*. *Tropenmed. Parasit.* 32: 31–34.
30. Jenkins, D. C., Rapson, E. B., Topley, P. 1986. The aggregation response of *Trichostrongylus colubriformis*: A basis for the rapid interpretation of *in vitro* anthelmintic screens. *Parasitology* 93: 531–537.
31. Kumar, S. 1987. The response of adult and free-living stages of *Necator americanus, in vitro,* to anthelmintics. *Rev. Biol. Trop.* 35: 73–76.
32. Leland, S. E., Jr., Ridley, R. K., Slonka, G. F., Zimmerman, G. Z. 1975. Detection of activity for various anthelmintics against *in vitro*-produced *Cooperia punctata*. *Am. J. Vet. Res.* 36: 449–456.
33. Lewis, J., Wu, C., Levine, J., Berg, H. 1980. Levamisole-resistant mutants of the nematode *Caenorhabditis elegans* appear to lack pharmacological acetylcholine receptors. *Neuroscience* 5: 967–989.
34. Marschke, C. K. Development of a radioimmunoassay for avermectin. Abstr. International Chemical Congress of Pacific Basin Societies, Honolulu, Hawaii, December 17–22, 1989.
35. Martin, P., LeJambre, L. 1979. Larval paralysis as an *in vitro* assay of levamisole and morantel tartrate resistance in *Ostertagia*. *Vet. Sci. Commun.* 3: 159–164.
36. Martin, R. J. 1982. Electrophysiological effects of piperazine and diethylcarbamazine on *Ascaris suum* somatic muscle. *Br. J. Pharmacol.* 77: 255–265.
37. Martin, R. J., Pennington, A. J. 1989. A patch-clamp study of effects of dihydroavermectin on *Ascaris* muscle. *Br. J. Pharmacol.* 98: 747–756.
38. Martlbauer, E., Gareis, M., Terplan, G. 1988. Enzyme immunoassay for the macrocyclic trichothecene roridin A: Properties and the use of rabbit antibodies. *Appl. Environ. Microbiol.* 54: 225–230.
39. Meyer, B. N., Ferrigni, N. R., Putnam, J. E., Jacobsen, L. B., Nichols, D. E., McLaughlin, J. L. 1982. Brine shrimp: A convenient general bioassay for active plant constituents. *Planta. Med.* 45: 31–34.
40. Michael, A. S., Thompson, C. G., Abramovitz, M. 1956. *Artemia salina* as a test organism for bioassay. *Science* 123: 464.
41. Pong, S. S., C. C. Wang. 1980. The specificity of high affinity binding of avermectin B_{1a} to mammalian brain. *Neuropharmacology* 19: 311–317.
42. Prior, M. G. 1979. Evaluation of brine shrimp (*Artemia salina*) larvae as a bioassay for mycotoxins in animal feedstuffs *Can. J. Comp. Med.* 43: 352–355.
43. Putter, I., MacConnell, J. G., Preiser, F. A., Haidri, A. A., Ristich, S. S., Dybas, R. A. 1981. Avermectins: Novel insecticides, acaricides and nematocides from a soil microorganism. *Experientia* 37: 963–964.

44. Rapson, E. B., Chilwan, A. S., Jenkins, D. C. 1986. Acetylcholinesterase secretion—a parameter for the interpretation of *in vitro* anthelmintic screens. *Parasitology* 92: 425–430.
45. Rapson, E. B., Jenkins, D. C., Topley, P. 1985. *Trichostrongylus colubriformis: In vitro* culture of parasitic stages and their use for evaluation of anthelmintics. *Res. Vet. Sci.* 39: 90–94.
46. Rew, R. S., Fetterer, R. H. 1989. Mode of action of antinematodal drugs. In *Chemotherapy of Parasitic Diseases*, eds. W. C. Campbell, R. S. Rew pp. 321–337. New York: Plenum.
47. Rew, R. S., Urban, J. F., Jr., Douvres, F. W. 1986. Screen for anthelmintics, using larvae of *Ascaris suum*. *Am J. Vet. Res.* 47: 869–873.
48. Dano, M., Terada, M., Ishii, A. I., Keno, H., Hayashi, M. 1981. Studies on chemotherapy of parasitic helminths (I). On the *in vitro* methods and paralyzing effects of avermectin B_{1a} on *Angiostrongylus cantonensis*. *Japan J. Parasitol.* 30: 305–314.
49. Schaeffer, J. M., Frazier, E. G., Bergstrom, A. R., Williamson, J. M., Liesch, J. M., Goetz, M. A. 1990. Cochlioquinone A, a nematocidal agent which competes for specific [^{3}H]-ivermectin binding sites. *J. Antibiot.* 43: 1179–1182.
50. Schaeffer, J. M., Haines, H. W. 1989. Avermectin binding in *Caenorhabditis elegans*: A two-state model for the avermectin binding site. *Biochem. Pharmacol.* 38: 2329–2338.
51. Simpkin, K. G., Coles, G. C. 1981. The use of *Caenorhabditis elegans* for anthelmintic screening. *J. Chem. Tech. Biotechnol.* 31: 263–265.
52. Takahashi, A., Kurasawa, S., Ikeda, D., Okani, Y., Takeuchi, T. 1989. Altmecidin. A new acaricidal and antitumor substance. *J. Antibiot.* 42: 1556–1561.
53. Takiguchi, Y., Mishima, H., Okuda, M., Terao, M., Aoki, A., Fukuda, R. 1980. Milbemycins, a new family of macrolide antibiotics: Fermentation, isolation and physico-chemical properties. *J. Antibiot.* 33: 1120–1127.
54. Thompson, D. P., Lee, B. L., Bowman, J. W. 1988. Effects of ivermectin and related compounds on stretcher muscle of the lined shore crab (*Pachygrapsus crassipes*): Electrophysiological responses and correlations with anthelmintic activity. *Pestic. Sci.* 24: 263–265.
55. Turner, M. J., Schaeffer, J. M. 1989. Mode of action of ivermectin. See Ref. 9, pp. 73–88.
56. VanFleteren, J. R., Roets, D. E. 1972. The influence of some anthelmintic drugs on the population growth of the free-living nematodes, *Caenorhabditis briggsae* and *Turbatrix asceti* (nematoda: rhabditida). *Nematologica* 18: 325–338.
57. Ward, J. B., Nobel, H. M., Porter, N., Fletton, R. A., Noble, D. 1986. UK Patent Application GB 2,166,436A.
58. Whaley, H. A., Wiley, V. H., Shilliday, F. B., Haber, C. L., Slechta, L., Mizsak, S. A. 1990. The coproduction of milbemycins and polyethers. *Planta Med.* 56: 500.
59. Yao, R. C., Mahoney, D. F. 1984. Enzyme-linked immunosorbant assay for the detection of fermentation metabolites: Aminoglycoside antibiotics. *J. Antibiot.* 37: 1462–1468.
60. Yao, R. C., Mahoney, D. F. 1989. Enzyme immunoassay for macrolide antibiotics: Characterization of an antibody to 23-amino-*O*-mycaminosyltylonolide. *Appl. Environ. Microbiol.* 55: 1507–1511.
61. Zufall, F., Franke, C., Hatt, H. 1989. The insecticide avermectin B_{1a} activates a chloride channel in crayfish muscle membrane. *J. Exp. Biol.* 142: 191–205.

15

Paraherquamide—A Novel Antiparasitic Agent; Production and Activity

Prakash S. Masurekar, Michel M. Chartrain, Kodzo Gbewonyo, Robert T. Goegelman, John G. Ondeyka, Margaret S. Sosa, Louis Kaplan, Dan A. Ostlind, Wesley L. Shoop
Merck and Co. Inc. P. O. Box 2000, Rahway, N.J. 07065

Maria T. Diaz-Matas
CIBE, Madrid, Spain

Penicillium charlesii *was found to produce paraherquamide and six other novel analogues. Fermentation processes for the production of paraherquamide in liquid and solid media were developed. In liquid medium the preferred carbon and nitrogen sources were glycerol and peptonized milk, respectively; whereas for the solid medium, these were glycerol and peptone, yeast extract and glutamine, respectively. For the liquid medium process, optimum seed concentration and the age of transfer respectively were 2.5% and 24 hrs. On solid medium, calcium stimulated sporulation and paraherquamide synthesis. For both processes the best yields were obtained at 25.5 C. Kinetics of production was similar in liquid and solid media with most of the synthesis occurring between 48 to 144 hrs. Scale-up to 14-L fermentors for the liquid and to large bioreactors for solid was successful. Paraherquamide was shown to be effective against* Trichostrongylus colubriformis *in the gerbil and against a spectrum of helminths in sheep including ivermectin-resistant* T. colubriformis *and* Haemonchus contortus.

Key words: *Penicillium charlesii*, anthelmintic, fermentation development

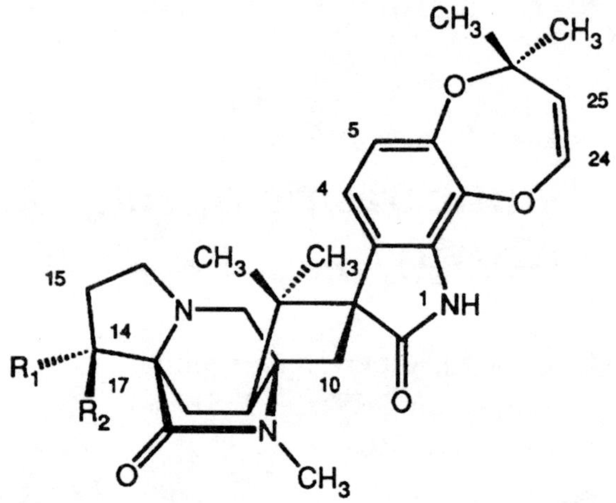

Analogs	R₁	R₂
1	OH	CH₃
B	H	H
C	—CH₂—	
D	—CH₂O—	
E	H	CH₃

Figure 15.1 / Paraherquamide and analogues B-E.

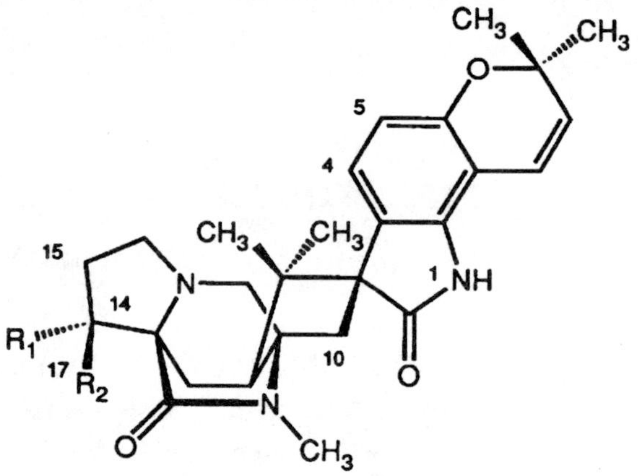

Analogs	R₁	R₂
F	H	CH₃
G	OH	CH3

Figure 15.2 / Paraherquamide analogues F and G.

INTRODUCTION

Throughout history, man has looked to nature for compounds to be used to meet such diverse needs as food, medicines, dyes and toxins. Today the pharmaceutical industry continues this tradition in search of new therapeutics. Since the discovery of the first microbially produced anthelmintic, avermectin, there has been considerable effort to discover other such compounds from microbial sources [2, 15, 20]. In our screen we found that a strain of *Penicillium*, which was later identified as *P. charlesii*, produced a compound which had anthelmintic activity [22]. The compound was isolated and identified to be paraherquamide (Figure 15.1) [21, 16]. Yamazaki et al. [29] had previously reported its production by *P. paraherquei*. These workers also determined its structure and noted its toxic properties [29]. The absolute stereochemistry was determined by Blizzard and coworkers [5]. Subsequently, in our laboratories six other analogues of paraherquamide were isolated from *P. charlesii* and their structures were determined (Figures 15.1 and 15.2) [21, 16].

In this paper we describe the chemistry, the development of fermentation processes and animal studies with paraherquamide.

CHEMISTRY OF PARAHERQUAMIDE

ISOLATION AND CHARACTERIZATION

As mentioned in the Introduction, *P. charlesii* produces paraherquamide and six structurally related analogues. These were isolated from solid medium fermentation. The isolation procedure from the small scale fermentation involved extraction of the fermentation mass with ethyl acetate, concentration of ethyl acetate extract, trituration with methanol and chromatography of the methanol extract on Sephadex LH-20 and silica-gel followed by reverse phase HPLC on a Magnum 20 column (Whatman ODS-3) [21]. Similar extraction procedure and chromatography on silica-gel was used for their isolation from the large scale fermentation [21]. UV spectra, mass spectra (electron impact (EI) and high resolution (HR)) and NMR spectra were used to determine the structure shown in figures 15.1 and 15.2 [21, 16]. The analogues B to E have different substituents on C-14 as compared to paraherquamide; the analogues F and G have a pyran ring in place of dioxypin of paraherquamide and furthermore, analogue F lacks the C-14 hydroxyl. The determination of the biological activity against *Caenorhabditis elegans* indicated that none of the analogues were as active as paraherquamide. The LD_{50} value for analogue E, the most potent one, was 6 µg/ml as compared to that for paraherquamide, which was 2.5 µg/ml. The other five were considerably less active, with the LD_{50} values ranging from 20 to 160 µg/ml [21]. Similarly, the potency of the 24,25-dihydro analogue prepared synthetically was very low. These results suggested that the alkyl substitution at C-14 and/or unsaturation in dioxypino ring is critical for the activity.

It was of interest to note that the scale of the fermentation affected the production of these analogues. Although paraherquamide was produced in shake flasks (small scale) and in trays (large scale), analogues B, C and D were the only ones produced in the shake flasks while E, F and G predominated at the large scale [21].

SEMI-SYNTHETIC DERIVATIVES OF PARAHERQUAMIDE

Blizzard and his coworkers [3, 4, 5, 6] have described preparation of about fifty derivatives of paraherquamide. In the first paper formation of a rearranged cyclic urethane by the reaction of paraherquamide with phosgene and its superhydride-mediated reductive cylization was reported [5]. In subsequent papers the preparation and the biological activity of these and other analogs against *C. elegans* was described. The first series of analogs included those where positions C-5, C-14 and C-17 were substituted [6]. The vinyl ether modified analogs and the 1-N substituted analogs were the two other types prepared [3, 4]. None of these was as effective as paraherquamide.

DEVELOPMENT OF FERMENTATION PROCESS

EFFECT OF CARBON SOURCE

Initial production of paraherquamide was done on solid medium. The inoculum was grown in a medium consisting of cerelose, oat flour, tomato paste, corn steep liquor and a trace element solution which contained $FeSO_4 \cdot 7H_2O$, $MnSO_4 \cdot 4H_2O$, $CuCl_2 \cdot 2H_2O$, $CaCl_2$, H_3BO_3, $(NH_4)_6MoO_2 \cdot 4H_2O$ and $ZnSO_4$. Generally, fifty ml of medium in a 250 ml unbaffled Erlenmeyer flask was inoculated with 1 ml of frozen vegetative mycelia. The flasks were shaken at 28°C and 220 RPM for 48 hrs. Ten ml of this seed was then used to inoculate the second stage seed consisting of 500 ml of medium in a 2-L unbaffled Erlenmeyer flask. The flasks were incubated at 28°C and 180 RPM for 48 hrs. Two ml of the second stage seed were used to inoculate the production medium which had a cracked corn base. The composition of the nutrient medium used to supplement the base was as

Table 15.1 / Effect of carbon source on the production of paraherquamide

	Liquid Medium		Solid Medium[a]	
Carbon Source	Conc. g/l	Yield mg/l	Conc. g/l	Yield mg/flask
Glycerol	50	11	52	18
Glucose	100	7	20	14
Starch	20	0	20	15
Corn oil	—[b]	—[b]	20[c]	13
Dextrin	100	10	—[b]	—[b]

[a] In solid medium the yield without any additions (basal medium) was 15 mg/flask
[b] Not tested
[c] ml/l

follows: glycerol 26.2 g/l, L-cysteine 6.8 g/l, yeast extract 33.4 g/l, cobalt chloride hexahydrate 0.134 g/l, sodium tartrate 0.68 g/l and ferrous sulfate heptahydrate 0.66 g/l. Ten grams of cracked corn and 15 ml of nutrient medium were added to 250 ml unbaffled Erlenmeyer flask and were inoculated as described. The flasks was statically incubated at 25°C for 14 days.

The liquid medium contained glycerol, peptonized milk and ardamine PH. The procedure used to prepare the seed was the same as that used for the solid medium fermentation. For production, the flasks were shaken at 220 RPM and at 28°C for 7 days.

The contents of the flask were extracted with ethyl acetate and the extract was dried. The residue was dissolved in methanol and analysed by HPLC.

In the first series of experiments the effect of various carbon sources on the production was studied. In solid medium only marginal improvement was seen with glycerol, while glucose, starch and corn oil had no effect (Table 15.1). These results indicate that the cracked corn was adequate as a carbon source. An increase in the concentration of glycerol up to 105 g/l showed no effect on the yield, however further increase to 209 g/l depressed the titer (data not shown). One possible explanation for this inhibition is that at this high concentration glycerol reduces the water activity of the substratum. On the other hand, in liquid fermentation glycerol, glucose and dextrin supported production (Table 15.1). The substrate concentrations at which the best yields were obtained are shown in Table 15.1. It was surprising that no paraherquamide was produced in the medium with modified starch since good production is obtained on the cracked corn. Another interesting observation was that while generally, glucose is not an ideal carbon source for the production of the secondary metabolites and that often slowly utilizable carbon sources are preferred [8, 11, 12, 18], in the case of paraherquamide, glucose did not seem to inhibit the synthesis.

EFFECT OF NITROGEN SOURCE

The additions of peptone and yeast extract improved the yields by 30% in solid medium (Table 15.2). Further increases in the concentration of peptone or yeast extract lowered the titer (data not shown). Similarly, production was stimulated when amino acids were added to the medium (Table 15.3). The highest increase of 65% was observed upon the supplementation with L-arginine, followed by that with L-histidine (48%), L-alanine (42%), L-serine (39%), L-glutamine (36%), L-orni

Table 15.2 / Effect of nitrogen source on the production of paraherquamide

Nitrogen Source	Liquid Medium[a]		Solid Medium[b]	
	Conc. g/l	Yield mg/l	Conc. g/l	Yield mg/flask
Peptonized mild	30	11	—[c]	—[c]
Peptone	—[c]	—[c]	10	19
Ardamine PH	4	3	—[c]	—[c]
Yeast extract	—[c]	—[c]	33	20
$(NH_4)_2SO_4$	—[c]	—[c]	5	15

[a]In liquid medium promosoy, lard water, corn steep liquor, and lexein supported good growth but not paraherquamide production

[b]In solid medium the yield without any additions (basal medium) was 15 mg/flask

[c]Not tested

Table 15.3 / Effect of amino acids on paraherquamide production in solid medium

Amino Acid Added[a]	Production mg/flask	Amino Acid Added[a]	Production mg/flask
None	15	L-Tyrosine	18
L-Arginine	24	L-Valine	21
L-Aspargine	19	L Glutamine	21
L-Aspartic acid	17	L-Glutamic acid	18
L-Glycine	20	L-Alanine	22
L-Histidine	23	L-Leucine	18
L-Lysine	18	L-Serine	21
L-Ornithine	21	L-Isoleucine	19
L-Proline	18	L-Phenyl alanine	16
L-Threonine	19		
L-Tryptophan	19		

[a]Amino acids were added at the final concentration of 7 g/l

thine (36%), L-valine (35%) and L-glycine (33%). Ammonium sulfate had no effect on the synthesis. These results indicated that the basal medium used to prepare the solid medium was nitrogen-limited and that the hydrolysis of the proteins and the peptides in the medium was probably a rate-limiting step.

In liquid fermentation peptonized milk and ardamine PH supported the synthesis of paraherquamide (Table 15.2). Although the culture grew well in the medium containing promosoy, pharmamedia, lard water, corn steep liquor or lexein, it did not produce any paraherquamide. Simple nitrogen sources like $(NH_4)_2SO_4$, $NaNO_3$ and urea were also ineffective. It is probable that various nitrogen compounds play a regulatory role in the biosynthesis of secondary metabolites. Since many of these metabolites contain nitrogen, the regulation of their nitrogen containing precursors may affect their synthesis. There are reports of repression by ammonia of tylosin, gilvocarcin and cephalosporin production as was observed in paraherquamide fermentation [7, 9, 28]. The effect of the complex nitrogen sources is more difficult

Table 15.4 / Effect of trace elements

Trace Element Solution[a] Added ml	Paraherquamide mg/l
None	15
10	19
20	23
50	20

[a]Composition of the trace element solution: per liter of 0.6 N HCl: $FeSO_4 \cdot 7H_2O$, 1.0 g; $MnSO_4 \cdot 4H_2O$, 1.0 g; $CuCl_2 \cdot 2H_2O$, 0.025 g; $CaCl_2$, 0.1 g; H_3BO_3, 0.056 g; $(NH_4)_6MoO_2 \cdot 4H_2O$, 0.01 g; $ZnSO_4 \cdot 7H_2O$, 0.2 g

Table 15.5 / Effect of aeration in solid medium fermentation

Conditions	Paraherquamide mg/flask
No aeration	14
Aeration 40 ml/min	18

to understand but there are examples of depression of antibiotic production upon their addition during the idiophase [19, 26]. It is indeed possible that certain amino acids, either by themselves or as precursors of ammonia, affect critical enzymes involved in the biosynthesis of the secondary metabolite. Complex nitrogen sources can also be inhibitors if they either supply specific amino acids too rapidly or at levels above critical concentration.

EFFECT OF OTHER VARIABLES

Trace elements had a beneficial effect on the production of paraherquamide on solid medium. The same trace element solution that was used to supplement the seed medium was used in this experiment. The optimum effect, an increase of 55%, was observed with 20 ml of the trace element solution (Table 15.4). The studies on the kinetics of fermentation (Figure 15.5) indicated a correlation between production and sporulation (see below). Therefore, addition of calcium, which has been reported to be a stimulator of fungal sporulation was tried. As shown in Figure 15.3 there was a stimulation of sporulation and paraherquamide produc-

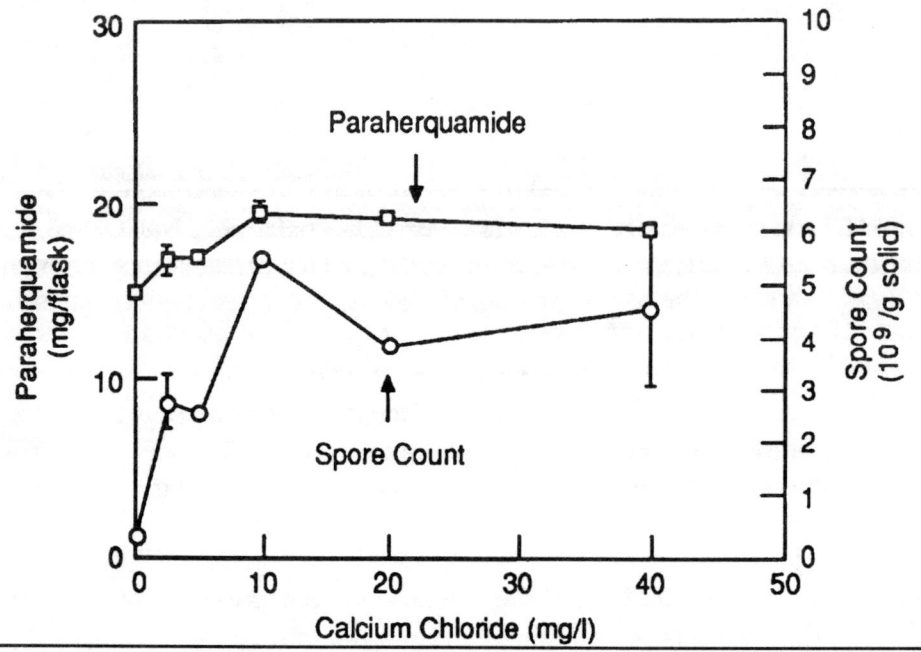

Figure 15.3 / Effect of calcium chloride on the sporulation and paraherquamide production in solid medium fermentation.

Figure 15.4 / Effect of environmental parameters on the production of paraherquamide in solution medium fermentation.

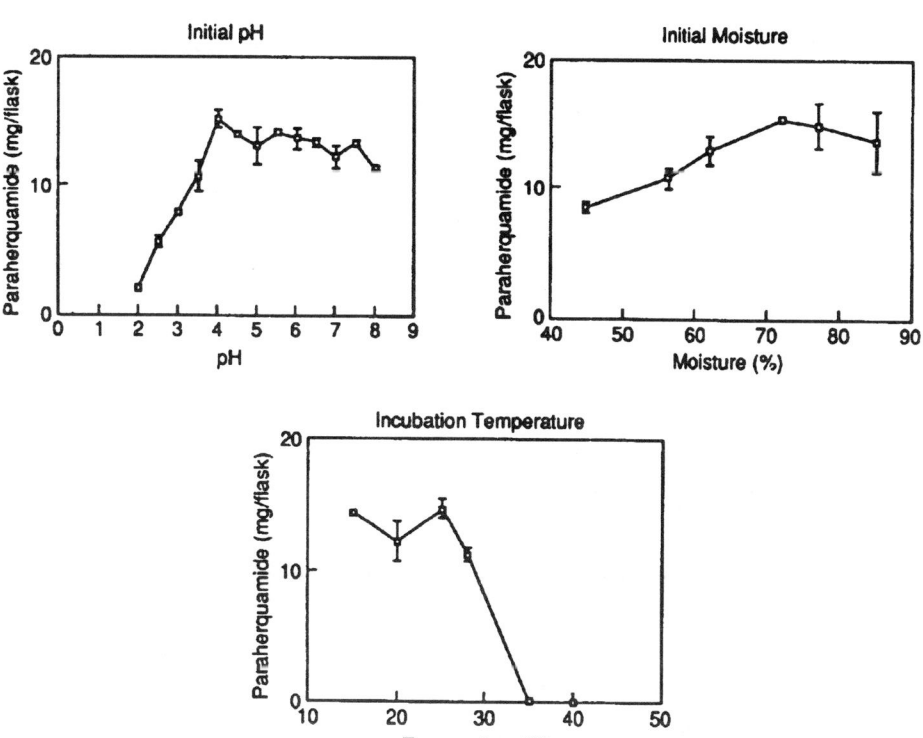

tion with increasing the $CaCl_2$ concentration up to 10 mg/ml. However, further increase had no effect on the yield or sporulation.

In liquid fermentation the addition of trace elements did not have any effect on production.

Aeration was found to have a positive effect on the synthesis in solid fermentation (Table 15.5). The flasks were aerated by passing 40 ml of air/min over the surface of the medium. We believe that reduction in the concentration of CO_2, which has been reported to be inhibitory in solid fermentation, was probably responsible for this improvement [1, 17]. A sharp reduction in the titer was observed as the initial pH of the nutrient medium was reduced below 4.0 (Figure 15.4a). However, there was no inhibitory effect of pH in the range 4 to 8. The optimum initial moisture concentration was 72% and that below 70% resulted in reduced yield (Figure 15.4b). About the same production was obtained in the range of the incubation temperature from 15 to 25°C and further increase in the temperature depressed synthesis (Figure 15.4c).

In liquid fermentation aeration had only a marginal effect and the best yields were obtained with 30 ml of the medium in 250 ml unbaffled Erlenmeyer flasks (data not shown). The optimum process temperature was 25.5°C and the yield was lower at 27.5°C as was observed in the case of solid fermentation (Table 15.6). In general very low titers were observed at 20°C. Two other variables studied were the age and the size of the inoculum. The age of the inoculum was varied from 24 to 72 hrs to determine the optimum time to transfer. In order to obtain the maximum yield it is essential to transfer the culture at the correct metabolic state. In many processes, but not in all, this is during the exponential growth phase [10]. The best production was obtained with a 24 hr old inoculum at 25.5° C (Table 15.6). The 48 hr old inoculum was less sensitive to the temperature between 25.5 to 27.5°C. A reduction in the yield was observed with

Table 15.6 / Effect of age of inoculum and fermentation temperature in liquid medium fermentation

Seed age Hrs	Temperature °C	Final pH	Paraherquamide mg/l
24	27.5	7.6	16
	25.5	7.6	27
	20.0	6.5	3
48	27.5	7.8	22
	25.5	7.9	24
	20.0	6.4	4
72	27.5	7.9	17
	25.5	7.8	21
	20.0	6.6	2

Table 15.7 / Effect of inoculum size in liquid medium fermentation

Size of Inoculum (%)	Paraherquamide mg/l
2.5	30
5.0	23
8.75	11
12.5	12
20.0	6

the 72 hr old inoculum. These results are consistent with the above mentioned requirement for the optimum production. The size of the seed was varied between 2.5 to 20%. Interestingly the lowest concentration supported the highest production (Table 15.7). Synthesis decreased as the size of the inoculum was increased. The optimum inoculum size is a function of the morphological characteristics of the microbial strain and of the design and the operation of the production vessel [14, 27]. In this case growth is heavy and finely pelleted and the 250 ml Erlenmeyer flasks used for the production might have been limited in the oxygen transfer capacity. These two factors probably favored the lower inoculum size.

KINETICS OF FERMENTATION

Kinetics of fermentation in solid medium are shown in Figure 15.5. It was found from the visual observation of the culture during the fermentation that after 72 hrs the surface of the corn was covered by a white-green mycelial mat. The formation of typical *Penicillium* conidiophores bearing spores of a diameter of 2–3 mm was also observed during this period. The glycerol was essentially exhausted in the first 48 hrs. During the same period the corn starch was hydrolysed as indicated by the increase in the level of glucose in the medium, which peaked at 48 hrs. The spore count reached the maximum value at 100 hrs and then dropped. The synthesis of paraherquamide began at 48 hrs and continued up to 144 hrs. During the linear phase of the synthesis there seemed to be a correlation between the production and sporulation. There was a small increased in the titer from 144 to 480 hrs. The pH remained steady around 6 throughout the fermentation (data not shown). These results suggest that glycerol is used to support the growth and probably the initial synthesis of paraherquamide and that glycerol did not repress the enzymes involved in the hydrolysis

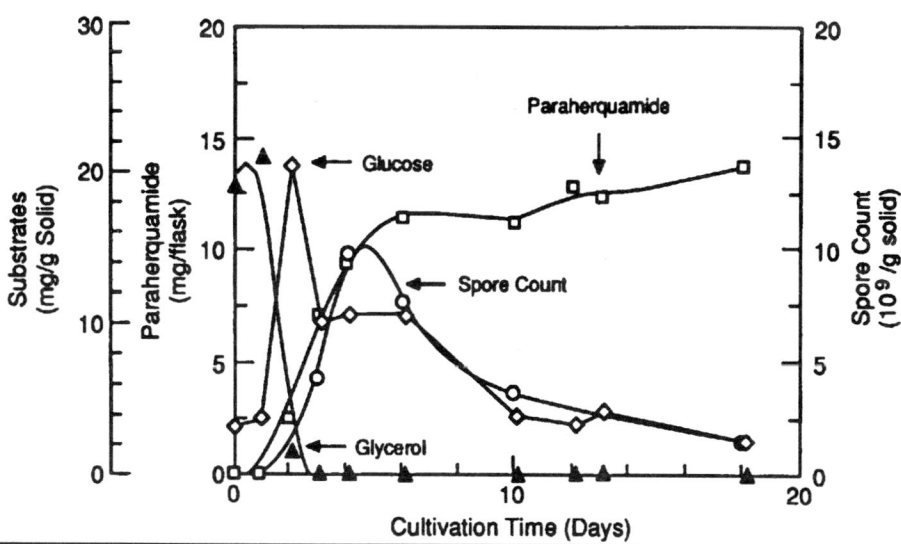

Figure 15.5 / Kinetics of fermentation in solid medium fermentation.

of the corn starch. Interestingly, glucose was the carbon source used during the production phase of the fermentation. The cause for the reduction in the concentration of glucose after 144 hrs is not clear. It is possible that it was due to the turnover of the hydrolytic enzymes. The loss of spore count is also puzzling. Indeed, these two events, namely the turnover of the enzymes and the reduction of the spore count may be related.

The growth in the liquid fermentation continued up to 192 hrs (Figure 15.6). The production of paraherquamide, as in the solid fermentation began at 48 hrs and reached the maximum value at 144 hrs. There was a considerable scatter in the dry cell weight and the yield data due to the sampling difficulties. The pH reduced to 5 in 72 hrs and remained low up to 120 hrs, after which it in-

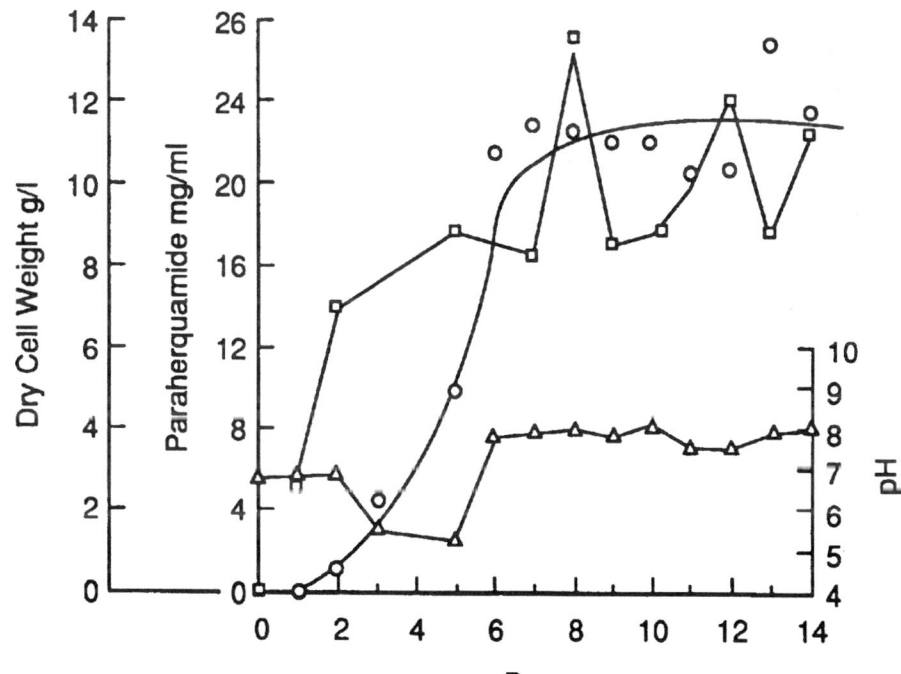

Figure 15.6 / Kinetics of fermentation in liquid medium in shake flasks.

Table 15.8 / Effect of improved medium on paraherquamide production in flasks and trays

Conditions	Paraherquamide		Specific Yield
	mg/Vessel	Percent of Control	mg/g of Corn
Basal Medium	14	100	1.4
Flasks (250 ml)[a] No Aeration	22	158	2.2
Flasks (250 ml)[a] With Aeration	24	173	2.4
Small Bioreactors[a]	290	NA	1.5
Large Bioreactors[a]	1,800	NA	0.9

NA: Not applicable
[a] Improved medium was tested in flasks and in the trays (bioreactors). It contained, in addition to the basal medium described in the text, the following supplements per liter: peptone, 10 g; yeast extract, 33 g; glutamine, 7g; glycerol, 26 g; calcium chloride, 10 mg; trace element solution, 20 ml

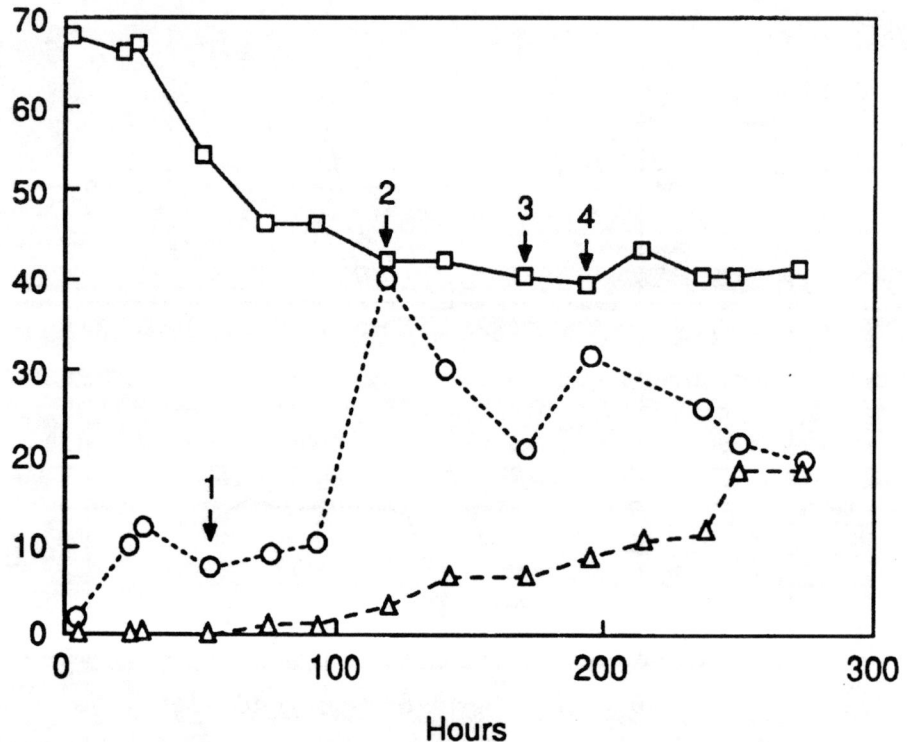

Figure 15.7 / Kinetics of fermentation in liquid medium in 14-L fermentor.

creased to 8 and was steady for the rest of the fermentation. These results suggested that glycerol was exhausted in the first 120 hrs. Lactose from peptonized milk probably supported the growth and the production beyond 120 hrs. This suggestion remains to be tested at this time.

SCALE-UP OF FERMENTATION

The optimized solid medium was tested both in the shake flasks and in the small and large bioreactors (trays). The dimensions of the small trays were 35 cm × 11 cm × 5 cm and those of the large were 80 cm × 55 cm × 5 cm. The capacity of these two types of trays was 200 g and 2000 g of cracked corn, respectively. The volume of the improved nutrient medium added was 300 ml and 3000 ml, respectively. The improved nutrient medium contained peptone, yeast extract, glutamine, glycerol, $CaCl_2$ and trace element solution. The seed volume was 40 ml and 400 ml, respectively. In the shake flasks, improved medium with and without aeration supported higher production (Table 15.8). Aeration increased the yields by 15% over those without aeration. Good production was obtained in both small and large trays. The specific yield (yield per g of cracked corn), however, was lower in the trays and decreased further as the size of the trays increased.

The fermentation in the liquid medium was scaled-up to 14-L fermentors. The medium (9 liters operating volume) contained glycerol, peptonized milk, ardamine PH, FD 62 (anti-foam) and glutamine. The inoculum was 5%. The rate of aeration was initially 4 l/min and was increased to 6 l/min at 126 hrs. Similarly, the rate of agitation was changed from initial 300 RPM to 700 RPM at the same time. The process temperature was 25°C. At 72, 126 and 14 hrs, 500 ml of 50% glycerol solution was added and at 192 hrs 1 liter of the complete medium was added. The packed cell column increased to 40% by 126 hrs and then declined to 20% by the end of the fermentation (Figure 15.7).

The pH decreased slowly to 4 and remained steady. The synthesis did not begin until 126 hrs and reached the maximum titer of 18 mg/l at 264 hrs. There were some striking differences in the kinetics in the shake flask and in the fermentor. The pH in the shake flask recovered after 120 hrs, whereas it remained low throughout the run in the fermentor. The cause for this may be the additions of glycerol during the fermentation. On the other hand, the reason for the slow rate of synthesis is not clear at this time.

ANIMAL STUDIES

EFFICACY IN THE GERBIL

The efficacy of paraherquamide against immature *Trichostrongylus colubriformis* was demonstrated by Ostlind et al. [22]. Five-week old male and female gerbils were each inoculated with 500–600 infective larvae of benzimidazole- and avermectin-sensitive *T. colubriformis*. On the sixth day of infections paraherquamide dissolved in an organic solvent was administered orally. The efficacy of paraherquamide was determined 72 hrs later from the worm count. Paraherquamide was found to be more than 98% effective at all dose levels above 1.56 mg/kg (Table 15.9).

ANTHELMINTIC ACTIVITY IN SHEEP

Preliminary laboratory studies showed that paraherquamide was effective against the common nematodes of the ovine gastrointestinal tract and therefore, it was titrated in helminth-free sheep which were experimentally infected with 6 species of nematodes [24]. The procedure involved inoculating helminth-free lambs with the infective larvae from monospecific cultures of *Haemonchus contortus*, *Ostertagia circumcincta*, *Trichostrongylus axei*, *Trichostrongylus colubriformis*, *Cooperia curticei* and *Oesophagostomum colum-*

Table 15.9 / Efficacy of paraherquamide against immature *Trichostrongylus colubriformis* in gerbils

Dosage (mg/kg)	Number of Animals	Efficacy (%)
Placebo	8	—[a]
200	3	100[b]
100	3	100[b]
50	3	100[b]
25	3	99.5[b]
12.5	3	100[b]
6.25	5	100[b]
3.12	4	99.7[b]
1.56	3	98.1[b]
0.78	3	96.5
0.39	3	66.3[b,c]

[a]Average worm burden was 286.7; range 199–370
[b,c]Each value showing different superscript is significant (P<0.05); Dunnett's t multiple comparison test

bianum; treating the lambs with either paraherquamide or the vehicle; and examining the lambs 7 days after treatment for residual worm burden. The strain of *H. contortous* used was ivermectin-resistant and that of *T. colubriformis* was ivermectin-and benzimidazole-resistant [13, 25]. The drug was given singly and orally as a drench in 1:7 dimethyl sulfoxide (DMSO): propylene glycol. The results showed that paraherquamide, even at the lowest dose of 0.25 mg/kg, is highly efficacious against the adult stages of 5 of the 6 nematodes tested (Tables 15.10 and 15.11). *O. columbianum* was the least sensitive with an ED_{95} approximately 18-fold higher than that of the other species tested

Table 15.10 / Anthelmintic efficacy of paraherquamide infected sheep (HS-5-87).

				Percent Efficacy						
Treatment	P.O. Dosage (mg/kg)	No. Sheep	*Haemonchus contortus*[a]	*Ostertagia circumcinta* L_4[b]	Adult	*Trichostrongylus* axei	*colubriformis*[c]	*Cooperia* spp, L_4[b]	*curtecei* Adult	*Oesophagostomum columbianum*
Control	—	3	(237)	(1600)	(65)	(4132)	(4321)	(96)	(2534)	(19)
L-668,422	2.0	2	>99**	96	99**	>99**	>99*	>99**	>99**	79
L-668,422	1.0	3	>99**	83	99**	>99**	>99**	>99**	>99**	37
L-668,422	0.5	3	>99**	98	99**	>99***	>99*	>99**	>99**	47
L-668,422	0.25	3	92*	79	95**	>99**	94	>99**	>99**	0

[a]Ivermectin resistant isolate.
[b]Inhibited developmental stage.
[c]Ivermectin and benzimidazole resistant isolate.
*,**,***Reduction from control level due to treatment; p<.05, p<.01, respectively.

Table 15.11 / Dose response estimates for paraherquamide against sheep gastrointestinal nematodes (HS-5-87)

Parasite	Regression Equation	$ED_{95} \pm$ S.E., mg/kg
Haemonchus contortus	None	~.295
Ostertagia circumcinta L_4	None	>2.0
Ostertagia circumcinta	None	~.250
Trichostrongylus axei	Log $\hat{Y}$ = −2.53050 Log X −.04247	.117 ± .227
Trichostrongylus colubriformis	Log $\hat{Y}$ = −3.08454 Log X +.43285	.242 ± .227
Cooperia curticei	None	<.250
Cooperia spp. L_4	None	<.250
Oesophagostomum columbianum	Log $\hat{Y}$ = −1.46715 Log X +.94825	4.659 ± 3.037 ɸ

ɸEstimate on extrapolation outside the range of dose levels tested.

(Table 15.11). The most striking observation was that paraherquamide was active against the ivermectin- and benzimidazole-resistant strains of *H. contortus* and *T. colubriformis*. These results suggest that the mode of action of paraherquamide is probably different from those of these two anthelmintic compounds. There were no adverse reactions in any sheep.

The limited animal studies described clearly indicate that though somewhat less potent as compared to the avermectins, paraherquamide due to its efficacy against the ivermectin and benzimidazole-resistant strains is a compound of great interest, especially since there are reports of benzimidazole- and ivermectin-resistant nematodes in sheep (cf. [24]). More studies are needed to broaden the spectrum of activity and to explore formulations which would offer increased activity in the lower bowel against *O. columbianum*.

MODE OF ACTION

The studies to determine the mode of action of paraherquamide were done in a free living nematode, *Caenorhabditis elegans*[23]. The specific and total binding of ³H-paraherquamide to isolated membrane from *C. elegans* along with its effect on the motility of the nematode was determined. It was found that paraherquamide bound specifically to the membranes and there was a positive correlation between the binding affinity and the biological activity of paraherquamide and its analogues [23]. This suggested that the interaction of this compound with the specific binding site is needed for its biological activity. Furthermore, there was a strong inhibition of binding by phenothiazine, a known anthelmintic, and a series of phenothiazine derivatives. However, none of the other anthelmintic compounds tested had any effect on the binding, which is consistent with the results obtained in sheep infected with the ivermectin- and benzimidazole-resistant nematodes [24]. Phenothiazine has been reported to be an antagonist of dopamine and to act on D-2 receptors, although it is not clear if that is the mechanism of its anthelmintic activity. It is possible that the paraherquamide effect is mediated in a similar manner. However, at this time the exact mechanism of the nematocidal action of paraherquamide remains to be elucidated.

CONCLUSIONS

Paraherquamide, an oxindole alkaloid, represents a new class of effective anthelmintic agents active against a number of common gastrointestinal nematodes in the gerbil and sheep. Fermentation processes to produce it on solid or in liquid medium were developed. A careful selection of carbon and nitrogen sources was required for good production. The structure-function relationship studies indicate that alkyl substitution at C-14 and/or unsaturation in dioxypino ring is critical for activity. It was found to bind specifically to membrane preparation of *C. elegans*, however at present, the exact mechanism of its action is not known. Since the animal studies have shown it to be efficacious against the nematodes resistant to ivermectin and thiabendazole, it merits careful scrutiny.

ACKNOWLEDGEMENTS

We thank Dr. James Schaeffer for making available a preprint of his paper on the binding of paraherquamide to *C. elegans* membranes.

REFERENCES

1. Bajracharya, R. and R. E. Mudgett. 1980. Effect of controlled gas environments in solid substrate fermentation of rice. *Biotech. Bioeng.* 22: 2219–2235.
2. Burg, R. W., B. M. Miller, E. E. Baker, J. Birnbaum, S. A. Currie, R. Hartman, Y. L. Kong, R. L. Monaghan, G. Olson, I. Putter, J. B. Tunac, H. Wallick, E. O. Stapley, R. Oiwa and S. Omura. 1979. Avermectins, new family of potent anthelmintic agents: Producing organism and fermentation. *Antimicrob. Agents Chemother.* 15: 361–367.
3. Blizzard, T. A., G. Margiatto, H. Mrozik, J. A. Schaeffer and M. H. Fisher. Chemical modifications of paraherquamide. III. Vinyl ether modified analogs. *Tetra. Lett.* In press.
4. Blizzard, T. A., G. Margiatto, H. Mrozik, J. M. Schaeffer and M. H. Fisher. Chemical modification of paraherquamide. IV. 1-N-substituted analogs. *Tet. Lett.* In press.
5. Blizzard, T. A., G. Marino, H. Mrozik, M. Fisher, K. Hoogsteen and J. Springer. 1989. Chemical modification of paraherquamide. 1. Unusual reactions and absolute stereochemistry. *J. Org. Chem.* 54: 2657–2663.
6. Blizzard, T. A., H. Mrozik, M. Fisher and J. M. Schaeffer. 1990. Chemical modification of paraherquamide. 2. Replacement of C-14 methyl group. *J. Oreg. Chem.* 55: 2256–2259.
7. Brana, A. F., S. Wolfe and A. L. Demain. 1985. Ammonium repression of cephalosporin production by *Streptomyces clavuligerus*. *Can. J. Microbiol.* 31: 736–743.
8. Buckland, B., K. Gbewonyo, T. Hallada, L. Kaplan and P. Masurekar. 1989. Production of lovastatin, an inhibitor of cholesterol accumulation in humans. In: Novel Microbial Products for Medicine and Agriculture (Demain, A. L., G. A. Somkuti, J. C. Hunter-Cevera and H. W. Rossmoore ed.), pp. 161–169, Elsevier, Amsterdam.
9. Byrne, K. M. and M. Greenstein. 1986. Nitrogen repression of gilvocarcin V production in *Streptomyces arenae* 2064. *J. Antibiot.* 39: 594–599.
10. Corbett, K. 1987. Production of Antibiotics. In: Basic Biotechnology (Bu'Lock, J. and B. Kristiansen ed.), pp. 425–448, Academic Press, London.
11. Davey, V. V. and M. J. Johnson. 1953. Penicillin production in corn steep medium with continuous carbohydrate addition. *Appl. Microbiol.* 1: 208–211.
12. Drew, S. W. and A. L. Demain. 1977. Effect of primary metabolites on secondary metabolism. *Ann. Rev. Microbiol.* 31: 343–356.
13. Egerton, J. R., D. Suhayda and C. H. Eary. 1988. Laboratory selection of *Haemonchus contortus* for resistance to ivermectin. *J. Parasitol.* 74: 614–617.
14. Hersbach, G. J. M., C. P. Van der Beek and P. W. M. Van Dijck. 1984. The penicillins: Properties, biosynthesis and fermentation. In: Biotechnology of Industrial Antibiotics (Vandamme, E. J. ed.), pp. 45–140, Marcel Dekker, New York.
15. Hood, J. D., R. M. Banks, M. D. Brewer, J. P. Fish, B. R. Manger and M. E. Poulton. 1989. A novel series of milbemycin antibiotics from *Streptomyces* strain E225. *J. Antibiot.* 42: 1593–1598.

16. Liesch, J. M. and C. F. Wichman. 1990. Novel antinematodal and antiparasitic agents from *Penicillium charlesii:* II Structure determination of paraherquamides B, C, D, E, F and G. *J. Antibiot.* 43:1380–1386.
17. Maheva, E., G. Djelveh, C. Larroche and J. B. Gros. 1984. Sporulation of *Penicillium roqueforti* in solid substrate fermentation. *Biotech. Lett.* 6:97–102.
18. Martin, J. F. and A. L. Demain. 1980. Control of antibiotic synthesis. *Microbiol. Rev.* 44:230–251.
19. Martin, J. F. and L. E. McDaniel. 1974. The submerged culture production of polyene antifungal antibiotics candidin and candihexin. *Dev. Ind. Microbiol.* 15:324–337.
20. Okazaki, T., M. Ono, A. Aoki and R. Fukuda. 1983. Milbemycins, a new family of macrolide antibiotics: Producing organism and its mutants. *J. Antibiot.* 36:438–441.
21. Ondeyka, J. G., R. T. Goegelman, J. M. Schaeffer, L. Keleman and L. Zitano. 1990. Novel antinematodal and antiparasitic agents from *Penicillium charlesii:* I. Fermentation, isolation and biological activity. *J. Antibiot.* 43:1375–1379.
22. Ostlind, D. A., W. G. Mickle, D. V. Ewanciw, F. J. Andriuli, W. C. Campbell, S. Hernandez, S. Mochales and E. Munguira. 1990. Efficacy of paraherquamide against immature *Trichostrongylus colubriformis* in gerbil (*Meriones unguiculatus*). *Res. Vet. Sci.* 48:260–261.
23. Schaeffer, J. M., T. A. Blizzard, J. Ondeyka, R. Goegelman and H. Mrozik. [3H] Paraherquamide binding to *Caenorhabditis elegans:* studies on a potent new anthelmintic agent. *J. Parasitol.* Submitted for publication
24. Shoop, W. L., J. R. Egerton, C. H. Eary and D. Suhayda. 1990. Anthelmintic activity of paraherquamide in sheep. *J. Parasitol.* 76:349–351.
25. Shoop, W. L., J. R. Egerton, C. H. Eary and D. Suhayda. 1990. Laboratory selection of a benzimidazole resistant isolate of *Trichostrongylus colubriformis* for ivermectin resistance. *J. Parasitol.* 76:186–189.
26. Smith, R. L., H. R. Bungay and R. C. Pittenger. 1962. Growth-biosynthesis relationships in erythromycin fermentation. *Appl. Microbiol.* 10:293–296.
27. Vardar, F. and M. D. Lilly. 1982. Effect of dissolved oxygen concentration on product formation in penicillin fermentation. *Europ. J. Appl. Microbiol. Biotech.* 14:203–211.
28. Vu-Trong, K. and P. P. Gray. 1987. Influence of ammonium on the biosynthesis of the macrolide antibiotic tylosin. *Enzyme Microb. Technol.* 9:590–593.
29. Yamazaki, M., E. Okuyama, M. Kobayashi and H. Inoue. 1981. The structure of paraherquamide, a toxic metabolite from *Penicillium paraherquei. Tetra. Lett.* 22:135–136.

16

Enzyme and Receptor Assays as Natural Product Screens

R. Murray Tait and Martyn N. Banks

Biotechnology, Microbiology Division, Glaxo Group Research Ltd, Greenford Road, Greenford, Middlesex, UK

Natural product screening against enzyme and receptor targets requires the development of simple, robust assay methodologies which are capable of operating at high throughput with crude test samples derived from microbial fermentations. The choice of assay method is an important determinant of success or failure in natural product drug discovery programmes. Two illustrative examples of novel screening assays, developed using conventional assay techniques, are described. In one, an assay for phosphoenolpyruvate carboxykinase, nucleoside diphosphatase and 5'-nucleotidase are used as coupling enzymes to convert GDP (a reaction product) to guanosine. A facile method for in situ separation of radiolabelled guanosine from unreacted substrate using an anion exchange resin completes the assay. In the other, an assay for tyrosine kinase activity of the epidermal growth factor receptor, a simple immunochemical assay was developed by employing poly(Glu: Tyr, 1:1) as a solid phase substrate and determining phosphorylation with the aid of a monoclonal antiphosphotyrosine antibody. In future, the application of new assay technologies, such as particle concentration fluorescence immunoassay, scintillation proximity assay and recombinant screening organisms, is likely to play a major role in the discovery of novel, biologically active compounds from natural sources. The potential impact of these techniques on natural product screening is discussed.

Key Words: Natural product screens; Particle concentration fluorescence immunoassay; Scintillation proximity assay; Recombinant screening organisms; Phosphoenolpyruvate carboxykinase; Epidermal growth factor receptor; Tyrosine kinase; Transforming growth factor-α

INTRODUCTION

Recent years have witnessed a growing interest in natural product screening against specific enzyme and receptor targets as a means of identifying new compounds with therapeutic potential for the treatment of human disease. The value of this approach to drug discovery has been amply demonstrated by the novel structures with potent biological activities which have already been described [for reviews, see 18, 19, 44]. It is particularly encouraging to note that potent and selective compounds have been discovered for a wide range of biological targets in many different therapeutic areas, and thus far there are no reasons to doubt that many more valuable compounds will continue to be identified. Indeed Williams et al [63] have recently argued that the energy expended by a microorganism for the biosynthesis of complex secondary metabolites implies specific survival functions for those natural products. They propose that such functions may be mediated via the action of natural products at specific receptor sites in competing organisms and conclude that potent and selective interactions of microbial metabolites with specific macromolecular targets may therefore be commonplace.

Whether this hypothesis is fully accepted or not, the evidence to date does inspire a belief that novel, biologically active natural products are likely to be found wherever they are sought—provided that the tools of discovery are a match for the magnitude of the task. Clearly, the assay methods employed for natural product screening are crucial components in the armament of drug discovery tools, and this article will focus primarily on issues relating to the development of assay methodologies for screening applications. Following a consideration of some general principles, two illustrative examples of assays developed specifically as screening tools are described. Finally, the potential of new assay technologies to yield novel and powerful screening systems is assessed.

GENERAL PRINCIPLES OF ASSAY DEVELOPMENT FOR NATURAL PRODUCT SCREENING

Too often, random natural product screening is regarded as a game of chance in which luck is the sole factor determining success or failure. While it would be foolish to deny the element of luck in a random screening approach to drug discovery, this view is simplistic and misses an important point; namely that successful natural product screening depends upon actively minimizing the inherent dependence on luck at all stages in the process. In particular, the choice of assay method for screening purposes is an important point at which control can be exercised in order to maximize the likelihood of success. In establishing a screening methodology for any biological target, the overriding premise must be that compounds of the desired potency and selectivity are not likely to be encountered commonly. Therefore, many thousands of samples may need to be screened before a compound with the "correct" characteristics is discovered. This fact immediately defines many of the ground rules for the development of assays as natural product screens.

Firstly, they must be able to handle a high throughput of test samples if interesting compounds are to be identified within realistic timescales. A natural product screen will typically be required to process upwards of 1,000 samples in a single experiment. Secondly, and as a direct consequence of the need for high throughput, assays should not be subject to excessive interference from common, uninteresting compounds which may be present in the fermentation broths of many different microbial species and/or be constituents of the fermentation media used for culturing those organisms. An assay which is sensitive to polyenes, for example, may have difficulty in identifying interesting non-polyene activities in fermentations of streptomycetes. Similarly, assays sensitive to non-specific interference by common salts, fatty acids, simple organic acids and carbohydrates, trace metals, detergents, etc. would be unlikely to operate effectively as natural product screens. Assays with any or all of these characteristics may be regarded as increasing, rather than decreasing, the dependence on luck in natural product screening, i.e., they are likely to detect interesting activities only in broth extracts which do not contain the offending substance(s), thus eliminating a significant proportion of test samples from useful study. Indeed, interference by such substances could mask the activities of other potentially interesting compounds which may be present simultaneously, leading to the unsettling conclusion that extracts containing the desired compound(s) may well be screened—and missed!

In practice there are few, if any, *in vitro* enzyme and/or receptor-ligand binding assays which are devoid of such problems. Difficulties are magnified substantially when intact mammalian cells are employed as a screening system, where the end point of the assay is some functional measure of cell behaviour, e.g., proliferation, surface antigen expression, etc. In these systems, dilution of broth extracts to a level at which the incidence of non-specific toxicity becomes "acceptable" is virtually mandatory, but it must be recognized that interesting activities may also be diluted beyond their detection limits. Dependence on luck therefore increases again and even more samples may need to be screened before a genuine "hit" is encountered. The specific problems of whole cell assays as natural product screens will not be addressed further in this article. The issue is raised to emphasize a third requirement of natural product screening which should be met if the chances of success are to be maximized—namely, that broth extracts should be tested at as high a concentration as the assay can reasonably tolerate so that potent, interesting compounds present in low titer can be detected. This must obviously be accomplished without inordinately increasing the negative effects of non-specific interference by common components in broth mixtures. The "fine-tuning" of an assay should aim to define an optimum balance between these two opposing influences. Clearly, each assay will have its own optimum setting.

Within the specifications imposed by these considerations, a wide range of assay techniques may be adapted or developed for the purposes of natural product screening. The choice of method for individual applications will be dependent to some extent on the intrinsic properties of the enzyme being targeted, but conventional techniques involving turbidimetric, colorimetric, fluorimetric and isotopic detection methods may all be used successfully in particular circumstances. Screen development should therefore begin with an appraisal of all possible ways in which a particular target could be assayed. The method most likely to result in a simple, rapid and robust screen, capable of high-throughput operations, is then selected for initial study. It is important to note here that the method most likely to yield the most suitable screen may not necessarily be the simplest option for the screen developer and difficult problems may need to be solved in order to achieve this goal. However, since the resulting screen may be used over a long period of time as the primary lead-generating tool in a new therapeutic target area, assay development time spent at the outset is normally a sound investment.

The following sections briefly describe two novel assay methods for therapeutically important target enzymes which illustrate some of these points. The first, an assay for phosphoenolpyruvate carboxykinase (PEPCK) illustrates the use of an isotopic technique in which facile separation of substrate and product was achieved through the use of coupling enzyme reactions. The second, an assay for tyrosine kinase activity of the epidermal growth factor (EGF)-receptor, exploits the techniques of immunochemistry to provide a simple colorimetric assay for this, and related enzymes.

A HIGH-THROUGHPUT SCREEN FOR INHIBITORS OF PEPCK

PEPCK (EC 4.1.1.32) catalyses the decarboxylation and phosphorylation of oxaloacetate (OAA) to yield phosphoenolpyruvate (PEP). GTP or ITP can act as the phosphate donor and the enzyme requires Mn^{2+} as a cation cofactor [2, 25, 42]. By blocking hepatic glucose production, inhibitors of PEPCK may have therapeutic value in the treatment of type II noninsulin dependent diabetes. The enzyme performs a regulatory function in controlling metabolic flux through the gluconeogenic pathway [57] and transcription of its mRNA is subject to multihormonal regulation [14, 22, 51, 52].

A variety of assay methods have been reported for the measurement of PEPCK activity. The reaction is readily reversible and may be assayed in either direction. In bicarbonate fixation methods the reaction is measured in the direction of OAA formation by following the incorporation of [^{14}C]bicarbonate into OAA. Since OAA is unstable, these assays are normally coupled either to malate dehydrogenase to produce [^{14}C]malate [4, 45] or to asparate-2-oxoglutarate aminotransferase to produce [^{14}C]asparate [5]. Coupling to malate dehydrogenase has also been employed to measure PEPCK activity spectrophotometrically by following NADH consumption at 340nm [46]. A similar method can be employed to measure PEPCK activity in the direction of PEP formation by coupling to pyruvate kinase and lactate dehydrogenase [37]. Some limitations of the spectrophotometric methods for determining PEPCK activity in crude tissue extracts have been described [21]. Wimmer [64] has recently reported an interesting variation of the pyruvate kinase coupling method in which ATP production from PEP and ADP is determined luminometrically using a luciferin/luciferase system. Finally, since OAA is phosphorylated in the course of the PEPCK reaction the enzyme can also be measured by employing [γ-^{32}P]GTP as substrate and determining formation of [^{32}P]PEP following removal of unreacted [γ-^{32}P]GTP on charcoal [60].

For the purposes of high-throughput screening in the presence of crude microbial extracts, our aim was to develop an assay which was not dependent upon spectrophotometric detection (to avoid problems of background absorption from broth samples), did not require the use of [^{32}P]labelled substrate (to enable convenient operation of the screen without the containment, disposal and worker protection problems associated with the use of this isotope) and which could be conducted within a microtiter plate format to facilitate high-throughput operation and automation. None of the methods could conveniently be adapted to meet these objectives and therefore a new approach to the problem of PEPCK assay was taken. The PEPCK reaction (in the direction of PEP formation) was coupled to nucleoside diphosphatase (EC 3.6.1.6) and 5′-nucleotidase (EC 3.1.3.5) to yield quantitative conversion of the GDP produced in the primary reaction to guanosine. By employing [8-^{14}C] GTP as substrate ([8-^{3}H] GTP may also be used), a facile separation of [8-^{14}C] guanosine from unreacted [8-^{14}C]GTP was then possible by adsorption of the latter onto a slurry of AG1-X8 anion exchange resin at pH 2.25. Similar anion exchange separations have been employed widely in assays of cyclic nucleoside phosphodiesterases [32, 53]. The key steps in this assay are therefore: 1) incubation of enzyme with OAA, [8-^{14}C]GTP and test sample; 2) addition of coupling enzyme mixture containing sufficient unlabelled GTP to "stop" the PEPCK reaction by isotope dilution, and sufficient quantities of nucleoside diphosphatase and 5′-nucleotidase to achieve quantitative conversion of [8-^{14}C]GDP produced in Step 1 to [8-^{14}C]guanosine during a subsequent incubation and 3) addition of a slurry of AG1-X8 anion exchange resin to stop the reaction and adsorb unreacted [8-^{14}C]GTP. After mixing and allowing the resin to settle under gravity, the supernatant can be removed for determination of [8-^{14}C] guanosine. For accurate quantitation of

Figure 16.1

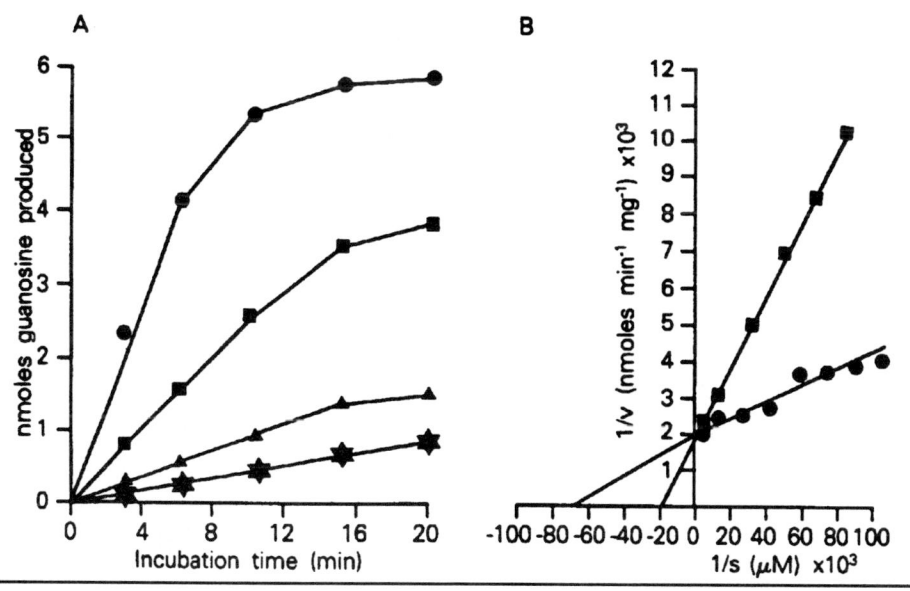

assays this may be conducted by liquid scintillation counting, but for primary screening purposes (where large numbers of samples are involved), it is more convenient to spot supernatants onto a solid phase (such as 20cm × 20cm silica gel TLC plates) for autoradiography.

The entire assay, including separation of substrate and product, is conducted *in situ* within the wells of microtiter plates. Liquid transfer operations are therefore minimal, the assay is amenable to automation, and upwards of 1,000 assays can be readily handled within a working day by one trained operator. Furthermore the assay may be used to generate quantitative data on PEPCK reaction kinetics. Figure 16.1 (panel A) illustrates a series of progress curves for different concentrations of PEPCK, showing linearity of reaction with respect to time and enzyme concentration. Panel B of Figure 16.1 shows double reciprocal plots under conditions where one substrate (GTP or OAA) was held constant and the other varied. The K_m values of 14 μM and 61 μM for OAA and GTP respectively are in good agreement with literature values obtained with conventional assay method [2, 23, 31, 37].

The assay method is simple, rapid and robust and is suitable for high-throughput natural product screening, but there are some intrinsic limitations which should be noted. Firstly the use of coupling enzymes in any screening assay automatically introduces the need for additional secondary testing of primary screen activities to eliminate those which inhibit the coupling enzymes and not the primary target. In the case of the PEPCK assay, this simply involves retesting positives against the second stage of the assay using exogenous [8-^{3}H]GDP as substrate. Secondly, since nucleoside diphosphatase is not commercially available, a suitable preparation must be generated before the PEPCK assay can be set up. A partially purified preparation obtained from frozen lamb liver by deoxycholate solubilisation, ammonium sulphate fractionation and anion exchange chromatography on DEAE-cellulose, similar to the procedures described by Kuriyama [36], was found to be suitable for use. Thirdly, the assay method may only be employed with PEPCK and coupling enzyme preparations which are devoid of contaminating "GTPase" activities. Cytosolic PEPCK purified to homogeneity from rat liver was used for the work and the nucleoside diphosphatase preparation after anion exchange chromatography was also free of contaminating phosphatases. Occasionally, however, phosphatase contamination was noted in some batches of 5'-nucleotidase from commercial sources. Finally, PEPCK is susceptible to inhibition by some simple organic simple acids, in particular oxalate [2, 23]. Uninteresting positives, due to oxalate or other simple acids, may therefore be encountered while screening against microbial fermentation extracts.

A HIGH-THROUGHPUT SCREEN FOR INHIBITORS OF EGF-RECEPTOR TYROSINE KINASE (ETK)

The EGF-receptor is a member of the growing list of signal transducing receptors and oncogenic proteins which modulate cellular behaviour by phosphorylating substrate proteins on tyrosine residues (EC 2.7.1.112) [for reviews, see 26, 58]. As in the case of PEPCK, we wished to develop a simple, robust assay to measure ETK activity in microtiter plate format, which could be employed as a high-throughput natural product screen. Tyrosine kinases are generally assayed by measuring the incorporation of [^{32}P]phosphate from [γ-^{32}P]ATP into either macromolecular substrates (e.g., casein, histones, random polymers of glutamate and tyrosine or the EGF-receptor itself) or low molecular weight substrates such as angiotensin II or src-related peptides [7, 10, 11, 12, 20, 35, 47, 49, 50, 61, 65]. Since we wished to avoid the use of [^{32}P]labelled substrate, we opted to establish a non-isotopic technique using an antiphosphotyrosine monoclonal antibody to detect phosphorylation of an exogenous macromolecular substrate. We found that poly(Glu:Tyr, 1:1), retained its substrate activity after immobilisation by passive adsorption onto polystyrene microtiter plate wells. This observation forms the basis of a simple assay method in which enzyme, ATP and cofactors are added to assay plates pre-coated with poly(Glu:Tyr, 1:1) to start the reaction. Reactions are stopped by aspiration and the amount of phosphotyrosine associated with the immobilised substrate is subsequently determined using monoclonal anti-phosphotyrosine antibody and an anti-mouse IgG-horseradish peroxidase conjugate as second antibody in traditional ELISA format.

Figure 16.2 illustrates some characteristics of the assay. Panel 16.2A shows the effect of varying the coating concentration of poly(Glu:Tyr, 1:1) on subsequent tyrosine phosphorylation by ETK in the presence of ATP. Maximum activity occurs at a coating concentration of ~2µg/ml. Other random polymers containing tyrosine, such as poly(Glu:Tyr, 4:1) and poly(Glu:Ala:Tyr, 1:1:1) produce similar results, but no reaction is observed when plates are coated with poly(Lys:Tyr, 1:1 or 4:1), poly(Glu:Lys:Tyr, 6:3:1) and poly (Lys:Ala:Glu:Tyr, 5:6:2:1). Phosphorylation of solid phase poly(Glu:Tyr, 1:1) is also dependent upon time and enzyme (A431 membrane protein) concentration (Panel B in Figure 16.2). Activity reaches a maximum at 25–50 µg/ml of membrane protein, and is significantly reduced at higher concentrations. Reasons for this are not fully understood but may be related to masking of potential phosphorylation sites by high concentrations of membrane protein. With 50 µg/ml of membrane protein in the assay, the ATP concentration required to produce half-maximal activity is about 2 µM and activity is near maximal at 45 µM-ATP. Higher ATP concentrations have no further effect (Fig. 16.2; panel C). Panel D in Figure 16.2 demonstrates that the assay can detect inhibition of ETK activity by erbstatin, a previously described natural product inhibitor of the enzyme [59]. The IC$_{50}$ of 29 µM is slightly higher than the values of about 1–5 µM reported using alternative assay methods [27, 29, 59]. This discrepancy may not be surprising in view of the very different assay conditions employed here (solid phase, macromolecular substrate) relative to the conventional assays (soluble substrates or autophosphorylation) used in the studies.

The assay method described may be adapted for the determination of other tyrosine kinase activities and is well suited for high-throughput operation as a natural product screen. Other macromolecular kinase substrates (e.g., histones)

Figure 16.2

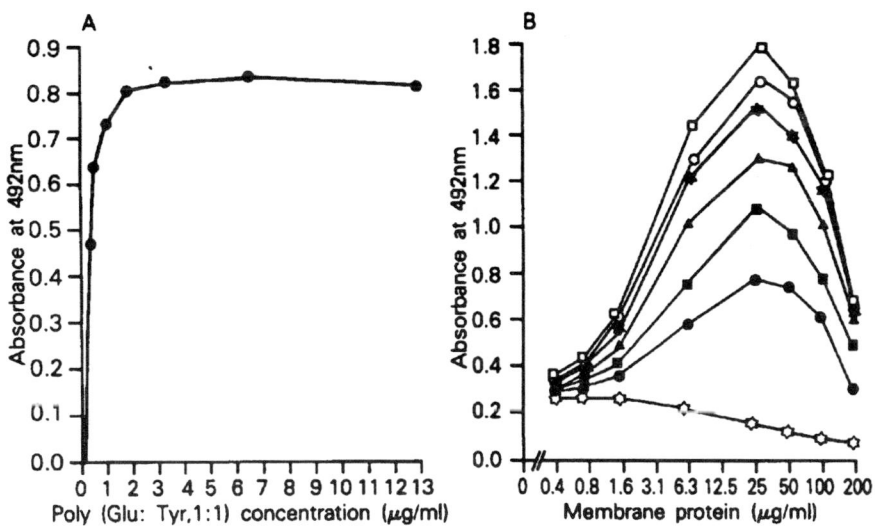

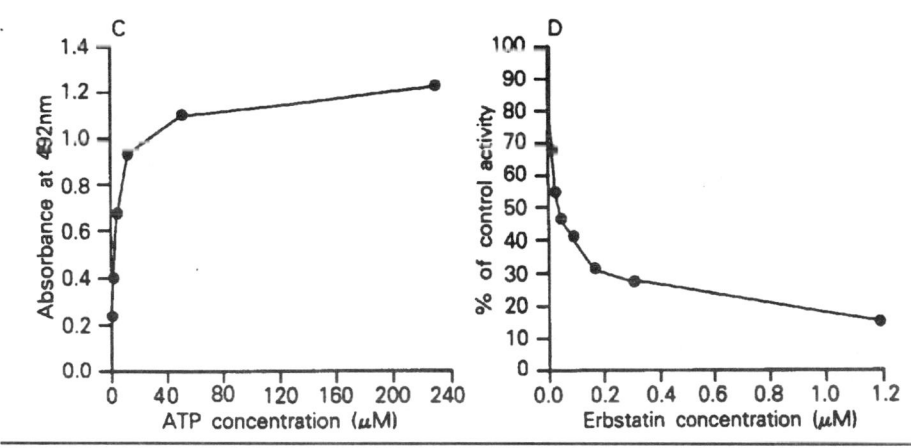

can also be adsorbed to microtiter plate wells and used successfully as solid phase substrates in the assay (results not shown). Rijksen et al. [49] have employed similar methodology in the development of a non-isotopic "dot-blot" assay for tyrosine kinases, but they used poly(Glu:Tyr, 4:1) as a substrate for phosphorylation in solution before adsorbing reaction mixtures onto polyvinylidene fluoride membranes and antibody detection of product using an immunogold staining procedure. Poly(Glu:Tyr, 4:1) has also been used as a solid phase substrate by others [50] although this method involved immobilization of the substrate in polyacrylamide gels and employed [γ-^{32}P]ATP as substrate.

IMPACT OF NEW TECHNOLOGIES ON SCREENING

The examples described briefly in the preceding sections are representative of a wide range of enzyme targets for which analogous assays, suitable for high-throughput natural product screening, may be developed. Simple separations of the type described for PEPCK, for example, can often be developed for other enzyme reactions based on isotopic detection, and the flexible techniques of immunochemistry can clearly be adapted to many different assay problems. A powerful extension of the latter which has emerged in recent years is the concept of particle concentration fluorescence immunoassay (PCFIA) first described by Jolley et al [30]. The system subsequently developed as the "Pandex Screen Machine" [39] has been designed to provide a high degree of automation and assay sensitivity in addition to high sample throughput. This is a significant development in an area where the attractions of fluorescence detection methods for screening applications have traditionally been compromised by the cumbersome nature of associated instrumentation, which has been inherently low-throughput.

Common to all assay formats in PCFIA is the use of latex microspheres as a solid phase for immobilization of proteins, peptides, nucleic acids or small molecules, and a detection system based on front surface fluorimetry as the end point of the assay. However, since intact cells may be used as the "particles," sensitive assays to detect surface antigen expression and cell proliferation may also be developed using the associated technique of cell concentration fluorescence assay (CCFA).

PCFIA and CCFA have already been used successfully in a wide range of assay applications [for examples, see 3, 13, 15, 16, 17, 28, 38, 40, 62] and, in principle at least, any assay which can be designed to incorporate an end point based on the measurement of a bound fluorescent signal could be adapted for use in the Pandex Screen Machine. This flexibility, coupled with high sensitivity, the capacity for high sample throughput and the facility to automate complex sequences of reagent additions, incubations, washings, filtrations, etc., make PCFIA/CCFA an attractive option for many natural product screening applications. It is interesting to note that the techniques of PCFIA and CCFA have been adapted from well-established, conventional fluorescence methodology. The major advance represented by the system lies not in the invention of new methodology *per se*, but in the careful consideration of instrument specifications and the needs of high-throughput screening.

Another emerging technique of major interest to the natural product screening laboratory is scintillation proximity assay (SPA). This technique is currently less well advanced than PCFIA although, surprisingly, the initial description of the concept pre-dated that of PCFIA by some five years. In 1979 Hart and Greenwald [24] reported the use of polystyrene microspheres to establish a homogeneous assay for quantifying human albumin. The method depended upon the use of two different microspheres incorporating either a tritium label or a scintillant, both of which were then coated with human albumin. Addition of rabbit anti-human albumin was shown to cross-link the two types of particles, thus bringing the tritium β-emission into close proximity with the scintillant and producing an elevated scintillation count rate. It was not until several years later that important advances with SPA began to be reported. In 1985, Udenfriend et al [55] demonstrated that labelled analytes could be determined directly using only antibody-coated scintillant microspheres and demonstrated that the Auger electron emission from $[^{125}I]$ was also of a suitable energy and range to be employed successfully in SPA.

The wider potential of SPA for application to the determination of binding interactions other than those between antibody and antigen was finally realized in 1987 when the first examples of receptor-ligand binding assays using the principles of SPA were published [43, 56]. In the report by Nelson [43], scintillant microspheres coated with membranes prepared from the electric organ of *Torpedo californica* were used to detect the binding of $[^{125}I]\alpha$-bungarotoxin to the acetylcholine receptor and the suitability of SPA for pharmacological studies of agonists and antagonists was demonstrated.

In principle, SPA may be used to monitor binding interactions between any cognate antigen/antibody or ligand/receptor pair provided that one (usually the antibody or receptor) can be successfully immobilized on fluorophoric microspheres with preservation of functional integrity and that the other (usually the antigen or ligand) can be radiolabelled with either tritium or [^{125}I]. The potential advantages of the technique relative to conventional binding methodologies are obvious. Primarily, since SPA is a homogeneous assay technique, requiring no separation of bound and free ligand, very simple assay protocols can be established. In most cases, the operator may need only mix two samples together in assay wells and wait until binding comes to equilibrium before determining light emission with suitable instrumentation. However, in order to realize the potential of SPA in practice, a great deal of refinement and modification of the basic techniques described in the early publications cited have been required. In recent years, the technology has been developed primarily in the laboratories of Amersham International plc who now market a range of antibody coated fluoromicrospheres for SPA method development. Additionally, specific SPA reagents in kit form for the determination of analytes traditionally measured by radioimmunoassay are also available and most recently, reagents for the measurement of specific receptor-ligand binding interactions have also been generated [9]. We have investigated SPA for the determination of [^{125}I]transforming growth factor-α (TGFα) binding to fluoromicrospheres coated with A431 cell membranes as a source of EGF-receptor (reagents supplied by Amersham). Figure 16.3 illustrates that binding of the labelled ligand was competitively displaced by unlabelled TGFα with half maximal displacement occurring at about 2ng/ml of unlabelled ligand. In our hands, binding equilibrium was attained after four hours incubation at room temperature and the resulting SPA signal was stable for at least 24 hours.

Techniques such as PCFIA/CCFA and SPA are likely to have a direct impact on natural product screening by allowing the development of much simpler, more easily automated assay methods and by opening up new target areas which have previ-

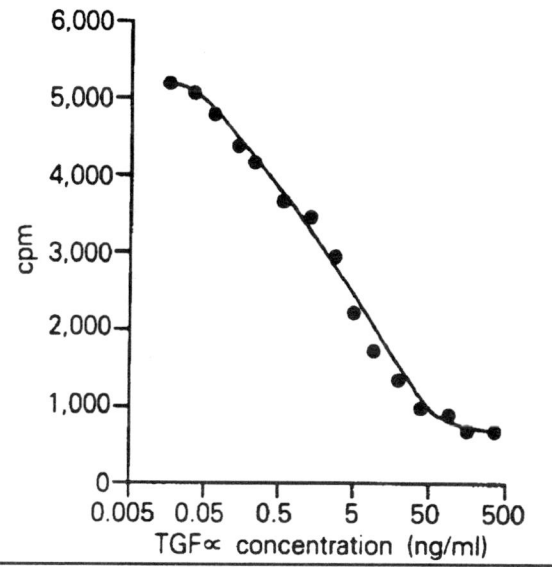

Figure 16.3

ously been less amenable to the development of high-throughput methodology. However, they are unlikely to be universally applicable, or even desirable, in all cases. For many enzymic transformations of simple organic molecules, traditional techniques such as those described earlier for PEPCK are likely to remain the methods of choice. Furthermore, and particularly in the case of SPA, the technology is new, and still developing, and it has not yet been fully exposed over long time periods to the rough and tumble of natural product screening. An element of caution should therefore be exercised until the technology has demonstrated robust characteristics in these situations.

IMPACT OF MOLECULAR BIOLOGY ON NATURAL PRODUCT SCREENING

No discussion of modern natural product screening methods would be complete without reference to the dramatic impact being made in this field by recent advances in molecular biology. Many enzyme and receptor drug targets are low abundance proteins in cells and tissues and the provision of sufficient quantities of recombinant proteins for high-throughput screening has therefore been instrumental in opening up new target areas for study. The ability to produce the relevant human proteins as biological targets for screening is also an important development.

However, the value of molecular biology in relation to natural product screening extends beyond a simple enabling role. In an analogous fashion to the technological innovations represented by techniques such as PCFIA and SPA, the construction of recombinant screening organisms by molecular biological techniques is likely to play a major role in the screening process itself. The introduction of human genes into *E. coli* or yeast hosts in such a way that the activity of the gene product influences the expression of an indicator gene has already been exemplified by the work of Metzger et al [41] who reported expression of the human oestrogen receptor in yeast. By simultaneously introducing an indicator gene (β-galactosidase) under the control of an oestrogen responsive element, expression of β-galactosidase in an oestradiol-dependent fashion was achieved. Thus, in the presence of a chromogenic substrate for β-galactosidase, the recombinant yeast colonies turned blue in response to addition of oestradiol. This, or a similar construction, could clearly form the basis of a very simple screening system to detect novel oestrogen antagonists. An added attraction of such an assay would be the ability to detect not only compounds which act as classical antagonists at the level of steroid-receptor binding, but also compounds which may specifically perturb binding of the activated steroid-receptor complex to its cognate recognition sequence on DNA.

In some cases, simple genetic constructions in which expression of a target protein may directly influence the behaviour of the host organism in an easily observable way could also form the basis of useful natural product screens. Recently, Baum et al [6] have illustrated this principle by demonstrating that expression of HIV protease in *E. coli* can produce a toxic effect leading to death of the host cells. Inhibitors of HIV protease should therefore reverse this toxicity and allow cell growth, providing a simple and convenient screening system.

Although there is clearly great potential for recombinant screening organisms as drug discovery tools, the limitations of these techniques should also be recognized. Screens based on recombinant organisms introduce a permeability barrier to the free diffusion of small molecules and therefore compounds which fail to penetrate the membranes and cell walls of host organisms will not be screened. Additionally, it seems likely that the behaviour of human proteins expressed in microorganisms may be subtly different to that observed in their normal cellular environments, and parameters such as ligand binding and/or efficacies, susceptibility to inhibition, etc, may well be influenced. Evidence for this has already been described in the study of human oestrogen receptor function in yeast [41]. Defining the limitations of recombinant screening organisms and assessing the extent to which subtle differences in functional parameters impact upon the validity of such screens will be important issues to address over the next few years.

Genetic manipulation of mammalian cells provides additional opportunities for the development of simple screens to detect compounds with novel mechanisms of action. For any protein target where there is reason to propose a therapeutic potential for specific inhibitors or antagonists, selective down-regulation of the target at the level of gene transcription may also be expected to result in the desired therapeutic effect. Advances in the construction and application of reporter genes to monitor the transcriptional activity of specific promoter and enhancer elements within relevant mammalian cell hosts have recently been reviewed by Alam and Cook [1]. By providing an easily measurable end point, these techniques open the way for more extensive natural product screening in models of transcriptional control and for the potential discovery of new classes of specific "transcriptional modifier" drugs. The immunosuppressant natural products cyclosporin A and FK506 are thought to act by blocking the transcription of specific T-cell genes required to propagate the immune response [48, 54]. Although their precise modes of action are still not understood in detail, they may be regarded as transcriptional modifiers and therefore as precedents for the discovery of other compounds with analagous mechanisms of action. It should nevertheless be borne in mind that the immunosuppressive actions of these important drugs were identified using conventional *in vitro* and *in vivo* models of the immune response [8, 33, 34] rather than engineered mammalian cell constructs. Thus, although the latter will undoubtedly aid the search for new transcriptional modifiers, conventional assay methodologies will also continue to play an important role in this area.

CONCLUSIONS AND FUTURE PROSPECTS

Natural product screening against enzyme and receptor targets has already led to the discovery of many important compounds, but the pool of interesting chemical entities has by no means been exhausted. Indeed, the prospects for novel drug discovery by natural product screening have never been more promising. To a large extent, optimism for the future is engendered by the dramatic innovations in assay technology which have already been made and which will continue to result in previously inaccessible targets becoming realistic candidates for high-throughput screening programmes. The history of natural product screening strongly indicates that new targets lead to the discovery of novel compounds and we may therefore expect to see a continuing expansion of the already impressive list of important new compounds isolated from natural sources.

REFERENCES

1. Alam, J. and Cook, J. L. 1990. Reporter genes: application to the study of mammalian gene transcription. *Anal. Biochem.* 188: 245–254.
2. Ash, D. E., Emig, F. A., Chowdhury, S. A., Satoh, Y. and Schramm, V. L. 1990. Mammalian and avian liver phosphoenolpyruvate carboxykinase. Alternate substrates and inhibition by analogues of oxaloacetate. *J. Biol. Chem.* 265: 7377–7384.
3. Averett, D. R. 1989. Anti-HIV compound assessments by two novel high capacity assays. *J. Virol. Methods* 23: 263–276.
4. Ballard, F. J. and Hanson, R. W. 1969. Purification of phosphoenolpyruvate carboxykinase from the cytosol fraction of rat liver and the immunochemical demonstration of differences between this enzyme and the mitochondrial phosphoenolpyruvate carboxykinase. *J. Biol. Chem.* 244: 5625–5630.
5. Barns, R. J. and Keech, D. B. 1968. The essential thiol group of sheep kidney mitochondrial phosphoenolpyruvate carboxykinase. *Biochim. Biophys. Acta* 159: 514–526.
6. Baum, E. Z., Bebernitz, G. A. and Gluzman, Y. 1990. Isolation of mutants of human immunodeficiency virus protease based on the toxicity of the enzyme in *Escherichia coli*. *Proc. Natl. Acad. Sci. USA.* 87: 5573–5577.
7. Bertics, P. J. and Gill, G. N. 1985. Self-phosphorylation enhances the protein tyrosine kinase activity of the epidermal growth factor receptor. *J. Biol. Chem.* 260: 14642–14647.
8. Borel, J. F., Feurer, C., Gubler, H. U. and Stahelin, H. 1976. Biological effects of cyclosporin A: a new antilymphocytic agent. *Agents Actions* 6: 468–475.
9. Bosworth, N. and Towers, P. 1989. Scintillation proximity assay. *Nature* 341: 167–168.
10. Braun, S., Raymond, W. E. and Racker, E. 1984. Synthetic tyrosine polymers as substrates and inhibitors of tyrosine-specific protein kinases. *J. Biol. Chem.* 259: 2051–2054.
11. Casnellie, J. E., Harrison, M. L., Pike, L. J., Hellstrom, K. E. and Krebs, E. G. 1982. Phosphorylation of synthetic peptides by a tyrosine protein kinase from the particulate fraction of a lymphoma cell line. *Proc. Natl. Acad. Sci. USA.* 79: 282–286.

12. Cassel, D., Pike, L. J., Grant, G. a., Krebs, E. G. and Glaser, L. 1983. Interaction of epidermal growth factor-dependent protein kinase with endogenous membrane proteins and soluble peptide substrate. *J. Biol. Chem.* 258: 2945–2950.
13. Chan, M. A. and Dosch, H-M. 1989. Human IgE response: virus-activated IgE secretors are interleukin-2 dependent cells. *Int. Arch. Allergy App. Immunol.* 89: 90–97.
14. Cimbala, M. A., Lamers, W. H., Nelson, K., Monahan, J. E., Yoo-Warren, H. and Hanson, R. W. 1982. Rapid changes in the concentration of phosphoenolpyruvate carboxykinase mRNA in rat liver and kidney. Effects of insulin and cyclic AMP. *J. Biol. Chem.* 257: 7629–7636.
15. Custer, M. C. and Lotze, M. T. 1990. A biologic assay for IL-4. Rapid fluorescence assay for IL-4 detection in supernatants and serum. *J. Immunol. Methods* 128: 109–117.
16. Damle, N. K. and Doyle, L. V. 1989. IL-2-activated human killer lymphocytes but not their secreted products mediate increase in albumin flux across cultured endothelial monolayers. Implications for vascular leak syndrome. *J. Immunol.* 142: 2660–2669.
17. Del Tito, B. J. Jnr., Zabriskie, D. W. and Arcuri, E. J. 1988. Detection of α_1-antitrypsin from recombinant *Escherishia coli* lysates utilising the particle concentration fluorescence immunoassay. *J. Immunol. Methods* 107: 67–72.
18. Demain, A. L. 1983. A new era of exploitation of microbial metabolites. *Biochem Soc. Symp.* 48: 117–132.
19. Deshpande, B. S., Ambedkar, S. S. and Shewale, J. G. 1988. Biologically active secondary metabolites from *Streptomyces. Enzyme Microb. Technol.* 10: 455–473.
20. Downward, J., Waterfield, M. D. and Parker, P. J. 1985. Autophosphorylation and protein kinase C phosphorylation of the epidermal growth factor receptor. Effect on tyrosine kinase activity and ligand binding affinity. *J. Biol. Chem.* 260: 14538–14546.
21. Duff, D. A. and Snell, K. 1982. Limitations of commonly used spectrophotometric assay methods for phosphoenolpyruvate carboxykinase activity in crude extracts of muscle. *Biochem J.* 206: 147–152.
22. Granner, D., Andreone, T., Sasaki, K. and Beale, E. 1983. Inhibition of transcription of the phosphoenolpyruvate carboxykinase gene by insulin. *Nature* 305: 459–551.
23. Guidinger, P. F. and Nowak, T. 1990. Analogs of oxaloacetate as potential substrates for phosphoenolpyruvate carboxykinase. *Arch. Biochem. Biophys.* 278: 131–141.
24. Hart, H. E. and Greenwald, E. B. 1979. Scintillation proximity assay (SPA)—a new method of immunoassay. *Mol. Immunol.* 16: 265–267.
25. Hebda, C. A. and Nowak, T. 1982. Phosphoenolpyruvate carboxykinase. Mn^{2+} and Mn^{2+} substrate complexes. *J. Biol. Chem.* 257: 5515–5522.
26. Hunter, T. 1989. Protein modification: phosphorylation on tyrosine residues. *Curr. Op. Cell Biol.* 1: 1168–1181.
27. Imoto, M., Umezawa, K., Isshiki, K., Kunimoto, S., Sawa, T., Takeuchi, T. and Umezawa, H. 1987. Kinetic studies of tyrosine kinase inhibition by erbstatin. *J. Antibiot.* 40: 1471–1473.
28. Islam, F., Urade, Y., Watanabe, Y. and Hayaishi, O. 1990. A particle concentration fluorescence immunoassay for prostaglandin D synthase in the rat central nervous system. *Arch. Biochem. Biophys.* 277: 290–295.
29. Isshiki, K., Imoto, M., Sawa, T., Umezawa, K., Takeuchi, T., Umezawa, H., Tsuchida, T., Yoshioka, T. and Tatsuta, K. 1987. Inhibition of tyrosine protein kinase by synthetic erbstatin analogues. *J. Antibiot.* 40: 1209–1210.
30. Jolley, M. E., Wang, C. H. J., Ekenberg, S. J., Zuelke, M. S. and Kelso, D. M. 1984. Particle concentration fluorescence immunoassay (PCFIA): a new rapid immunoassay technique with high sensitivity. *J. Immunol. Methods* 67: 21–35.
31. Jomain-Baum, M. and Schramm, V. L. 1978. Kinetic mechanism of phosphoenolpyruvate carboxykinase (GTP) from rat liver cytosol. Product inhibition, isotope exchange at equilibrium, and partial reactions. *J. Biol. Chem.* 253: 3648–3659.
32. Kincaid, R. L. and Manganiello, V. C. 1988. Assay of cyclic nucleotide phosphodiesterase using radiolabelled and fluorescent substrates. *Methods Enzymol.* 159: 457–470.

33. Kino, T., Hatanaka, H., Hashimoto, M., Nishiyama, M., Goto, T., Okuhara, M., Kohsaka, M., Aoki, H. and Imanaka, H. 1987. FK506, a novel immunosuppressant isolated from a streptomyces. I. Fermentation, isolation and physico-chemical and biological characteristics. *J. Antibiot. 40:* 1249–1255.
34. Kino, T., Hatanaka, H., Miyata, S., Inamura, N., Nishiyama, M., Yajima, T., Goto, T., Okuhara, M., Kohsaka, M., Aoki, H. and Ochiai, T. 1987. FK506, a novel immunosuppressant isolated from a streptomyces. II. Immunosupprssive effect of FK506 *in vitro. J. Antibiot. 40:* 1256–1265.
35. Kruse, C. H., Holden, K. G., Pritchard, M. L., Field, J. A., Rieman, D. J., Greig, R. G. and Poste, G. 1988. Synthesis and evaluation of multisubstrate inhibitors of an oncogene-encoded tyrosine-specific protein kinase. *J. Med. Chem. 31:* 1762–1767.
36. Kuriyama, Y. 1972. Studies on microsomal nucleoside diphosphatase of rat hepatocytes. Its purification, intramembranous localization and turnover. *J. Biol. Chem. 247:* 2979–2988.
37. Lee, M. H., Hebda, C. A. and Nowak, T. 1981. The role of cations in avian liver phosphoenolpyruvate carboxykinase catalysis. Activation and regulation. *J. Biol. Chem. 256:* 12793–12801.
38. Liebert, M., Laino, L. and Wahl, R. L. 1987. A semi-automated fluorescent (SAF) assay using viable whole cells for screening hybridoma supernatants. *J. Immunol. Methods 101:* 85–90.
39. MacCrindle, C., Schwenzer, K. and Jolley, M. E. 1985. Particle concentration fluorescence immunoassay: a new immunoassay technique for quantification of human immunoglobulins in serum. *Clin. Chem. 31:* 1487–1490.
40. Magus, J., Macke, K., Shackelford, P., Kim, J. and Nahm, M. 1986. Human IgG subclass assays using a novel assay method. *J. Immunol. Methods 88:* 65–73.
41. Metzger, D., White, J. H. and Chambon, P. 1988. The human oestrogen receptor functions in yeast. *Nature 334:* 31–36.
42. Miller, R. S., Mildvan, A. S., Chang, H. C., Easterday, R. L., Maruyama, H. and Lane, M. D. 1968. The enzymatic carboxylation of phosphoenolpyruvate IV. The binding of manganese and substrates by phosphoenolpyruvate carboxykinase and phosphoenolpyruvate carboxylase. *J. Biol. Chem. 243:* 6030–6040.
43. Nelson, N. 1987. A novel method for the detection of receptors and membrane proteins by scintillation proximity radioassay. *Anal. Biochem. 165:* 287–293.
44. Nisbet, L. J. and Porter, N. 1989. The impact of pharmacology and molecular biology on the exploitation of microbial products. In: Symposium of the Society for Microbiology, 44. Microbial products: New Approaches. (Baumberg, S., Hunter, I. and Rhodes, M., eds.), pp. 309–342, Cambridge University Press, New York, USA; Cambridge, England.
45. Nolte, J., Brdiczka, D. and Pette, D. 1972. Intracellular distribution of phosphoenolpyruvate carboxylase and (NADP) malate dehydrogenase in different muscle types. *Biochim. Biophys. Acta 284:* 497–507.
46. Opie, L. H. and Newsholme, E. A. 1967. The activities of fructose 1,6-diphosphatase, phosphofructokinase and phosphoenolpyruvate carboxykinase in white muscle and red muscle. *Biochem. J. 103:* 391–399.
47. Pike, L. J., Gallis, B., Casnellie, J. E., Bornstein, P. and Krebs, E. G. 1982. Epidermal growth factor stimulates the phosphorylation of synthetic tyrosine containing peptides by A431 cell membranes. *Proc. Natl. Acad. Sci. USA. 79:* 1443–1447.
48. Randak, C., Brabletz, T., Hergenrother, M., Sobotta, I. and Serfling, E. 1990. Cyclosporin A suppresses the expression of the interleukin 2 gene by inhibiting the binding of lymphocyte-specific factors to the IL-2 enhancer. *EMBO J. 9:* 2529–2536.
49. Rijksen, G., van Oirschot, B. A. and Staal, G. E. J. 1989. A non-radioactive dot-blot assay for protein tyrosine kinase activity. *Anal. Biochem. 182:* 98–102.
50. Sahal, D. and Fujita-Yamaguchi, Y. 1989. Solid-phase tyrosine-specific protein kinase assay in multiwell substrate-immobilised polyacrylamide gel. *Anal. Biochem. 182:* 37–43.
51. Sasaki, K., Cripe. T. P., Koch, S. R., Andreone, T. L., Petersen, D. D., Beale, E. G. and Granner, D. K. 1984. Multihormonal regulation of phosphoenolpyruvate carboxykinase gene transcription. The dominant role of insulin. *J. Biol. Chem. 259:* 15242–15251.
52. Sasaki, K. and Granner, D. K. 1988. Regulation of phosphoenolpyruvate carboxykinase gene transcription by insulin and cAMP: Reciprocal actions on initiation and elongation. *Proc. Natl. Acad. Sci. USA. 85:* 2954–2958.

53. Thompson, W. J. and M. M. Appleman. 1971. Multiple cyclic nucleotide phosphodiesterase activities from rat brain. *Biochemistry 10:* 311–316.
54. Tocci, M. J., Matkovich, D. A., Collier, K. A., Kwok, P., Dumont, F., Lin, S., Degudicibus, S., Siekierka, J. J., Chin, J. and Hutchinson, N. I. 1989. The immunosuppressant FK506 selectively inhibits expression of early T-cell activation genes. *J. Immunol. 143:* 718–726.
55. Udenfriend, S., Gerber, L. D., Brink, L. and Spector, S. 1985. Scintillation proximity radioimmunoassay utilizing ^{125}I-labelled ligands. *Proc. Natl. Acad. Sci. USA. 82:* 8672–8676.
56. Udenfriend, S., Gerber, L. and Nelson, N. 1987. Scintillation proximity assay: a sensitive and continuous isotopic method for monitoring ligand/receptor and antigen/antibody interactions. *Anal. Biochem. 161:* 494–500.
57. Ui, M., Claus, T. H., Exton, J. H. and Park, C. R. 1973. Studies on the mechanism of action of glucagon on gluconeogenesis. *J. Biol. Chem. 248:* 5344–5349.
58. Ullrich, A. and Schlessinger, J. 1990. Signal transduction by receptors with tyrosine kinase activity. *Cell 61:* 204–212.
59. Umezawa, H., Imoto, M., Sawa, T., Isshiki, K., Matsuda, N., Uchida, T., Iinuma, H., Hamada, M. and Takeuchi, T. 1986. Studies on a new epidermal growth factor receptor kinase inhibitor, erbstatin, produced by MH435-hF3. *J. Antibiot. 39:* 170–173.
60. Vandewalle, A., Wirthensohn, G., Heidrich, H. G. and Guder, W. G. 1981. Distribution of hexokinase (EC 2.7.1.1) and phosphoenolpyruvate carboxykinase (EC 4.1.1.32) along the rabbit nephron. *Am. J. Physiol. 240:* F492–F500.
61. Weber, W., Bertics, P. J. and Gill, G. N. 1984. Immunoaffinity purification of the epidermal growth factor receptor. Stoichiometry of binding and kinetics of self-phosphorylation. *J. Biol. Chem. 259:* 14631–14636.
62. Wierda, W. G., Mehr, D. S. and Kim, Y. B. 1989. Comparison of fluorochrome-labelled and ^{51}Cr-labelled targets for natural killer cytotoxicity assay. *J. Immunol. Methods 122:* 15–24.
63. Williams, D. H., Stone, M. J., Hauch, P. R. and Rahman, S. K. 1989. Why are secondary metabolites (natural products) biosynthesized? *J. Nat. Prod. 52:* 1189–1208.
64. Wimmer, M. 1988. A bioluminescent assay for the determination of phosphoenolpyruvate carboxykinase activity in nanogram-sized tissue samples. *Anal Biochem. 170:* 376–381.
65. Wong, T. W. and Goldberg, A. R. 1983. In vitro phosphorylation of angiotensin analogues by tyrosyl protein kinases. *J. Biol. Chem. 258:* 1022–1025.

17

TAN-950 A, a Novel Glutamate Receptor Agonist Produced by *Streptomyces platensis* A-136

Seiji Hakoda, Sigetoshi Tsubotani, Tenichi Tamura, Toshi Iwama, Setsuo Harada and Takashi Iwasa

Research and Development Division, Takeda Chemical Industries, Ltd., Juso, Yodogawa-ku, Osaka 532, Japan

In the course of screening for new antifungals, a new amino acid antibiotic, TAN-950 A was found. The producing strain (A-136) was identified as Streptomyces platensis. *The chemical structure of TAN-950 A was determined to be (S)-2-amino-3-(2, 5-dihydro-5-oxo-4-isoxazolyl)propanoic acid. This antibiotic showed* in vitro *and* in vivo *activities against* Candida albicans *and had low toxicity. Antagonism of TAN-950 A by L-glutamate and L-aspartate and its unique structure led to the idea that this compound shows affinity for excitatory amino acid (EAA) receptors in the central nervous system (CNS). It bound strongly to these receptors and also elicited the firing of rat hippocampal CA1 neurons* in vitro. *The methyl analog of TAN-950 A had a slightly reduced affinity for the three EAA receptors but increased potency to elicit the neuronal firing. The (R)-enantiomer of TAN-950 A had increased selectivity for the N-methyl-D-aspartate(NMDA) receptor subtype. The most selective NMDA agonist was (R)-3-methyl-Nor-TAN-950 A.*

Key Words: TAN-950 A, *Candida albicans, Streptomyces platensis,* antifungal, excitatory amino acid, hippocampus, kainate, quisqualate, NMDA, glutamate

INTRODUCTION

In the course of screening for the new antifungals, a new amino acid antibiotic, TAN-950 A was found. This antibiotic showed *in vitro* and *in vivo* activities against *Candida albicans*. TAN-950 A also showed affinity for excitatory amino acid receptors in the central nervous system. The *in vitro* antifungal activity of TAN-950 A was antagonized by glutamate. Glutamate is known as a major excitatory amino acid transmitter [3,6,7]. Excitatory amino acid transmitters participate in normal synaptic transmission in the central nervous system [4,10]. The excitatory amino acid receptor agonists and antagonists have recently been of major interest as potential drugs for the CNS disorders [5]. Therefore, the relationship between TAN-950 A and excitatory amino acid receptors was investigated.

TAXONOMIC STUDIES ON THE PRODUCING ORGANISM

Methods adopted by the International Streptomyces Project (ISP) [9, 13, 15] were used for taxonomic characterization and carbohydrate utilization studies. Aerial mycelia were found abundantly on some media such as glycerol-asparagine agar, inorganic salts-starch agar and yeast extract-malt extract agar, and they were in the gray series. A trace light brown, soluble pigment was produced only with yeast extract-malt extract agar [16].

Strain A-136 grew between 10 to 36°C and the optimum growth temperature was 28 to 32°C on yeast extract-malt extract agar. Brown melanoid pigment was not produced. This strain was positive for starch hydrolysis and gelatin liquefaction and was negative for milk peptonization and nitrate reduction. Analysis of whole cell hydrolysates revealed the presence of LL-diaminopimelic acid. This strain could utilize D-glucose, D-fructose, raffinose, L-arabinose, inositol and mannitol [16].

Aerial mycelia were simply branched and the chains of spores formed spirals. Mature spore chains had 10 or more spores per chain. The spores were elliptical with a smooth surface and measured 0.8 to 1.3 µm in size. Sporangia, flagellated spores or sclerotic granules were not observed.

From these taxonomic properties it was concluded that strain A-136 belongs to the genus *Streptomyces*, and that it fits the description of *Streptomyces platensis*. This strain was designated as *Streptomyces platensis* A-136.

FERMENTATION

The fermentation medium contained 0.5% glucose, 5% dextrin and 3.5% soy bean meal. Fermentation was carried out at 28°C for 5 days. Antibiotic production reached about 750 µg/ml (Fig. 17.1).

ISOLATION OF TAN-950 A

Trichostatin-A [14], a known antifungal antibiotic present in the culture broth, was removed by ethylacetate extraction. The resulting aqueous layer was chromatographed on columns of cation exchange resin, adsorption resin, anion exchange Sephadex and microcrystalline cellulose. Finally, TAN-950 A was isolated as the sodium salt.

In the QAE-Sephadex column isolation procedure, TAN-950 B, C, D and E fractions were detected. These fractions showed antifungal activity. However, due to isomerization, these components could not be isolated individually and a mixture of TAN-950 A, B, C, D and E was obtained [16].

The ion-paired method using tetrabutylammonium hydroxide as the mobile phase separated these components very well [17].

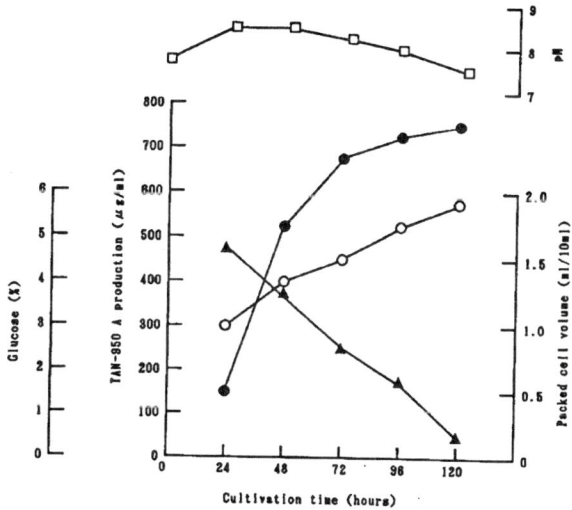

Figure 17.1 / Time course of TAN-950 A production. Glucose (▲), potency (●), packed cell volume (○) and pH (□).

Table 17.1 / Physico-chemical properties of TAN-950 A sodium salt.

Appearance	White powder
Nature	Water-soluble amphoteric substance
Optical rotation	$[\alpha]_D^{23}$ −69.5° (c 0.52, H_2O)
SI-MS: m/z	195 $(M + H)^+$, 217 $(M + Na)^+$
Molecular formula	$C_6H_7N_2O_4Na(H_2O)$
Elemental analysis	Found: C 33.64, H 4.31, N 12.72, Na 11.0
	Calcd: C 33.97, H 4.28, N 13.21, Na 10.66
UV: $\lambda_{max}^{H_2O}$ nm(ε)	253 (8,060)
IR: ν_{max}^{KBr} cm^{-1}	3430, 1640, 1500, 1410, 1350

PHYSICO-CHEMICAL PROPERTIES OF TAN-950 A

TAN-950 A is a white, water-soluble amphoteric substance. The optical rotation is −69.5 degree. The molecular formula is determined to be $C_6H_7N_2O_4$•Na by secondary-ion-mass spectrometry and elemental analysis (Table 17.1).

STRUCTURE DETERMINATION OF TAN-950 A

The ^{13}C NMR data suggested the presence of two carbonyl groups, an olefin, a methine and a methylene. A coupling constant analysis of the ^1H NMR spectral data showed the methine and methylene to be bound to each other. The long range ^1H-^{13}C correlation spectroscopy data showed that the olefin and carbonyl moieties adjoin the methylene. Therefore the structure of TAN-950 A is revealed except for the linkage of NH and O (Table 17.2).

Table 17.2 / ^{13}C and ^1H NMR spectral data of TAN-950 A in D_2O.

Position	^{13}C NMR	^1H NMR
6	180.58 (s)	
1	177.09 (s)	
5	155.54 (d)	7.98 (1H, s)
4	82.85 (s)	
2	58.49 (d)	3.86 (1H, dd, J=4.4, 7.3)
3	26.51 (t)	2.80 (1H, dd, J=4.4, 15.6)
		2.70 (1H, dd, J=7.3, 15.6)

TAN-950 A gave an N-benzoyl derivative. On treatment with diazomethane, the N-benzoate gave two dimethyl compounds; one was an N-methyl compound and the another was an O-methyl compound. In the ^1H-NMR spectrum of the N-methyl derivative, the nuclear Overhauser and exchange effects were found between an N-methyl and an olefinic proton. This shows that the

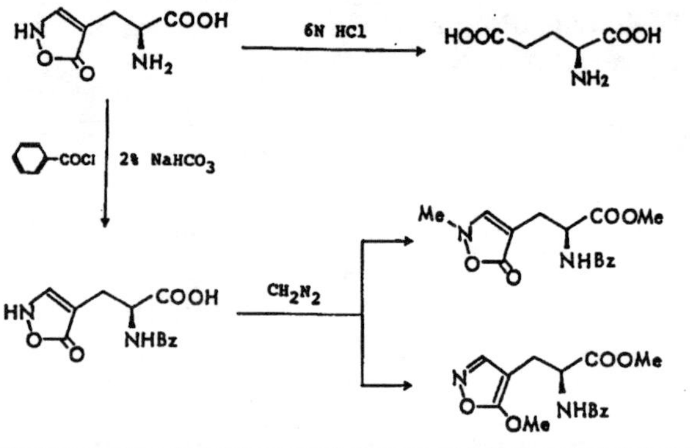

Figure 17.2 / Structure determination of TAN-950 A.

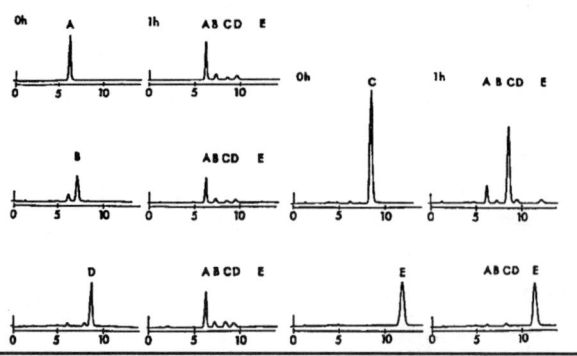

Figure 17.3 / HPLC patterns of TAN-950 A-E after preparative HPLC. [ODS, 0.25mM n-Bu₄N⁺OH⁻/0.02M p.b. (pH 6)].

nitrogen atom is bound to the olefin. From these findings the structure of TAN-950 A was deduced as shown in Fig. 17.2.

The absolute configuration was determined to be (S), as the acidic hydrolysis of TAN-950 A gave L-glutamate.

Furthermore, this structure was confirmed by X-ray crystallographic analysis of the N-benzoyl derivative.

ISOMERIZATION OF TAN-950 A, B, C, D, AND E

Each component was detected by HPLC at the time just after the separation by preparative HPLC and after allowing to stand at 60°C for an hour. The isomerization velocity of TAN-950 B and D was faster than those of C and E (Fig. 17.3).

The structures of TAN-950 B, C, D and E were assigned as shown in Fig. 17.4 from the physicochemical and NMR data of their derivatives and elucidated with the synthetic studies. The configuration of the oxime group is assigned to be *syn* in B and D, because the hydroxyl group of the oxime can attach easily to the amide carbonyl group.

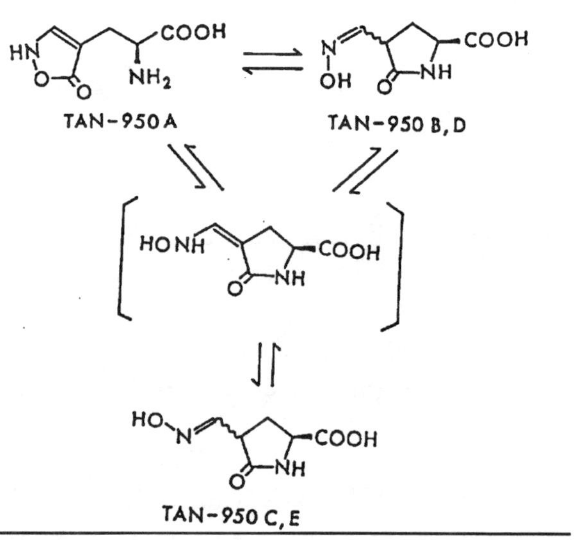

Figure 17.4 / Equilibrium mechanism of TAN-950 A-E.

BIOLOGICAL PROPERTIES OF TAN-950 A

ANTIFUNGAL ACTIVITY AND ACUTE TOXICITY

The antifungal activity was determined by an agar dilution method on Yeast Nitrogen Base supplemented with 2% glucose and 1.5% agar. TAN-950 A had a highly specific activity against yeasts, especially *Candida* and *Saccharomyces* (Table 17.3).

Table 17.3 / Antifungal activities of TAN-950 A.

Test organism		MIC (µg/ml)
Candida albicans	IFO 0583	3.13
Candida parakrusei	IFO 0640	3.13
Candida parapsilosis	IFO 1396	1.56
Candida tropicalis	IFO 0006	0.78
Torulopsis grabrata	IFO 1085	>100
Cryptococcus neoformans	IFO 0410	>100
Saccharomyces cerevisiae	IFO 0209	1.56
Aspergillus fumigatus	IFO 6344	>100
Trichophyton rubrum	IFO 5467	>100
Microsporum gypseum	IFO 6075	>100

Medium : yeast nitrogen base (Difco) + 2% glucose + 1.5% agar
Incubation temperature and time : 28 °C, 2 ~ 3 days

TAN-950 A showed strong activity against candidiasis with ED_{50} values of 35.5 mg/kg by subcutaneous administration and 100 mg/kg by oral. TAN-950 A was as effective as miconazole. TAN-950 A did not show any toxicities at a dose of 4 g/kg in the two weeks after dosing intraparenterally, intravenously, subcutaneously and orally to mice [16].

As the antifungal activity of TAN-950 A was shown only with the synthetic test medium, the effect of several amino acids on the antifungal activity of this antibiotic was examined. Glutamate greatly reduced the antifungal activity of TAN-950 A against Saccharomyces cerevisiae (Table 17.4).

EXCITATORY AMINO ACID RECEPTOR AGONIST ACTIVITY

Two assays were carried out for this screen. The first was a binding assay for excitatory amino acid receptor subtypes [18], and the second was a electrophysiological assay for neuronal activity.

Table 17.4 / Effect of amino acids on the antifungal activity of TAN-950 A.

Amino acid (10 mM)	Paper disc assay(mm)	
	100 µg/ml	1000 µg/ml
Ala	13	24
Val	13	22
Leu	+	+
Ile	12.5	23
Ser	13	25
Thr	13	23
Met	13	23
Phe	15	24
Trp	13	23
Pro	14	22.5
Asp	0	+
Asn	12	25
Glu	0	0
Gln	13	24
Lys	14	23
none	13	22.5

Test organism : Saccharomyces cerevisiae IFO 0209
Medium : yeast nitrogen base + 2% glucose + 1.5 % agar

THE PROCEDURES FOR THE RECEPTOR BINDING ASSAY

A crude synaptic membrane prepared from the forebrain of Wister rats was washed with Tris buffer and incubated with a radioactive ligand and a test compound. For the K site binding assay, ^{3}H-kainate was added. For the Q site binding assay, ^{3}H-AMPA was added. For the N site binding assay, ^{3}H-CPP was added [11, 12]. After incubation, the bound ligand was separated either by vacuum filtration or by centrifugation, and was counted by liquid scintillation. Nonspecific binding was determined by the addition of 1 mM glutamate (Fig. 17.5).

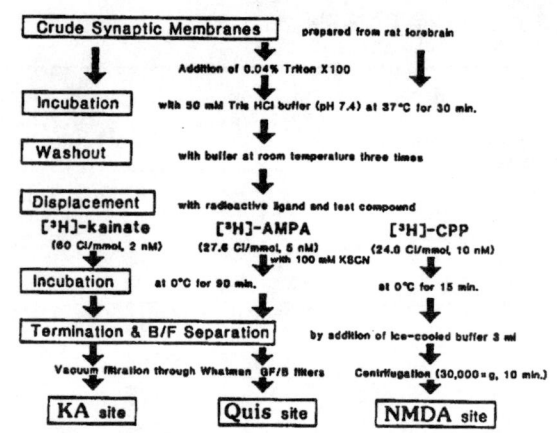

Figure 17.5 / Receptor binding assay.

THE SCHEME USED TO INVESTIGATE NEURONAL ACTIVITY ELECTROPHYSIOLOGICALLY

A slice of hippocampus from a brain of a Wister rat, incubated for 2 hour at 32°C, was submerged into a recording chamber perfused with Krebs Ringer solution at a rate of 2 ml/min. A bipolar stimulating electrode was placed in the Schaffer collaterals. A glass electrode having 2 M sodium chloride was placed in the pyramidal cell layer of the CA1 area and the evoked potential was recorded [18].

A TYPICAL EVOKED POTENTIAL AND A TYPICAL PATTERN OF THE FIRING RATE

1. When the recording electrode was moved carefully, a spike from a single neuron was obtained. The number of spontaneous spikes were counted using a spike counter and recorded sequentially [18].
2. Two criteria were set. First, only neurons with a spontaneous firing rate of less than one hertz were used. And second, neurons which responded to less than 0.1 mM glutamate, or those which did not respond to more than 1 mM glutamate were excluded [18].

Test compounds were added to the bath solution, and the concentration was sequentially increased approximately three fold each time. The minimum effective concentration of each test compound was determined, and defined as the concentration which elicited a firing rate of more than one hertz.

THE RECEPTOR BINDING AFFINITIES AND ELECTROPHYSIOLOGICAL PROFILES OF AUTHENTIC EXCITATORY AMINO ACIDS AND TAN-950 A

Glutamate has relatively high affinities for all three sites, but has a low potency to excite neurons. This action seems to be through mainly Q/K sites. Kainate is a potent K agonist and quisqualate is a potent Q agonist. On the other hand, NMDA has little affinity for K or Q sites and excites neurons via N sites. TAN-950 A has binding affinity for all three subtypes and acts as a Q/K agonist. The potency to excite neurons is as weak as that of glutamate (Table 17.5).

Table 17.5 / Receptor binding affinities and electrophysiological profiles of authentic excitatory amino acid and TAN-950 A.

Compound	Binding study			Excitation of neurons	
	K	Q	N	MEC	subtype
	(IC_{50}: µM)			(µM)	
TAN-950 A	3.6	0.28	19	300	Q/K
L-Glutamate	0.11	0.24	0.98	300	Q/K(N)
Kainate	0.0026	5.7	>100	0.1	Q/K
Quisqualate	0.028	0.0086	26	1	Q/K
AMPA	87	0.036	>100	1	Q/K
NMDA	>100	>100	14	10	N
DNQX	0.85	0.21	44		
CPP	>100	>100	0.20		
AP7	>100	>100	0.93		

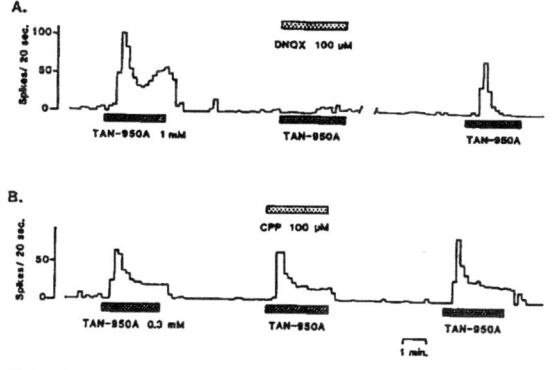

Figure 17.6 / Excitatiory effects of TAN-950 A on the rat hippocampal CA1 neurons *in vitro*. A. Tan-950 (1mM excited the neuron, and this effect was antagonized by DNQX (100 μM). B. The excitation elicited by TAN-950 A (0.3 mM) was not antagonized by CPP (100 μM).

THE TYPICAL EFFECTS OF TAN-950 A ON THE RAT HIPPOCAMPAL CA1 NEURONS

TAN-950 A excites the neurons at a relatively high concentration. This effect is antagonized by DNQX [8], but not by CPP (Fig. 17.6).

SUMMARY ON THE RECEPTOR BINDING AND ELECTROPHYSIOLOGICAL PROFILES OF TAN-950 A AND ITS DERIVATIVES

The results show three characteristics. First, in order to have affinity for amino acid receptors, it is necessary that both the carboxyl group and the amino group of TAN-950 A are free. Second, alkylation at the R1 position results in a decrease in the binding affinity relative to that of TAN-950 A but greatly increases the potency to excite neurons. For this alkylation, a methyl group was found to be most suitable. Third, the R enantiomers were found to have neuronal activity at N sites, although the original enatiomer is a Q/K agonist. The most potent NMDA agonist is (R)-Me-Nor-TAN-950 A (Table 17.6).

DISCUSSION

TAN-950 A shows *in vitro* and *in vivo* activity against *Candida albicans* and has low toxicity. The antifungal activity of TAN-950 A was reduced by glutamate. The mode of action of TAN-950 A has not yet been clarified.

Glutamate is known as a major excitatory amino acid transmitter in the central nervous system. The pharmacological effects of TAN-950 A on the excitatory amino acid (EAA) receptors in the rat brain were investigated. The receptors can be divided into at least three subtypes: kainate (K), quiasqualate (Q) and NMDA (N) subtypes [1, 2]. TAN-950 A had high affinity for all three receptor subtypes. TAN-950 A effectively elicited the firing of the hippocampal CA1 neurons *in vitro*, and this action was antagonized by a selective Q/K antagonist, DNQX. Alkylation, especially methylation at the 3-position of the isoxazole ring, slightly reduced the affinity for the three subtype receptors but increased the effect activating the hippocampal neurons; this activating effect was also antagonized by DNQX. The enantiomer of TAN-950 A had less affinity for Q/K subtypes but more affinity for NMDA subtypes than TAN-950 A itself. In hippocampal slices, it showed a receptor activation which was antagonized by AP7. One derivative of enatiomer, (R)-2-amino-2-(2,5-dihydro-3-methyl-5-oxo-4-isoxazolyl)acetic acid, was the most potent and selective MNDA agonist, and it had a relatively rigid structure. These derivatives of TAN-950 A seem to be good tools for investigating the pharmacological and physiological roles of EAA receptors, and to be potential drugs for CNS disorders.

ACKNOWLEDGEMENT

The authors wish to thank Drs. Y. Nakao and H. Okazaki for their encouragement throughout this work.

Table 17.6 / Receptor binding and electrophysiological profiles of TAN-950 A and its derivatives.

Compound	Chemical structure		Binding study (IC$_{50}$: μM)			Excitation of neurons MEC (μM)	subtype
	R$_1$	R$_2$	K	Q	N		
TAN-950 A	-H	-CH$_2$CHNH$_2$COOH	3.6	0.28	19	300	Q/K
	-H	-CH$_2$CHNH$_2$COOCH$_3$	>100	>100	>100	-	
	-H	-CH$_2$CH(N(CH$_3$)$_2$)COOH	>100	>100	>100	-	
Me-TAN-950 A	-CH$_3$	-CH$_2$CHNH$_2$COOH	17	0.30	40	1	Q/K
	-CH(CH$_3$)$_2$	-CH$_2$CHNH$_2$COOH	67	3.8	>100	3	Q/K
	-C$_6$H$_5$	-CH$_2$CHNH$_2$COOH	26	15	11	30	Q/K
	-H	-CH$_2$C*HNH$_2$COOH	12	3.6	4.9	100	N
	-H	-C*HNH$_2$COOH	>100	>100	5.8	30	N
(R)-Me-Nor-TAN-950 A	-CH$_3$	-C*HNH$_2$COOH	>100	>100	8.3	10	N

REFERENCES

1. Collingridge, G. L.; S. J. Kehl and H. McLennan *J. Physiol.*, 344, 19–31, 1983.
2. Collingridge, G. L.; S. J. Kehl and H. McLennan *J. Physiol.*, 334, 33–46, 1983.
3. Cotman, C. W.; D. T. Monaghan, O. P. Ottersen and J. Storm-Mathisen *TINS*, 10, 273–279, 1987.
4. Cotman C. W.; T. Monaghan and A. H. Ganong *Ann. Rev. Neurosci.*, 11, 61–80, 1988.
5. Deutsch S. I. and J. K. Morihisa *Clin. Neuropharmacol.*, 11, 18–35, 1988.
6. Fagg, G. E. and A. C. Foster *Neurosic.*, 9, 701–703, 1988.
7. Fonnu, F. *J. Neurochem.*, 42, 1–11, 1984.
8. Honore, T.; S. N. Davies, J. Drejer, E. J. Fletcher, P. Jacobsen, D. Lodge and E. Nielsen *Science*, 241, 701–703, 1988.
9. Lechevalier, M. P. and H. A. Lechevalier (1980) SIM special Publication No. 6. Eds., A. Dietz & D. W. Thayer, pp. 277–291, Society for Industrial Microbiology, Arlington.
10. Morris, R. G. M.; E. Anderson, G. S. Lynch and M. Baudry *Nature* 319, 774–776, 1986.
11. Murphy, D. E; E. W. Snowhill and M. Williams *Neurochem. Res.* 12, 775–782, 1987.
12. Murphy, D. E.; J. Schneider, C. Boehm, J. Lehmann and M. Williams *J. Pharmcol. Exp. Ther.* 240, 778–784, 1987.
13. Shirling, E. B. & D. Gottlieb *Int. J. Syst. Bacteriol.* 16, 313–340, 1966.
14. Tsuji, N.; M. Kobayashi, K. Nagashima, Y. Wakisaka and K. Koizumi *J. Antibiotics* 29, 1–6, 1976.
15. Waksman, S. A. (Ed.) (1961) The Actinomycetes. Vol. 2.
16. Hakoda S.; S. Tsubotani, T. Iwasa, M. Suzuki, M. Kondo & S. Harada *J.Antibiotics* submitted.
17. Tsubotani S.; Y. Funabashi, M. Takamoto, S. Hakoda & S. Harada *Tetrahedron Lett.* submitted.
18. a) Iwama T.; Y. Nagai, S. Harada, K. Itoh & A. Nagaoka *Europ. J. Pharmacol.* 1990, 183, 471 (brief communication). b) Iwama T.; Y. Nagai, N. Tamura, S. Harada & A. Nagaoka *Europ. J. Pharmacol* in press.

18

Inhibitors of Proline Specific Peptidases and Their Possible Applications in Medicine

Takaaki Aoyagi, Yasuhiko Muraoka and Tomio Takeuchi
Institute of Microbial Chemistry, Tokyo, Japan

In order to elucidate the cause of diseases, it seems to be useful to analyze the change in various enzymatic activities as the function of time and follow them for a rather long period. It is also important to study the dynamics of the activities of multiple enzymes and to examine the data with multivariate analysis. In this way, a certain enzyme can be possibly identified to be most essential in the disease process. Once such an enzyme is found, it will be useful to search for a specific inhibitor against such an enzyme and to study its effect on animal models of the disease process.

Searching for inhibitors in culture filtrate of microbes, many substances which specifically inhibit various enzymes such as serine-, cysteine-, aspartic, and metal-proteinases, aminopeptidases, dipeptidyl aminopeptidases, carboxypeptidaes, dipeptidyl carboxypeptidases, and so forth were discovered. Inhibitors of proline specific peptidase such as poststatin, ebelactones A and B, dioctatins A and B, and diprotins A and B, and so forth were discovered.

Thus the studies on inhibitors of proline specific peptidase may well afford important keys to understand various aspects of biological phenomena and diseases: inflammation, immune response, autoimmune diseases, demyelination diseases, Alzheimer's disease and so forth. Because of their interesting pharmacological activities, some of the inhibitors are now under clinical evaluation for their uses as drugs. Enzyme inhibitors seem to propose a new promising field of science.

INTRODUCTION

It is known that the peptide bonds involving the imino residue or the carboxy residue of proline are hard to be hydrolyzed with usual peptidases.

Recently, proline specific peptidases directed to such peptide bonds have attracted researcher's attentions. In 1971, Walter and his colleagues [1] found an endopeptidase with a specificity for the cleavage of proline containing peptides at the carboxyl side of a proline residue. This enzyme, named prolyl endopeptidase (PEP) or post-proline cleaving enzyme, hydrolyzes several biologically active peptides which contain proline residue. Vasopressin is one of such peptides and is known to regulate peripheral as well as central functions. Another proline specific enzyme, dipeptidyl peptidase IV (DPP-IV), has been shown to be localized in T-cells of the immune system and is in fact a good T-cell marker [2,3]. In the serum from systemic lupus erythematosus (SLE) and rheumatoid arthritis (RA) patients, DPP-IV activity was reported to be markedly decreased, while DPP-II is increased [4]. Similarly, in the plasma from New Zealand Black (NZB/W) mice, DPP-IV activity is decreased in contrast with the marked increase in DPP-II and also in PEP [5].

Because of the wide distribution of proline specific endo- and exo-peptidases in the animal tissues, the inhibitors of such enzymes are expected to influence physiologic functions in the body. In particular, our final goal is to contribute to the therapeutics of various diseases including intractable ones. With this goal in mind, we analyze diseases, find out the key enzyme which is the most relevant to the disease cause and examine effects or roles of specific inhibitors against the key enzyme. In the present study, our focus was placed on inhibitors of proline specific peptidases and their possible applications in medicine.

ENZYMATIC CHANGES IN VARIOUS PATHOLOGIC STATES

1. Abnormality of the PEP activity with SLE-like syndrome. Previous studies indicated the importance of hydrolytic enzymes in the pathogenesis of autoimmune diseases. We examined the activities of such enzymes in various organs of the hybrids of the New Zealand Black and New Zealand White (NZB/W) mouse as a laboratory model of human SLE. Of the 18 enzymatic activities tested the activities of PEP showed a paticular behavior in the spleen of NZB/W mouse.

Table 18.1 compares the splenic enzyme activities between the NZB/W mice and their controls (DBA/Z mice). At 10 weeks of age, chymotrypsin (Chy-try)-like, elastase-like, glucosidase, and mannosidase showed significantly increased activities, whereas leucine aminopeptidase (Leu-AP) and PEP activities were decreased in NZB/W mice. At 20 weeks of age, the activities of AP-A, AP-B, proline iminopeptidase (Pro-IP), formyl-methionine aminopeptidase (fMet-AP), glucosidase, and mannosidase were significantly increased, while those of Gly-Pro-Leu-AP, Chy-try-like, and elastase-like were decreased. At 30 weeks of age, the activities of AP-B, Pro-IP, fMet-AP, PEP, glucosidase, and mannosidase were all significantly increased, while those of trypsin-like and elastase-like were lower.

Of such extensive changes in enzyme networks, the most interesting were the changes in the activity of PEP. It catalyzes hydrolysis of peptides at the carboxyl side of prolyl residues [6-9] and is known to catalyze hydrolysis of several biological active peptides, including oxytocin, vasopressin, thyroliberin (TRH), substance P, neurotensin, angiotensin [8], and A_4 amiloid peptide [10]. In this regard it is to be noted that the activity of this enzyme paralleled that of Pro-IP in most organs of control animals. It is probable that the activi-

Table 18.1 / Enzymatic changes in spleen of NZB/W and control mice

Enzyme	Specific activity (nmol/min/mg prot.) n = 5					
	10 W		20 W		30 W	
	Control	NZB/W	Control	NZB/W	Control	NZB/W
AP-A	0.90 ± 0.04	0.95 ± 0.19	1.01 ± 0.07	1.28 ± 0.14^b	0.77 ± 0.16	0.59 ± 0.13
AP-B	8.13 ± 1.30	9.17 ± 1.03	7.81 ± 0.93	10.63 ± 1.44^b	7.33 ± 1.09	9.69 ± 1.04^a
Pro-IP	1.35 ± 0.29	1.38 ± 0.16	1.14 ± 0.13	1.57 ± 0.23^b	0.97 ± 0.09	1.25 ± 0.16^a
Leu-AP	8.75 ± 0.77	7.58 ± 0.69^a	8.64 ± 0.63	9.31 ± 0.84	7.25 ± 0.62	8.80 ± 1.51
fMet-AP	11.51 ± 0.87	11.67 ± 2.01	8.03 ± 1.57	12.24 ± 0.81^c	10.14 ± 1.06	12.57 ± 1.04^a
Gly-Pro-Leu-AP	2.12 ± 0.24	1.85 ± 0.28	2.39 ± 0.32	1.78 ± 0.12^b	1.78 ± 0.15	1.58 ± 0.19
PEP	3.39 ± 0.44	1.94 ± 0.37^c	2.89 ± 0.34	2.95 ± 0.44	2.40 ± 0.39	3.36 ± 0.41^b
Trypsin-like	86.04 ± 7.35	92.99 ± 26.52	101.83 ± 16.43	95.42 ± 11.35	82.13 ± 23.90	48.85 ± 10.51a
Chy-try-like	70.32 ± 8.11	97.79 ± 16.62^b	121.72 ± 10.61	78.13 ± 28.12^a	87.18 ± 19.47	59.73 ± 17.29
Elastase-like	55.94 ± 10.98	76.79 ± 15.11^a	97.51 ± 13.74	71.64 ± 5.84^b	69.32 ± 13.52	46.58 ± 14.33^a
Glucosidase	1.28 ± 0.11	1.62 ± 0.19^b	1.20 ± 0.13	1.93 ± 0.07^c	0.99 ± 0.09	1.46 ± 0.19^b
Mannosidase	0.37 ± 0.03	0.54 ± 0.07^c	0.22 ± 0.03	0.34 ± 0.05^b	0.23 ± 0.03	0.54 ± 0.12^c

^{a} $p<0.05$, ^{b} $p<0.01$, ^{c} $p<0.001$
NS: Cathepsin D, N-Ac-β-D-Glucosaminidase, Phosphatase, Arylesterase

ties of both enzymes correlate well with the general metabolism of the proline-containing peptides in various organs and tissues.

Because of these diversities of enzymatic changes with increasing age of the animals. A multivariate analysis was performed to assess the role of each enzyme in the pathophysiology of this SLE model. Since a particular role of PEP was indicated by previous studies, we focused our attention on this enzyme in the present study.

The activity of PEP showed a highly positive correlation with that of Pro-IP in all the organs tested from the control animals. In the spleen of control animals, there was a significant correlation (r = 0.85) between the two enzymatic activities, but the correlation was not significant (r = 0.03) in NZB/W mice. Furthermore, it is to be noted that the relationship of these two enzymes to the aging process completely changed in the pathologic model in comparison with the control. Namely, it is clear that both of the enzymatic activities decreased with increasing age in the control animals. In NZB/W mice, however, the activity of PEP increased, rather than decreased, with age; and no significant trend was seen for Pro-IP (Table 18.1). This phenomenon was not found in any organ other than the spleen. This may suggest an important pathogenetic role for the PEP in immunological disturbances in this model animal.

2. Abnormality of kallikrein and PEP activities in cerebral tissue of patients with Alzheimer's disease. Recently, several reports were published which suggested the accumulation of abnormal protein in the brains of patients with Alzheimer's disease [11–14]. Also, as more direct evidence, injection of a protease inhibitor into animal brains was shown to induce the formation of lysosome-associated granular aggregates (dense bodies) which closely resem-

Table 18.2 / Changes in protease activities in brain of patients with Alzheimer's disease

Enzyme	Substrate	Specific Activity (Mean ± SD, nmol/min/mg protein)	
		Normal (N = 6)	Alzheimer (n = 7)
AP-A	Glu•NA	0.93 ± 0.26	1.02 ± 0.39
AP-B	Arg•NA	17.65 ± 10.52	19.97 ± 8.71
Pro-IP	Pro•NA	10.61 ± 4.20	12.99 ± 4.33
DPP-I	Gly-Arg•NA	0.35 ± 0.16	0.22 ± 0.13
DPP-II	Lys-Ala•NA	7.09 ± 2.55	5.47 ± 2.40
DPP-III	Arg-Arg•NA	7.65 ± 1.07	7.95 ± 2.02
DPP-IV	Gly-Pro•NA	0.48 ± 0.22	0.61 ± 0.25
PEP	Z•Gly•Pro•NA	2.31 ± 1.52	4.68 ± 1.39[a]
Cathepsin B	Z•Arg•Arg•NA	21.38 ± 7.72	14.21 ± 5.37
Kallikrein	Pro•Phe•Arg•MCA	0.037 ± 0.012	0.019 ± 0.003[b]
Plasmin	Boc•Val•Leu•Lys•MCA	0.015 ± 0.006	0.009 ± 0.004
Thrombin	Boc•Val•Pro•Arg•MCA	0.026 ± 0.022	0.015 ± 0.001
Trypsin	Boc•Gln•Ala•Arg•MCA	0.061 ± 0.061	0.023 ± 0.011
Chy-try	Suc•Leu•Leu•Val•Tyr•MCA	0.014 ± 0.001	0.014 ± 0.003
Elastase	Suc•Ala•Pro•Ala•MCA	0.132 ± 0.117	0.26 ± 0.11

[a] $p<0.05$, [b] $p<0.01$
Normal: 80.5 ± 11.0 (Mean ± SD) years old, Alzheimer: 80.6 ± 12.0 (Mean ± SD) years old

bled the ceroid-lipofuscin that accumulates in certain disease states and during aging [15]. These reports prompted us to study the protease changes in the brain of patients with Alzheimer's disease.

Table 18.2 compares the intracerebral activities of 12 kinds of proteases between the two groups, control and pathologic. The activity of kallikrein-like enzyme was significantly decreased, while that of PEP was increased in the patient group when compared with the control group. Besides the kallikrein-like activity, we found weak cleavage of the substrates specific to various endopeptidases including elastase, trypsin and chymotrypsin. Since it is unlikely that these particular enzymes occur in brain tissue in active forms, it seems that some endopeptidases of lysosomal origin were responsible for these endopeptidase activities.

In order to examine the properties of the kallikrein-like activity in the brain, the inhibition patterns of several inhibitors were tested. The 50% inhibition concentration (IC_{50}) of leupeptin was 3×10^{-9} M, whereas that of aprotinin was 2×10^{-7} M. SBTI showed no inhibition within the same range of concentration, excluding the possibility that the kallikrein-like activity came from the plasma kallikrein in cerebral blood vessels.

To elucidate the meaning of the multiple enzymatic changes in the disease, a multivariate study using discriminant function analysis [16] was performed. The highest correlation, a negative correlation of −0.78, was seen with the activity of kallikrein-like enzyme. The activity of PEP showed the second highest, in this case positive, correlation of 0.63. In addition, the correlation between the discriminant function score and the length of the time which lapsed from death to the postmortem

Figure 18.1 / Structure of various inhibitors.

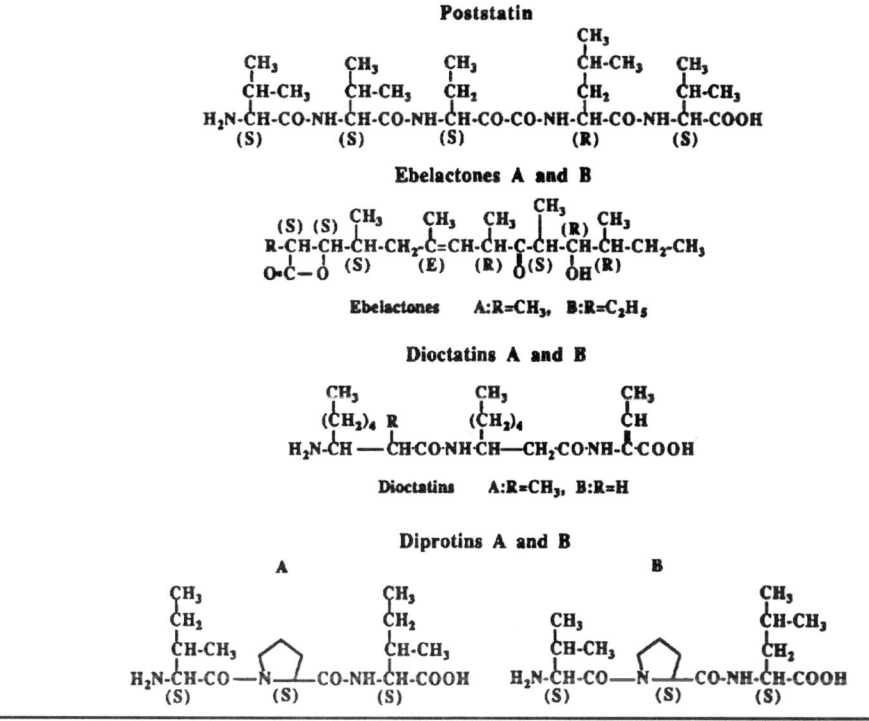

autopsy was calculated. This correlation was only 0.06.

The changes in the intracerebral activities, at the time of postmortem autopsy, in patients with Alzheimer's disease was examined. When compared with the control group, the activity of kallikrein-like enzyme was significantly decreased, while prolyl endopeptidase activity increased, in the patients group. Taken together with the results of a multivariate study, these findings may indicate that intracerebral kallikrein deficiency and the increase in PEP activity play an important role in the pathogenesis of Alzheimer's disease.

INHIBITORS AGAINST PROLINE SPECIFIC PEPTIDASES

Inhibitors of proline specific peptidases such as PEP, Pro-IP, DPP-II, DPP-IV and so on are of considerable interest because these enzymes are involved in kinin formation, autoimmune disease, Alzheimer's disease, and so forth. The inhibitors of such enzymes may have applications to studies on the related diseases.

Screening on culture filtrates of microbes for inhibitory activities against PEP, Pro-IP, DPP-II, and DPP-IV led to the discovery of the following inhibitors; poststatin [17, 18], ebelactones [19, 20], dioctatins A and B [21] and diprotins A and B [22]. Poststatin, inhibiting PEP strongly and cathepsin B weakly; ebelactones A and B, inhibiting Pro-IP, formylmethionine aminopeptidase (fMet-AP), esterase and lipase strongly; dioctatins A and B, inhibiting DPP-II; diprotins A and B, inhibiting DAP-IV have all been found from culture filtrates of *Streptomyces* strains. Chemical structures of these inhibitors are shown in Fig. 18.1.

The type of inhibition of these inhibitors against various enzymes is shown in Table 18.3.

Table 18.3 / Kinetic constants of various inhibitors

Inhibitor	Enzyme	Ki (10^{-7} M)	Type of inhibition
Poststatin	PEP	0.56	Competitive
Ebelactone A	fMet-AP	1.73	Non-competitive
Ebelactone B	fMet-AP	0.63	Non-competitive
Dioctatin A	DPP-II	2.0	Competitive
Dioctatin B	DPP-II	13.0	Competitive
Diprotin A	DPP-IV	22.0	Competitive
Diprotin B	DPP-IV	76.0	Competitive

PEP: prolyl endopeptidase, fMet-AP: formylmethionine aminopeptidase
DPP: dipeptidyl peptidase

PHARMACOLOGIC APPLICATIONS OF INHIBITORS OF PROLINE SPECIFIC PEPTIDASES

Of the low-molecular-weight enzyme inhibitors discovered in our institute, those acting against the enzymes located on cellular surfaces, such as aminopeptidases, carboxypeptidases, esterase, alkaline phosphatase, and so forth, proved to have immunomodifying effects [23]. Thus it seemed worthwhile to test the effects of immunomodifiers on the development of experimental allergic encephalomyelitis (EAE), which is known to be a useful model for studies of cell-mediated autoallergic immune diseases. Suppression of the development of autoimmune diseases in NZB/W as a laboratory model of human systemic lupus erythematosus was tried. The effects of inhibitors of proline specific peptidases together with immunomodyfying agents for suppression of autoimmune diseases were tested. The most effective inhibitor used was poststatin and dioctatins A and B.

Finally, the biological and therapeutic effects of enzyme inhibitors found in our institute are summarized (Table 18.4). The actions of immunopotentiation were found to be possessed by actinonin, amastatin, arphamenines A and B, benzylmalic acid, bestatin, dioctatins A and B, diprotins A and B, ebelactones A and B, esterastin, forphenicinol, forphenicine, foroxymithine, histargin, poststatin, and probestin. Fertilization is known to be suppressed by leupeptin, antipain, and chymostatin. Analgesic action is exerted by actinonin, amastatin, arphamenines A and B, and bestatin. Carrageenin edema is suppressed by antipain, chymostatin, elastatinal, esterastin, leupeptin, phosphoramidon and pepstatin, whereas ascites and pleural fluid by pepstatin. Blister formation after burns is suppressed by leupeptin. The serine proteinases released by pancreatitis can be inhibited by leupeptin. The essential and renal vascular hypertension are suppressed by bestatin, foroxymithine, histargin and pepstatin. Malignant diseases in man can be suppressed by bestatin, forphenicine, and forphenicinol. SLE and EAE can be suppressed by bestatin, poststatin and dioctatins A and B.

We would like to close this section hoping that the application of these enzyme inhibitors will help to elucidate the cause of diseases and even to cure the diseases.

Table 18.4 / Biological and therapeutic effects of enzyme inhibitors

Immunopotentation	Ac, Am, Ar(A,B), Bm, Bs, Do(A,B), Dp(A,B), Eb(A,B), Es, Fl, Fn, Fo, Hs, Po, Pr
Fertilization	Ap, Cs, Lp
Analgesic action	Ac, Am, Ar(A,B), Bs
Inflammation	Ap, Cs, El, Es, Lp, Ph, Ps
Ascites and pleural fluid	Ps
Chemical carcinogenesis	Ap, Lp
Burn	Lp
Pancreatitis	Lp
Hypertension	Bs, Fo, Hs, Ps
Malignant diseases	Bs, Fl, Fn
Autoimmune diseases	Be, Do(A,B), Po
Alzheimer's disease	Po

Am: amastatin, Ac: actinonin, Ap: antipain, Ar: arphamenine, Be: benastatin, Bm: benzylmalic acid, Bs: bestatin, Cs: chymostatin, Do: dioctatin, Dp: diprotin, Eb: ebelactone, El: elastatinal, Es: esterastin, Fl: forphenicinol, Fn: forphenicine, Fo: foroxymithine, Lp: leupeptin, Hs: histargin, Ph: phosphoramidon, Po: poststatin, Pr: probestin, Ps: pepstatin

REFERENCES

1. Walter, R., H. Shlank, J. D. Glass, I. L. Schwartz and T. D. Kerenyi. 1971. Leucylglycinamide released from oxytocin by human uterine enzyme. *Science* 173: 827–829.

2. Ansorge, S. and E. Schon. 1986. Dipeptidyl peptidase IV in human T lymphocytes: an approach to the function of this peptidase in the immune system. In: *Peptides and Proteases: Recent Advances* (Schowen, R. L. and A. Barth eds.), pp. 3–10, Advances in the Biosciences, Vol. 65, Pergamon Press, Oxford•New York.

3. Aoyagi, T., T. Wada, F. Kojima, S. Yoshida, S. Sasakawa, M. Tamura, and H. Umezawa. 1989. Multiple enzymes related to differentiation of lymphocyte subpopulations in man. *Biochem. Int.* 18: 383–389.

4. Iwase-Okada, K., T. Nagatsu, K. Fujita, K. Torikai, T. Hamamoto, T. Shibata, Y. Maeno and S. Sakakibara. 1985. Serum collagenase-like peptidase activity in rheumatoid arthritis and systemic lupus erythematosus. *Clinica Chimica Acta* 146: 75–79.

5. Hagihara, M., M. Ohhashi and T. Nagatsu. 1987. Activities of dipeptidyl peptidase II and dipeptidyl peptidase IV in mice with lupus erythematosus-like syndrome and in patients with lupus erythematosus and rheumatoid arthritis. *Clinical Chemistry* 33, 1463–1465.

6. Yoshimoto, T., M. Fischl, R. C. Orlowski and R. Walter. 1978. Post-proline cleaving enzyme and post-proline dipeptidyl aminopeptidase. *J. Biol. Chem.* 253: 3708–3716.

7. Koida, M. and R. Walter. 1976. Post-proline cleaving enzyme. *J. Biol. Chem.* 251: 7593–7599.

8. Walter, R. and T. Yoshimoto. 1978. Postproline cleaving enzyme: kinetic studies of size and stereospecificity of its active site. *Biochemistry* 17: 4139–4144.

9. Yoshimoto, T., K. Tsukumo, N. Takatsuka and D. Tsuru. 1982. An inhibitor for post-proline cleaving enzyme; distribution and partial purification from porcine pancreas. *J. Pharm. Dyn.* 5: 734–740.

10. Ishiura, S., T. Tsukahara, T. Tabira, T. Shimizu, K. Arahata and H. Sugita. 1990. Identification of a putative amyloid A4-generating enzyme as a prolyl endopeptidase. *FEBS Letters* 260: 131–134.

11. Ihara, Y., N. Nukina, R. Miura and M. Ogawara. 1986. Phosphorylated tau protein is integrated into paired helical filaments in Alzheimer's disease. *J. Biochem.* 99: 1807-1810.

12. Selkoe, D. J., D. S. Bell, M. B. Podlisny, D. L. Price and L. C. Cork. 1987. Conservation of brain amyloid proteins in aged mammals and humans with Alzheimer's disease. *Science* 235: 873–877.
13. Perry, G., R. Friedman, G. Shaw and V. Chau. 1987. Ubiquitin is detected in neurofibrillary tangles and senile plaque neurites of Alzheimer disease brains. *Proc. Natl. Acad. Sci. USA* 84: 3033–3036.
14. Mori, H., J. Kondo and Y. Ihara. 1987. Ubiquitin is a component of paired helical filaments in Alzheimer's disease. *Science* 235: 1641–1644.
15. Ivy, G. O., F. Schottler, J. Wenzel, M. Baudry and G. Lynch. 1984. Inhibitors of lysosomal enzymes: accumulation of lipofuscin-like dense bodies in the brain. *Science* 226: 985–987.
16. Kendall, M. 1975. Principle components. In: *Multivariate Analysis*, pp. 13–29, Griffin, London.
17. Aoyagi, T., M. Nagai, K. Ogawa, K. Okada, T. Ikeda, M. Hamada and T. Takeuchi. 1990. Poststatin, a new inhibitor of prolyl endopeptidase, produced by *Streptomyces viridochromogenes* MH534–30F3. I. Taxonomy, production, isolation, physicochemical properties and biological activities. *J. Antibiot.*, in press.
18. Nagai, M., K. Ogawa, Y. Muraoka, H. Naganawa, T. Aoyagi and T. Takeuchi. 1990. Poststatin, a new inhibitor of prolyl endopeptidase, produced by *Streptomyces viridochromogenes* MH534–30F3. II. Structure determination. *J. Antibiot.*, in press.
19. Umezawa, H., T. Aoyagi, K. Uotani, M. Hamada, T. Takeuchi and S. Takahashi. 1980. Ebelactone, an inhibitor of esterase, produced by actinomycetes. *J. Antibiot.* 33: 1594–1596.
20. Uotani, K., H. Naganawa, S. Kondo, T. Aoyagi and H. Umezawa. 1982. Structural studies on ebelactone A and B, esterase inhibitors produced by actinomycetes. *J. Antibiot.* 35: 1495–1499.
21. Aoyagi, T., K. Ogawa, H. Iinuma, S. Harada, C. Imada, Y. Okami, T. Takeuchi and T. Nagatsu. 1990. Dioctatins A and B: new inhibitors dipeptidyl peptidase II produced by *Streptomyces* sp. SA-2581. I. Taxonomy, production, isolation, physicochemical properties and biological activities. *J. Antibiot.*, in press.
22. Umezawa, H., T. Aoyagi, K. Ogawa, H. Naganawa, M. Hamada and T. Takeuchi. 1984. Diprotins A and B, inhibitors of dipeptidyl aminopeptidase IV, produced by bacteria. *J. Antibiot.* 37: 422–425.
23. Aoyagi, T. 1990. Small molecular protease inhibitors and their biological effects. In: *Biochemistry of Peptide Antibiotics* (Kleinkauf, H. and H. Dohren. eds.), pp. 311–363, Walter de Gruyter, Berlin•New York.

19

Screening, Isolation and Subsequent Chemical Modifications of Low Molecular Weight Cysteine Proteinase Inhibitors from Micro-Organisms

M. Saito, N. Higuchi and T. Tanaka

Institute for Fundamental Research, Suntory Ltd., Shimamoto-cho, Mishima-gun, Osaka, 618, Japan

New types of cysteine proteinase inhibitors, strepin P-1, staccopin P-1, and P-2 are low molecular weight peptide derivatives isolated from micro-organisms. They strongly inhibited cysteine proteinases like papain, calpains and cathepsins. But the specificity toward cysteine proteinases is not very high. We attempted to synthesize several di- and tri-peptidyl aldehydes, with the aim for obtaining derivatives to fulfill the specificity requirements of different cysteine proteinases. Some of them modulate the signal transmission on neuromuscular junction of insects and prevent the Ca^{2+}-ionophore induced degradation of actin binding protein and P235 in intact platelets. We also synthesized various acyl-peptidyl inhibitores containing proline residue and have found that some prevent experimentally-induced amnesia in rats.

INTRODUCTION

Proteinous inhibitors acting on proteolytic enzymes have been isolated from the tissues of a variety of animals and plants and products of microbes, and their biological significances, chemical properties, and distribution have been widely studied (1, 2). Most of the inhibitors discovered earlier were high molecular weight (>5,000) peptides or glycopeptides. Many proteinase inhibitors of microbial origin were originally developed with the pioneering works by Umezawa during the

1970's (3, 4). These inhibitors are low molecular weight (<1,000) peptide. They are being widely used in biological and biochemical studies (5, 6). It is further expected that these proteinase inhibitors can be used as valuable tools in medical research, because of their unique pharmacological properties and potential clinical applications. In an attempt to isolate new specific inhibitors to the different cysteine proteinase from microbes, strepin P-1 (7), staccopins P-1 and P-2 (8) were isolated from new strains of *Streptomyces* and *Staphylococcus*, respectively. The staccopins are the first isolated peptidyl cysteine proteinase inhibitors with low molecular weights from the culture fluid of a bacterium.

However, the inhibition by these microbial inhibitors of cysteine proteinase was not as specific as we had expected. We thought that the specificity of inhibitors toward cysteine proteinase should be dependent on their unique chemical structures. If different peptidyl inhibitors could be synthesized each specifically acting on one of the different cysteine proteinases in a discriminative fashion, they would be useful tools for investigating the physiological roles of cysteine proteinase in tissue. Several di- and tri-peptidyl aldehyde derivatives were synthesized and tested for their inhibitory effects on several serine and cysteine endopeptidases. These compounds showed almost no inhibition of trypsin and chymotrypsin and some cathepsins, while they exhibited marked inhibition of cathepsin, calpain and papain, specifically (9). A further step was undertaken to obtain cell-penetrative agents by modifying the peptidyl inhibitors. We found that one of the newly synthesized dipeptidyl aldehydes (calpeptin) possessed selective inhibition to calpain and cell-permeability by employing platelets as representative of cells (10). This inhibitor also modulates the signal transmission on neuromuscular junction of insects (11). We have synthesized various acyl-peptidyl-prolinal derivatives as potential post-proline cleaving enzyme inhibitor based on the information of structure activity relationship of cysteine proteinase inhibitors and clarified both the relationship between structure and inhibitory activity in vitro and between inhibitory activity and preventing experimentally-induced amnesia in rats (12).

SCREENING AND ISOLATION OF CYSTEINE PROTEINASE INHIBITOR FROM MICROBE

The primary object of this study was to isolate some novel inhibitors acting specifically on cysteine proteinase as calpains and cathepsins from the culture fluid of certain microbes. The screening methods used for the inhibitors are the test tube enzyme assay and TLC assay. Test tube enzyme assays have played major roles in the monitoring of isolation and purification procedures, the determination of specificity on the different enzymes, and the quantitation of inhibitors. The test tube enzyme assay is sensitive, reproducible, and simple to operate on a routine basis. TLC assay played a role in the search characterization of new compounds. After enzyme assay, active culture fluid was applied to column of HP-20. The column was washed with water and active fraction was eluted with methanol. The active fractions were concentrated and the sample was hydrolyzed in 6N HCl at 105°C for 48hr. The hydrolyzate was developed on TLC of cellulose glass plate with n-butanol, pyridine, acetic acid, water (15:10:3:12,v/v). Amino acids on TLC were stained with ninhydrin, Pauly, Sakaguchi, Shiff and Tollens' reagents (13). Further analysis of amino acid was carried out by dansyl methods (14). Strepin P-1, staccopins P-1 and P-2, which have structures unprecedented among the known proteinase inhibitors from microbes were isolated (7, 8). Respectively, the staccopins are the first isolated peptidyl cysteine proteinase inhibitors with low molecular weight from the culture fluid of a bacterium, *Staphylococcus* sp.. The purification procedures for these inhibitors as summarized in Table 19.1 and 19.2 included HP-20

Table 19.1 / Purification of Strepin P-1 from Culture Fluid

Step	Dry weight (mg)	Total activity (units)	Yield (%)	Specific activity (units/mg)	Purification (-fold)[c]
Culture fluid[a]	n.d.[b]	3840000	100		
HP-20	13400	2800000	73	209	1.0
DEAE-cellulose	5710	1600000	42	280	1.3
Amberlite CG-50	263	800000	21	3040	14.5
Sephadex LH-20	53	620000	16	11700	56.0
Sephadex G-25	12	490000	13	40500	194.0

[a]From 8 liters of culture fluid
[b]Total dry weight of this fraction could not be determined.
[c]A specific activity of 209 units/mg found at 2nd step was taken as unity.

chromatography, DEAE-cellulose, CM-cellulose, and Sephadex Gel column chromatographies, and HPLC. With the use of amino acid analysis, field adsorption mass spectra (FD-MS), and NMR spectra, the chemical structures of strepin P-1, staccopins P-1 and P-2 were identified as N-isovaleryl-tyrosyl-valyl-argininal, H-valyl-valyl-valyl-valyl-phenylalaninal, and H-valyl-valyl-valyl-valyl-tyrosinal, respectively (Fig. 19.1).

The purified strepin P-1, staccopin P-1 and P-2 showed strong inhibition against calpain I, and calpain II, and papain (Table 19.3). The specificity spectra of staccopins toward cysteine and serine proteinase were different from those of leupeptin, antipain, and strepin P-1 found in *Streptomyces*. Staccopins inhibited cysteine proteinases very strongly, with no action on non-cysteine proteinase including serine proteinases like trypsin and chymotrypsin. The free amino N-terminal valine residues in staccopins are unique among naturally occurring inhibitors.

Figure 19.1

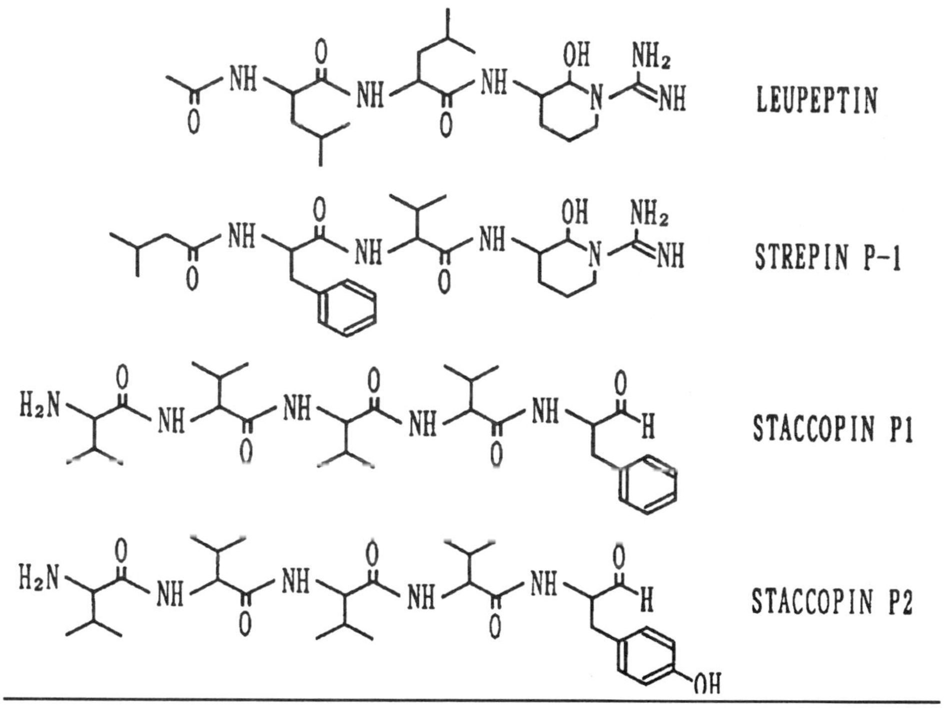

Table 19.2 / Purification of Staccopins P1 and P2 from Culture Fluid

Step	Dry weight (mg)	Total activity (units)	Yield (%)	Specific activity (units/mg)	Purification (-fold)[c]
Culture fluid[a]	n.d.[b]	690000			
HP-20	n.d.	780000	100		
DEAE-cellulose	17100	580000	75	34	1.0
1) Staccopin P1					
CM-cellulose	516	370000	48	721	21.1
Sephadex LH-20 and HPLC	492	60000	33	5100	150.0
2) Staccopin P2					
CM-cellulose	197	130000	17	699	19.7
Sephadex LH-10 and HPLC	18	90000	9	5000	147.0

[a]From 20 liters of culture fluid
[b]Total dry weight of this fraction could not be determined.
[c]A specific activity of 34 units/mg found at 2nd step was taken as unity.

Table 19.3 / Inhibitory Activity of Proteinase Inhibitors

Inhibitors	IC_{50} (µM)		
	Papain	Calpain I	Calpain II
Leupeptin	1.3 (1.0)[a]	0.30 (1.0)	0.15 (1.0)
Antipain	4.5 (0.3)	2.6 (0.1)	0.60 (0.3)
Strepin P-1	0.20 (6.5)	0.30 (1.0)	0.40 (0.4)
Staccopin P1	0.20 (6.5)	0.60 (0.5)	0.20 (0.8)
Staccopin P2	0.14 (9.3)	1.0 (0.3)	0.70 (0.2)
Ac[b]-Leu-Leu-Met-H	2.5 (0.5)	0.17 (1.8)	0.090 (1.7)
Ac-Leu-Leu-Nle[c]-H	2.0 (0.7)	0.14 (2.1)	0.080 (1.9)
Ac-Leu-Leu-Leu-H	10 (<0.1)		

[a]Relative potency, [b]Acetyl, [c]Norleucine

Table 19.4 / Inhibition of Calpains and Cathepsins by Di- and Tripeptidyl Aldehydes

Inhibitors	Ki (µ M)				
	Calpain I	Calpain II	Cathepsin L	Cathepsin B	Cathepsin H
Z^a-Leu-Nle^b-H	0.067	0.062	0.0034	0.13	>10
PB-^{c}Leu-Nle-H	0.065	0.068	0.0033	0.22	>10
Ac^d-Leu-Leu-Nle-H	0.19	0.22	0.00050	0.15	>10
Z-Leu-Met-H	0.036	0.068	0.013	0.25	>10
PB-Leu-Met-H	0.036	0.050	0.0045	0.16	>10
Ac-Leu-Leu-Met-H	0.12	0.23	0.00060	0.10	>10
Z-Leu-Phe-H	0.060	0.10	0.020	2.40	>10
PB-Leu-Phe-H	0.038	0.078	0.0014	0.43	>10

^{a}Benzyloxycarbonyl, ^{b}Norleucine, ^{c}4-Phenylbutanoyl, ^{d}Acetyl

CHEMICAL MODIFICATION OF MICROBIAL INHIBITORS

Many cysteine proteinase inhibitors from microbes were characterized, but the inhibitory spectra of the different cysteine proteinase inhibitors was not as expected. Attempts were made to synthesize many di- and tri-peptidyl aldehydes, aiming at obtaining some derivatives which would fulfill the specificity requirement of calpains and cathepsins. As shown in Table 19.3, the results indicate that the P^2 residue of the inhibitor, using the subsite notation originated by Schechter and Berger (15), is important in matching with the specificity of the enzyme. Papain favors a valine at the P_2 position over a leucine residue, while the reverse is true with calpains, both I and II. Thus, strepin P-1, staccopin P-1 and P-2, and antipain, which all have a valine residue at P_2 position, inhibited papain more strongly than calpain, while leupeptin and synthetic analogues with a leucine at the P_2 position inhibited calpains more strongly than papain. All of the 8 compounds containing a leucine at P_2 position in Table 19.4 were found, however, not to be specific for calpains, but to quite strongly inhibit also the cathepsins (9). From direct comparison of K_i values, the strength of affinity of cysteine proteinases tested with these peptidyl aldehydes generally increases in the following order: cathepsin H < < cathepsinB < calpain II ≤ calpain I < cathepsin L. The data shown in Table 19.4 also give some information on the subsite specificity of cathepsins L and B. With a Met or Phe residue at P_1 position, the nature of the P_3 residue showed a profound influence of K_i, while such an influence was not seen when P_1 was Nle. Elongation from di- to tri-peptide analogs by inserting one Leu residue at P_3 position strongly increased the affinity of the peptide analogs to cathepsin L, but not to cathepsin B. All of these lines are in accord with the known importance of the P_3 residue in determining the specificity of these cathepsins. The nature of the P_3 residue also influenced calpain I and II, but not always in the same ways as those for cathepsins. The presence of a Leu at P_3 always resulted in a remarkable decrease in affinity of inhibitor to calpains.

From information based on the structure activity relationship of cysteine proteinase inhibitors, several prolinal derivatives were synthesized and examined for their inhibitory activity on post-proline cleaving enzymes from *Flavobacterium meningosepticum* and bovine brain. One of the most potent post-proline cleaving enzyme inhibitors has a structure of Y-X-Pro-H, where Y is a usual amino-protecting group and X is an amino acid residue. Some Y and X groups were examined earlier for designing inhibitors of post-proline cleaving enzyme and it was found that Z-Val-Pro-H, and Boc-Ile-Pro-H exhibited strong inhibition toward both ascidian and *Flavobacterium* enzymes with IC_{50} value of 10~30nM, Z-Val-Pro-H, Z-pGlu-Pro-H, and Boc-Pro-Pro-H inhibited bovine brain enzyme with K_i values of 0.5~1.8nM, and Z-Pro-Pro-H also inhibited bacteria and bovine enzymes with low K_i values of 0.2~0.5nM. Therefore, the Y-X-Pro-H type compounds listed in Table 19.5 were designed and synthesized, and their inhibitory activities toward *Flavobacterium* and bovine enzymes were compared. P_2 subsite specificity studies were also carried out for various acyl moieties. According to the studies of cysteine proteinase inhibitors, the compounds with valine or proline in the P_2 subsite showed higher activity further than other amino acid residues. All the proline containing compounds in the P_2 subsite have much higher activity for the bacterial enzyme. On the contrary, some valine containing compounds showed the highest activity for the bovine enzyme. These facts indicate that proline and valine are suitable for the S_3 site of both enzymes but there are some differences around the S_2 site.

EFFECTS ON TISSUES AND CLINICAL APPLICATION OF PROTEINASE INHIBITOR

APPLICATION OF CALPAIN INHIBITORS

Cysteine endopeptidases, which include papain, calpains and cathepsins, are known to be strongly inhibited by leupeptin, antipain and strepin P-1. These streptomyces products have been frequently used as a calpain inhibitor to demonstrate the possible participation of calpains in platelet activation, fibrinogen receptor expression and serotonin release by thrombin. Recently two reports (16, 17) successively showed that these inhibitors exerted their effect primarily on thrombin rather than on endogenous calpain activity because actinomycetes inhibitors abolished amidolytic and fibrinolytic activity of thrombin. The peptidyl aldehydes synthesized were found to inhibit calpains strongly and selectively. Benzyloxycarbonyl-Leu-Nle-H (calpeptin) was most potent to calpain I and showed cell permeability (10). By incubating platelets with 100 μg/ml of calpeptin for 30 min at 37°C, intrinsic calpain activity was completely abolished after separation from the inhibitor. Acceleration of 20K protein phosphorylation was one of interesting calpain dependent intracellular events in platelets.

Phosphorylation of 20K protein was inhibited in the presence of calpeptin in platelets stimulated by thrombin, ionomycin or collagen. However, streptomyces inhibitors inhibited only thrombin induced 20K phosphorylation which was presumably due to direct inactivation of thrombin. Calpeptin seems to be a useful cell penetrative calpain inhibitor (Fig. 19.2).

Calpeptin was found to regulate the signal transmission of neuromuscular junction of insect hindgut.

Domoic acid isolated from a red seaweed, *Condria armata*, shows strong insecticidal activity toward various insects and also causes the contraction of hindgut excised from cockroaches (11, 18). The contraction of the hindgut did not occur when calcium ions were absent or when a calcium channel blocker was added to the incubation medium. The addition of calpain inhibitor, calpeptin, blocked the domoic acid activity. Incubation of the hindgut with calpeptin decreased the contractile response to a level close to that elicited by glutamic acid alone. Since calpeptin did not affect the sensitivity to the released glutamic acid but lowered the effect of domoic acid, it seems probable that the normal glutamic acid receptor sites are left intact. The effect of domoic acid at postsynaptic sites is that which activates Ca^{2+}-dependent neutral cysteine proteinases, possibly

Table 19.5 / Inhibition of Post-Proline Cleaving Enzymes from *Flavobacterium* and Bovine Brain by Acyl Pepteidyl Prolinal Derivatives

Inhibitors	IC_{50} (nM)	
	Flavobacterium	Bovine brain
1 4-Phenylbutanoyl-Ala-Pro-H	57	140
2 4-Phenylbutanoyl-Val-Pro-H	0.87	4.2
3 4-Phenylbutanoyl-Leu-Pro-H	1.2	12
4 4-Phenylbutanoyl-Nle-Pro-H	1.7	250
5 4-Phenylbutanoyl-Met-Pro-H	2.7	79
6 4-Phenylbutanoyl-Phe-Pro-H	1.1	22
7 4-Phenylbutanoyl-Pro-Pro-H	0.87	8.7
8 Phenylacetyl-Pro-Pro-H	4.7	4.7
9 3-Phenylpropionyl-Pro-Pro-H	1.8	4.5
10 5-Phenylpentanoyl-Pro-Pro-H	0.84	4.2
11 StBu[a]-Val-Pro-H	0.25	19
12 StBu-Leu-Pro-H	6.2	32
13 StBu-Phe-Pro-H	0.27	20
14 StBu-Pro-Pro-H	0.23	6.5
15 BzBu[b] Val-Pro-H	6.7	2.0
16 BzBu-Leu-Pro-H	6.5	65
17 BzBu-Phe-Pro-H	6.2	62
18 BzBu-Pro-Pro-H	0.20	6.7
19 IqBu[c]-Val-Pro-H	3.5	3.0
20 IqBu-Leu-Pro-H	3.5	6.0
21 IqBu-Nle-Pro-H	3.5	69
22 IqBu-Phe-Pro-H	2.3	97
23 IqBu-Pro-Pro-H	1.1	7.2
24 Oleoyl-Ala-Pro-H	52	
25 Oleoyl-Val-Pro-H	2.2	320
26 Oleoyl-Leu-Pro-H	2.5	3200
27 Oleoyl-Nle-Pro-H	2.2	6200
28 Oleoyl-Phe-Pro-H	1.2	45000
29 Oleoyl-Lys (Z)-Pro-H	1.4	6000
30 Oleoyl-Glu (H)-Pro-H	0.94	190
31 Oleoyl-Pro-Pro-H	0.97	65
32 Linoleoyl-Leu-Pro-H	1.9	320
33 Stearoyl-Leu-Pro-H	3.0	170
34 Palmitoyl-Leu-Pro-H	2.3	6500

[a] 4-(2-Styrylphenoxy)butanoyl, [b] 4-(4-Benzylphenoxy)butanoyl, [c] 4-(5-Isoquinolyloxy)butanoyl

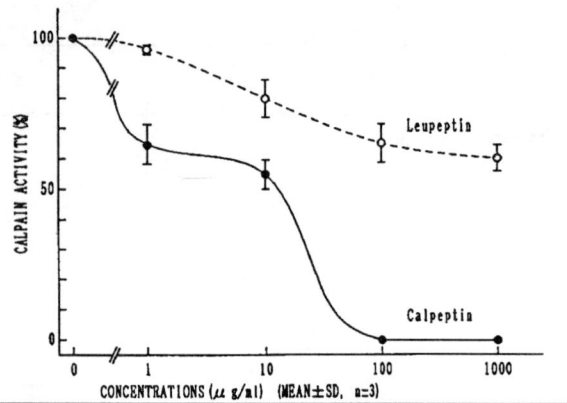

Figure 19.2

calpain, that unmask the latent glutamic acid receptors. A low concentration of domoic acid enhanced the sensitivity of the insect neuromuscular junctions to glutamic acid. These results suggest that domoic acid not only affects calcium ion transport and stimulates a calcium-dependent process that regulates the release of glutamic acid from presynaptic nerve endings, but also activates a calcium-dependent proteinase leading to generation of glutamic acid receptors on neuromuscular postsynaptic membrane from the latent state. Proctolin, a neuropeptide of insects, also caused contraction of the hindgut. Proctolin is probably a natural regulator of signal transmission of neuromuscular junctions of the insect hindgut, and domoic acid may bind to proctolin receptors to act.

From these reports, the following sequence of events is proposed to be caused by domoic acid at neuromuscular junction (Fig. 19.3). In both places of the pre- and the postsynaptic terminals it activates the opening of calcium channels. Calcium influx into the presynaptic nerve endings released glutamic acids from synaptic vesicle, and calcium influx into postsynaptic terminal activates calpain, which generates more glutamic acid receptors. This results in a potent muscle contraction of insect hindgut (11).

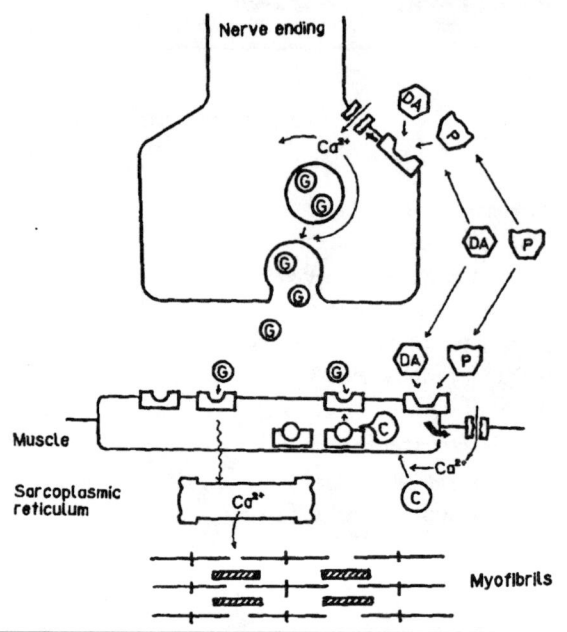

Figure 19.3

CLINICAL APPLICATION OF POST-PROLINE CLEAVING ENZYME INHIBITORS

Various prolinal derivatives with potent inhibitory activity toward post-proline cleaving enzyme were examined for their anti-amnesic effect of scopolamine-induced amnesic rats in the passive avoidance learning test (Table 19.6). Pretreatment with compounds, which have potent in vitro inhibitory activities toward the enzymes, prevented the induction of anmesia by scopolamine in the passive avoidance learning test at the doses of 10~25 µg/Kg (12). It can be stated that Y-Pro-Pro-H derivatives were effective in the retention test but no clear difference was observed among the compounds with different P_2 structures. We have determined the relationship between enzyme inhibitory activity and preventing in vivo methods of amnesic effects on animals such as passive avoidance learning in rats, active avoidance response in mice, one trial passive-active avoidance response in mice, and Y-maze learning in mice. From these results it was concluded that the pro-

Table 19.6 / Preventive Action of Post-Proline Cleaving Inhibitors toward Scopolamine-Induced Retrograde Amnesia

Compounds (i. p.)	Drug administered after training (i. p.)	Training (Means ± S.E.)[a]		Retention test[b]		Amnesia (%)
		No. of total rats	Stepdown latency	No. of decending	No. of amnesic rats/No. of total rats	
saline	saline	9	6.2 ± 1.7	2.3 ± 0.6	3/9	33
saline	scopolamine (3mg/kg)	10	2.7 ± 0.6	2.9 ± 0.3	8/10[c]	80
1 (250 µg/kg)	scopolamine (3mg/kg)	10	5.2 ± 1.1	3.0 ± 0.8	0/10[e]	0
2 (25 µg/kg)	scopolamine (3mg/kg)	10	1.9 ± 0.6	2.2 ± 0.6	1/10[d]	10
3 (25 µg/kg)	scopolamine (3mg/kg)	10	2.7 ± 0.9	2.4 ± 0.8	0/10[e]	0
4 (20 µg/kg)	scopolamine (3mg/kg)	10	1.7 ± 0.6	2.3 ± 0.4	0/10[e]	0
7 (20 µg/kg)	scopolamine (3mg/kg)	10	3.1 ± 0.7	2.4 ± 0.8	0/10[e]	0
14 (200 µg/kg)	scopolamine (3mg/kg)	5	3.3 ± 0.6	2.3 ± 0.6	3/5	60
25 (100 µg/kg)	scopolamine (3mg/kg)	10	1.8 ± 0.4	1.6 ± 0.2	1/10[d]	10
26 (10 µg/kg)	scopolamine (3mg/kg)	5	4.7 ± 0.7	2.0 ± 0.9	2/5	40
27 (25 µg/kg)	scopolamine (3mg/kg)	10	3.7 ± 0.3	2.1 ± 0.5	0/10[e]	0
28 (25 µg/kg)	scopolamine (3mg/kg)	10	2.7 ± 0.5	2.4 ± 0.7	0/10[e]	0
31 (10 µg/kg)	scopolamine (3mg/kg)	10	2.2 ± 0.8	2.4 ± 0.9	0/10[e]	0
33 (100 µg/kg)	scopolamine (3mg/kg)	10	4.7 ± 0.9	2.0 ± 0.5	0/10[e]	0
34 (100 µg/kg)	scopolamine (3mg/kg)	10	2.2 ± 0.3	3.0 ± 0.2	0/10[e]	0

[a]Total time for the training, 110 sec. (7, 14); 100 sec (1, 2, 3, 4, 25, 26, 27, 28, 31, 33, 34)
[b]The step down latency was measured for 300 sec.
[c]Significantly different from the saline group ($P<0.05$, x^2-analysis).
[d,e]Significantly different from the scopolamine group ([d]$P<0.005$, [e]$P<0.001$)

linal derivatives, which have potent inhibitory activity, showed strong anti-amnesic activities. In this study, however, compounds which inhibited only the bacterial enzyme showed as strong an anti-amnesic activity as those which inhibited both enzymes. The compounds showed the maximum effective dose, the so-called bell-shape dose dependency. The reason and mechanism for all these facts cannot presently be explained but at least the anti-amnesic activity of the compounds is due not only to enzyme inhibitory activity but also to other factors such as membrane permeability, hydrophobicity and inhibitory activity toward other related enzymes. From the results of dose dependency and other experiments described, it seem that the mechanism of the anti-amnesic affects of these compounds is complicated. However, the main mechanism is probably due to the inhibitory activity of the compounds toward the post-proline cleaving enzyme in the brain which is thought to cleave some special peptides, vasopressin, thyrotropin releasing hormone (TRH), substance P and neurotensin (19, 20) by acting directly on memory and related processes. Therefore the post-proline cleaving enzyme is thought to play an important role in the regulation of learning and memory consolidation in brain and inhibitors of this enzyme are suggested as possible candidates for an anti-amnesic drug.

REFERENCES

1. Vogel, R., I. Trautschold and E. Werle. 1968. *Natural Proteinase Inhibitor.* Academic Press, New York.
2. Birk, Y. 1987. Proteinase Inhibitors. In: *Hydrolytic Enzymes, New Comprehensive Biochemistry,* (Neuberger, A. and K. Brocklehurst, eds.), Vol. 16, pp. 257–306, Elsevier Amsterdam.
3. Umezawa, H. 1976. Enzyme Inhibitors Produced by Microorganisms. In: *Method in Enzymology.* (Lorand, L. ed.), Vol. 45, pp. 678–698, Academic Press, New York.
4. Umezawa, H. 1989. Enzyme Inhibitors Produced by Microorganisms. In: *Natural Products Isolation.* Journal of Chromatography Library (Wagman, G. H. and R. Cooper, eds.), Vol. 43, pp. 481–538. Elsevier, Amsterdam.
5. Aoyagi, T. 1978. Bioactive Peptides Produced by Microorganisms, (Umezawa, H., T. Shibata and T. Takita, eds.) pp. 129 Kodansha Scientific Press, Tokyo.
6. Libby, P. and A. L. Goldberg, 1978. *Science, 199:* 534
7. Ogura, K., M. Maeda, M. Nagai, T. Tanaka, K. Nomoto and T. Murachi. 1985. Purification and Structure of a Novel Cysteine Proteinase Inhibitor, Strepin P-1. *Agric. Biol. Chem. 49(3):* 799–805.
8. Saito, M., N. Kawaguchi, M. Hashimoto, T. Kodama, N. Higuchi, T. Tanaka, L. Nomoto and T. Murachi. 1987. Purification and Structure of Novel Cysteine Proteinase Inhibitors, Staccopins P1 and P2, from *Staphylococcus tanabeensis. Agri. Biol. Chem. 51(3):* 861–868.
9. Sasaki, T., M. Kishi, M. Saito, T. Tanaka, N. Higuchi, E. Kominami, N. Katunuma and T. Murachi. 1990. Inhibitory Effect of Di- and Tripeptidyl Aldehydes on Calpains and Cathepsins. *J. Enzyme Inhibition. 3:* 195–201.
10. Tsujunaka, T., Y. Kajiwara, J. Kambayashi, M. Sakon, N. Higuchi, T. Tanaka and T. Mori 1988. Synthesis of A New Cell Penetrating Calpain Inhibitor. *Biochem. Biophys. Res. Commun. 153(3):* 1201–1208.
11. Maeda, M., T. Kodama, M. Saito, T. Tanaka, H. Yoshizumi L. Nomoto and T. Fujita. 1987. Neuromuscular Action of Insecticidal Domoic Acid on the American Cockroach. *Pestic. Biochem. Physiol. 28:* 85–92.
12. Saito, M., M. Hashimoto, N. Kawaguchi, H. Fukami, T. Tanaka and N. Higuchi. 1990. Synthesis and Inhibitory Activity of Acyl-Peptidyl-Prolinal Derivatives toward Post-Proline Cleaving Enzyme as Nootropic Agents. *J. Enzyme Inhibition. 3:* 163–178.
13. Tanaka, T., H. Kita, T. Murakami and K. Narita. 1977. Purification and Amino Acid Sequence of Mating Factor from *Saccharomyces cerevisiae. J. Biochem. 82:* 1689–1693.
14. Hartlay, B. S. 1970. *Biochem. J. 119:* 805–822.
15. Schechter, I. and A. Berger. 1967. *Biochem. Biophys. Res. Commun. 27:* 157–162

16. Maeda, M., T. Kodama, T. Tanaka, Y. Ohfune, K. Nomoto, K. Nishimura and T. Fujita. 1984. Insecticidal and Neuromuscular Activities of Domoic Acid and Its Related Compounds. *J. Pesticide Sci.* 9:27–32.
17. Ruda, E. M. and M. C. Scrutlon. 1987. *Thromb. Res.* 47:611–619.
18. Ruggiero, M. and E. G. Lapetina. 1986. *Proc. Natl. Acad. Sci.* 83:3456–3459.
19. DeWied, D. and J. M. van Ree. 1982. Neuropeptides, Mental Performance and Aging. *Life Sciences,* 31:709–719.
20. Burbach, J. P., G. L. Kovacs, D. De Wied, J. W. Van Nispen and H. M. Greven. 1983. A Major Metabolite of Arginine Vasopressin in the Brain Is a Highly Potent Neuropeptide. *Science* 23:1310–1312.

20

Microbial Metabolites as Gastric H⁺-K⁺-ATPase Inhibitors

Jon S. Mynderse, Dennis R. Berry, Rosanne Bonjouklian, Otis W. Godfrey, Frederick P. Mertz, Walter N. Nakatsukasa, Raymond C. Yao, Ann H. Hunt, Jack B. Deeter, Jaswant S. Gidda, and Anne H. Dantzig

Lilly Research Laboratories, Lilly Corporate Center, Eli Lilly and Company, Indianapolis, IN, 46285 U.S.A.

INTRODUCTION

The stomach produces copious amounts of acid achieving a pH of about 1.0 (8). The production of acid is thought to play an important role in the digestion of food substances, and is bacteriostatic to microorganisms that would thrive in the nutrient-rich environment of the intestine. The overproduction of acid, however, is contributory to the development of duodenal ulcers. The clinical healing benefits of acid suppression are well established (18). Oversuppression of acid secretion can result in chronic hypergastrinemia and has the potential for carcinoid tumor formation (3, 22). Thus acid secretion must be properly controlled.

OVERVIEW OF GASTRIC ACID SECRETION

Acid is secreted by parietal cells located in the epithelium of gastric (oxyntic) glands of the stomach mucosa (13). These specialized cells possess the H^+-K^+-ATPase (or "proton pump"), a membrane-bound enzyme that pumps hydrogen ions extracellularly in exchange for potassium ions across the luminal membrane. As illustrated in Fig. 20.1, the parietal cell is a polarized cell that possesses receptors for the secretagogues, acetylcholine, gastrin, and histamine on the basolateral membrane and possesses an apical membrane that contains the proton pump when the cell is activated. The cell can exist in two morphological states as either a resting (non-secreting) cell or as a stimulated (acid secreting) cell. The series of events involved in the activation of the cell is somewhat complex. In the resting state, the H^+-K^+-ATPase

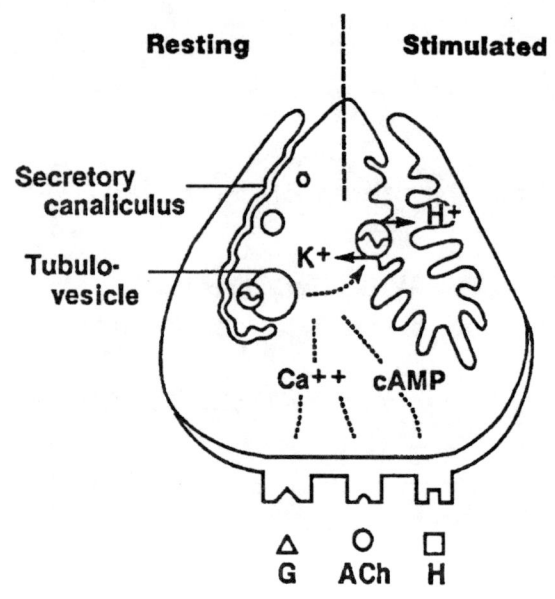

Figure 20.1 / Schematic representation of a parietal cell showing the conversion of a resting cell to a stimulated cell initiated by the binding of either acetylcholine (ACh), gastrin (G), or histamine (H) to specific receptors. Modified from Sachs et al. (23).

resides in the membranes of tubulovesicles distributed throughout the cytoplasm and the apical membrane infolds into the cytoplasm forming intracellular canaliculi (13). When one of the secretagogues binds to its receptor, the cell is converted morphologically as well as metabolically to an activated state. This can be achieved by two distinct pathways. Binding of histamine to the H_2-receptor results in an increase in cAMP production, whereas binding of the other two secretatogues to their receptors increases cytosolic calcium. Both pathways result in the fusion of tubulovesicles to the apical membrane, increasing dramatically the surface area of the secretory canaliculus (8). Simultaneously, H^+-K^+-ATPase is inserted into the canalicular membrane and activated to secrete acid into the lumen of the stomach.

CONTROL OF ACID PRODUCTION

Over the years there has been a variety of medications developed that intervene in the activation of the parietal cell (1, 21, 24). The H_2 receptor antagonists have been the most widely accepted and have provided a safe and an effective means of reducing acid secretion and healing duodenal ulcers. In many cases, however, more complete control of acid secretion is needed to heal gastric ulcers and to treat gastroesophagael reflux disease (heartburn). A more direct point for intervention by antisecretory drugs is the proton pump itself, the H^+-K^+-ATPase.

Omeprazole (Fig. 20.2) is an irreversible inhibitor of H^+-K^+-ATPase that heals duodenal and benign gastric ulcers (27) as well as provides some relief in reflux esophagitis (26). In the acid environment of the parietal cell, omeprazole undergoes chemical rearrangement to species that react covalently with H^+-K^+-ATPase sulfhydryl groups. Covalent modification of the enzyme can be prevented (*in vitro*) by the presence of sulfhydryl-reducing agents such as dithiothreitol.

Long-acting antisecretory agents, whether "super" H_2 antagonists or proton pump inhibitors, can lead to hypergastrinemia and carcinoid tumor formation (3). Drug discovery programs have been initiated at a number of pharmaceutical companies to develop reversible inhibitors of the proton pump that may lack this potential side effect. SCH 28080 (Fig. 20.2), a well-studied synthetic H^+-K^+-ATPase inhibitor lead, does not react covalently with the proton pump; its effect is not reversed by the presence of sulfhydryl-containing compounds. Instead, SCH 28080 competes for the high affinity potassium site on the luminal side of the enzyme. With this as a background, we initiated a drug discovery program aimed at developing reversible inhibitors of the proton pump from fermentation-derived leads.

Figure 20.2 / Structures of the synthetic H⁺-K⁺-ATPase inhibitors, omeprazole and SCH 28080.

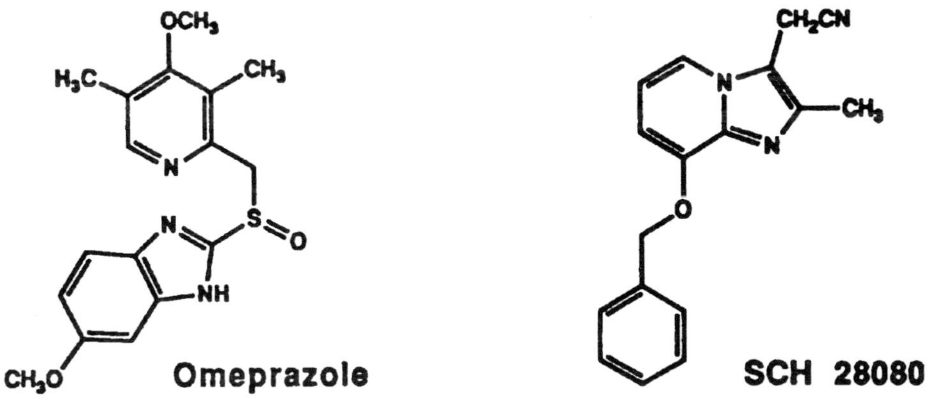

STRATEGIES FOR IDENTIFYING NEW MICROBIALLY-PRODUCED PROTON PUMP INHIBITORS

Soil microorganisms have long been recognized as an important source of novel compounds. In the search for new microbially-produced bioactive compounds two approaches can be used. Fermentation products previously isolated for an unrelated activity may be evaluated for their efficacy at a new pharmacological target. Alternatively fermentation broths may be screened directly for the activity of interest. Novel lead structures are subsequently isolated using the target assay. Both approaches have resulted in the discovery of novel natural product inhibitors of H⁺-K⁺-ATPase.

H⁺-K⁺-ATPase INHIBITORS: ISOLATED MICROBIAL METABOLITES

H⁺-K⁺-ATPase ASSAY

A standard assay for the proton pump was adapted to a 96-well microtiter dish (7). Gastric vesicles containing H⁺-K⁺-ATPase were prepared from the fundic mucosa of stomachs obtained from freshly slaughtered hogs. Typically, test compounds and enzyme were coincubated for a period of time before starting the reaction by addition of substrate. Routinely, enzyme activity (K⁺, Mg^{2+} dependent) was evaluated with the artificial substrate, p-nitrophenyl phosphate (p-NPP). During the assay, the colorless substrate p-NPP was hydrolysed to the yellow product, p-nitrophenol (p-NP). Product formation was detected colorimetrically at 405 nm using a microtiter dish reader.

A80915 H⁺-K⁺-ATPase INHIBITORS

Previously purified fermentation products were evaluated for inhibition of H⁺-K⁺-ATPase. Compound A80915A, a novel semi-naphthoquinone antibiotic produced by *Streptomyces aculeolatus*, was identified as a potent inhibitor of H⁺-K⁺-ATPase. Four members of the A80915 antibiotic complex (factors A-D, Fig. 20.3) as well as the known, related structure napyradiomycin B1 (25) were isolated from culture A80915 on the basis of their activity against Gram positive bacteria (9). Under modified fermentation conditions a new compound A80915G, as well as other known compounds SF2415A2, SF2415B1, and SF2415B2 (10) (Fig. 20.4) were produced (9).

The inhibition of H⁺-K⁺-ATPase by *Streptomyces aculeolatus*-produced antibiotics is shown in Table 20.1. With the exception of A80915 factor C and the diazo-containing factors B and D, the compounds were more potent than the synthetic inhibitors omeprazole and SCH 28080. The presence of the reducing agent, DTT, during the coincubation period had no significant effect on inhibition of the H⁺-K⁺-ATPase by any of the

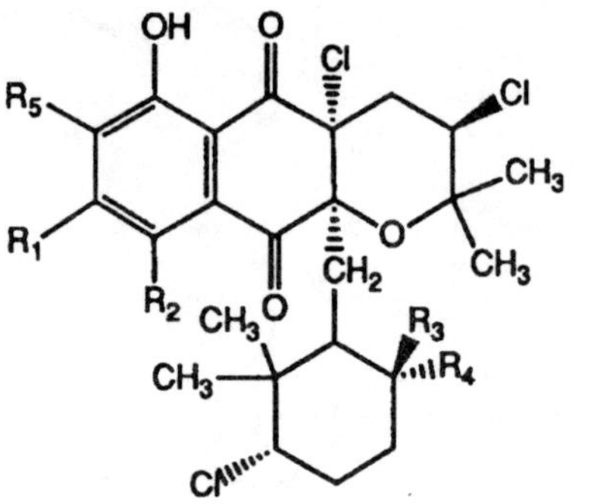

	R₁	R₂	R₃	R₄	R₅
A80915A	OH	H	=CH₂	-	CH₃
A80915B	O⁻	N≡N⁺	=CH₂	-	CH₃
A80915C	OH	H	OH	CH₃	CH₃
A80915D	O⁻	N≡N⁺	OH	CH₃	CH₃
Napyradiomycin B1	OH	H	=CH₂	-	H

Figure 20.3 / Semi-napthoquinone antibiotics A80915A-D and napyradiomycin B1 isolated from culture A80915 (9).

Table 20.1 / Inhibition of gastric H^+-K^+-ATPase by *S. aculeolatus*-produced antibiotics and reversal by reducing agent.

Compound	IC$_{50}$ (μM)	Reversal of inhibition by DTT[a]
Omeprazole	3.8	+
SCH 28080	14.0	-
A80915A	1.9	-
A80915B	16.0	-
A80915C	5.5	-
A80915D	8.7	-
A80915G	2.4	-
SF2415A2	1.3	-
SF2415B1	2.0	-
SF2415B2	0.8	-
Napyradiomycin B1	1.9	-

[a]K^+-stimulated *p*-NPPase activity was inhibited by 39 to 94% at the test concentration by preincubation of the enzyme with the inhibitor for 20 min in the absence of dithiothreitol, DTT. Reversal of inhibition by the addition of 5 mM DTT to the incubation mixture is denoted by plus (+) and no significant effect of DTT is denoted by a minus (−). Compounds were tested at 1 to 10 μg/ml final concentration in the assay mixture (19).

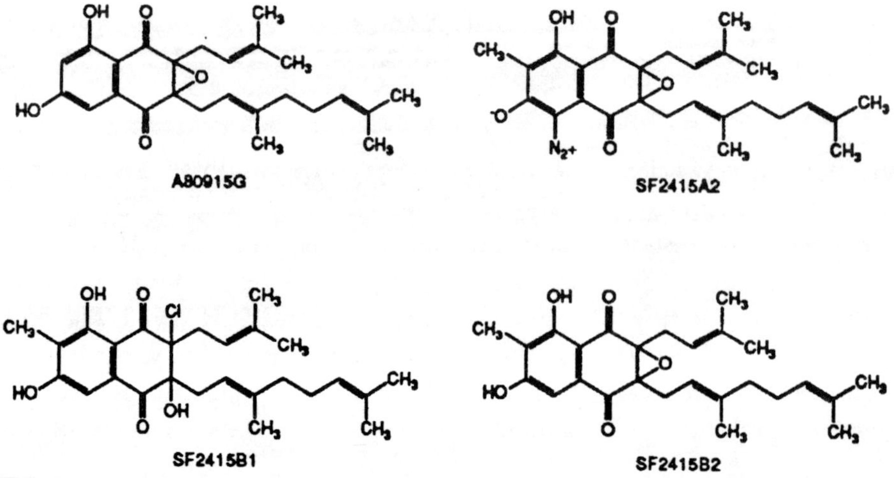

Figure 20.4 / A80915G and related antibiotics isolated from culture A80915 fermented in alternate medium (9).

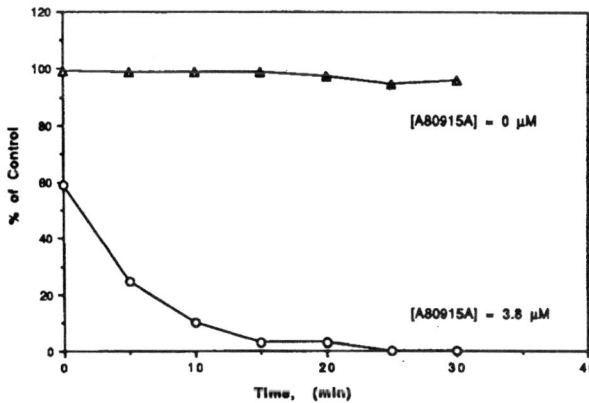

Figure 20.5 / Time-dependence of H^+-K^+-ATPase inhibition by A80915A. K^+-stimulated p-NPPase activity was determined after incubation of enzyme in the absence and presence of A80915A for increasing time periods.

A80915 produced antibiotics or by SCH 28080; whereas inhibition by omeprazole, which binds to enzyme thiol groups, was prevented (7, 19).

MECHANISM OF INHIBITION BY A80915A

The mechanism of action of H^+-K^+-ATPase inhibition by compound A80915A was chosen for further study (7). Enzyme inhibition was time-dependent (see Fig. 20.5). Incubation of enzyme with 3.8 µM A80915A resulted in approximately 90% inhibition within 10 minutes, whereas, the same incubation conditions in the absence of inhibitor had no effect. Inhibition by omeprazole, an irreversible inhibitor, also shows time-dependence (12, 14), whereas inhibition by the reversible inhibitor SCH 28080 is not time dependent (15).

Table 20.2 demonstrates the reversible nature of inhibition of H^+-K^+-ATPase by A80915A. IC_{50}'s were determined by incubation of enzyme with increasing drug concentrations, followed by no dilution or a 10-fold dilution of the preincubation mixture prior to measuring enzyme activity. A reversible inhibitor (such as SCH 28080) should give the same IC_{50} in both protocols, whereas an irre-

Table 20.2 / Reversibility of H^+-K^+-ATPase inhibition by A80915A.[a]

Compound	IC_{50} (µM)	
	No dilution	Diluted 10-fold
Omeprazole	4.0	0.3
SCH 28080	1.8	1.8
A80915A	1.5	2.0

[a]Enzyme was incubated with increasing drug concentrations for 30 min. Subsequently K^+-stimulated p-NPPase activity was measured after a 10-fold dilution of the incubation mixture or without dilution. Values listed are the IC_{50}'s determined by these two methods (7).

Table 20.3 / Effect of cations and sucrose on inhibition of H^+-K^+-ATPase by A80915A.[a]

Addition	Reduced Inhibition
KCl	+
RbCl	+
NH_4Cl	+
NaCl	−
LiCl	−
Choline Cl	−
Sucrose	−

[a]Enzyme was incubated 20 min with 3.8 µM A80915A in the presence of 20 mM of each of the salts or in the presence of 40 mM sucrose prior to measuring K^+-stimulated p-NPPase activity. Plus (+) denotes that the level of inhibition was significantly reduced and minus (−) denotes that the addition had no significant effect (7).

versible inhibitor (such as omeprazole) should give an apparent 10-fold lowering of IC_{50} after dilution. These two methods resulted in the same IC_{50} for A80915A, consistent with the action of a reversible inhibitor.

The presence of potassium was observed to slow the rate of inhibition of H^+-K^+-ATPase by A80915A (data not shown, (7)). The effect of various cations and sucrose on inhibition is shown in Table 20.3. Addition of osmotically equivalent concentrations of potassium chloride, rubidium chloride, or ammonium chloride during the coincubation of enzyme with inhibitor significantly reduced the level of inhibition by A80915A, whereas the addition of sodium chloride, lithium chloride, choline chloride, or sucrose did not significantly reduce the level of inhibition by A80915A. Thus the effect of potassium appears to be a direct effect on the gastric H^+-K^+-ATPase itself rather than an osmotic effect on the gastric membrane vesicles.

Taken together, these data suggest that the mode of action of A80915A is distinct from that of omeprazole and SCH 28080. The mode of action of A80915A appears similar to that of melittin, a polypeptide inhibitor present in bee venom (6). Melittin binds to the gastric H^+-K^+-ATPase at a site distinct from the luminal binding site of SCH 28080 (6). Binding prevents the conversion of the enzyme to a conformer that transports hydrogen ion through the membrane (4, 5).

IN VIVO ANTISECRETORY EVALUATION OF A80915A

The antisecretory activity of A80915A was evaluated in two *in vivo* tests and the results are shown in Table 20.4. In the pylorus ligated rat model, i.p. administration of A80915A inhibited basal acid secretion collected over a 4 hr time period in a dose dependent manner. In the gastric fistula dog model, intravenous administration of A80915A showed no significant antisecretory activity (19).

PUMILACIDINS/DAITOCIDINS

Other microbially-produced natural products have been found to be inhibitors of gastric H^+-K^+-ATPase. For example, pumilacidins A and B (Fig. 20.6), were isolated for antiviral activity against herpes simplex virus type I. Upon broader testing they were found to be more potent inhibitors of hog gastric H^+-K^+-ATPase than is omeprazole. Pumilacidin B afforded significant protection from gastric ulcer formation in the

Table 20.4 / Antisecretory activity of A80915A in *in vivo* rat and dog models.

Pylorus Ligated Rat Model[a]:

Drug	Dose (µmole/kg)	% Inh. of Acid Secretion
A80915A	50	78
	25	23
Omeprazole	50	95
	5	50

Gastric Fistula Dog Model[b]:

Drug	Dose (µmole/kg)	% Inh. of Acid Secretion
A80915A	0.64	NS
Omeprazole	1.0	94

[a]Drugs were given intraperitoneally and basal acid secretion was monitored for 4 hours with 6 rats per dose group (19).
[b]Drugs were given intravenously in 3 dogs (19).

Shay (pylorus ligated) rat model at a dose of 100 mg/kg s.c. (20). The pumilacidins are thought to be the same as the daitocidins (16), also produced by a *Bacillus* sp., which were found to be inhibitors of phospholipase A_2 as well as H^+-K^+-ATPase.

H^+-K^+-ATPase INHIBITORS: DIRECTED FERMENTATION SCREENING

The screening of fermentation broths and cyanophyte extracts was accomplished in our laboratory by increasing the buffering capacity of the H^+-K^+-ATPase assay. 12,821 broths and extracts were screened; of these 130 cultures were identified as producing inhibitors of H^+-K^+-ATPase. Inhibitory activity was produced by cultures of actinomycetes, molds, and cyanophytes (19).

Figure 20.6 / Structures of the pumilacidins (daitocidins) (20, 16).

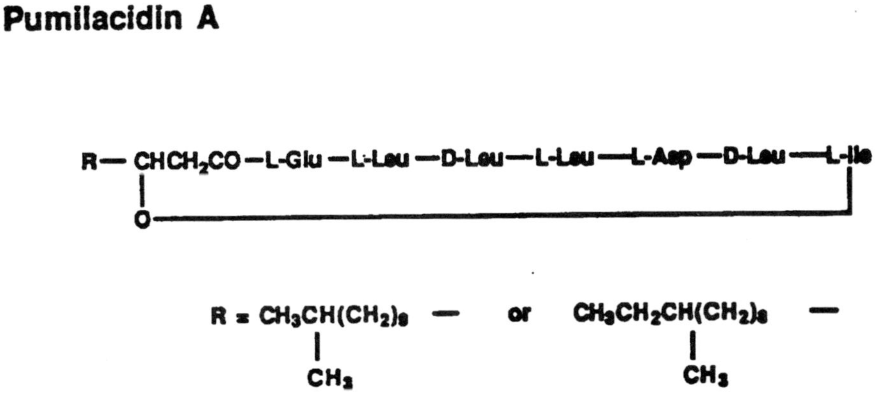

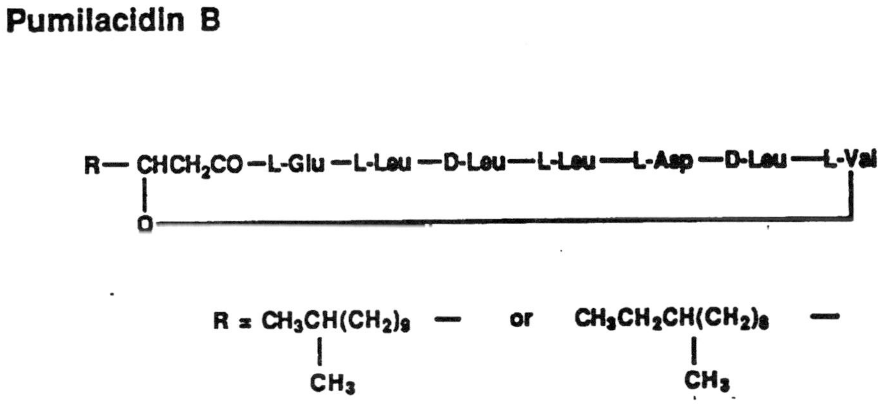

Pumilacidin A

R—CHCH$_2$CO—L-Glu—L-Leu—D-Leu—L-Leu—L-Asp—D-Leu—L-Ile
|
O——

R = CH$_3$CH(CH$_2$)$_8$—   or   CH$_3$CH$_2$CH(CH$_2$)$_8$—
         | |
        CH$_3$ CH$_3$

Pumilacidin B

R—CHCH$_2$CO—L-Glu—L-Leu—D-Leu—L-Leu—L-Asp—D-Leu—L-Val
|
O——

R = CH$_3$CH(CH$_2$)$_9$—   or   CH$_3$CH$_2$CH(CH$_2$)$_8$—
         | |
        CH$_3$ CH$_3$

ASSAY-GUIDED ISOLATION OF A88696 INHIBITORS

The isolation of H$^+$-K$^+$-ATPase inhibitors from culture A88696, *Streptomyces sclerotialus*, serves to illustrate the utility of the microtiter dish-based K$^+$-stimulated *p*-NPPase assay for the assay-guided purification of novel inhibitors. As shown in Figure 20.7, the activity was extracted from the mycelial mass and then fractionated by column chromatography on Sephadex LH-20. The assay had both the sample capacity and the reproduceability to permit separation of multiple factors. Figure 20.7 illustrates the sharp resolution of peaks that was obtained. Preparative RPHPLC final purification under unbuffered conditions yielded A88696 factors A, B, C, and D. Using an acidified mobile phase, RPHPLC purification of mycelial extract yielded factors C and F (19).

STRUCTURES AND ACTIVITIES OF A88696 FACTORS

A88696 "factors" A and B were shown to be the tautomeric salts of factors F and C respectively. Factors C and D crystallized and their structures were solved by x-ray crystallography. The structures of the A, B, C, and F factors were determined by chemical and spectroscopic means (2). The structures of A88696 factors C, D, and F and their potency as inhibitors of gastric H$^+$-K$^+$-ATPase are shown in Figure 20.8. The factors contain a spirotetronic acid within a macrocyclic ring, as do

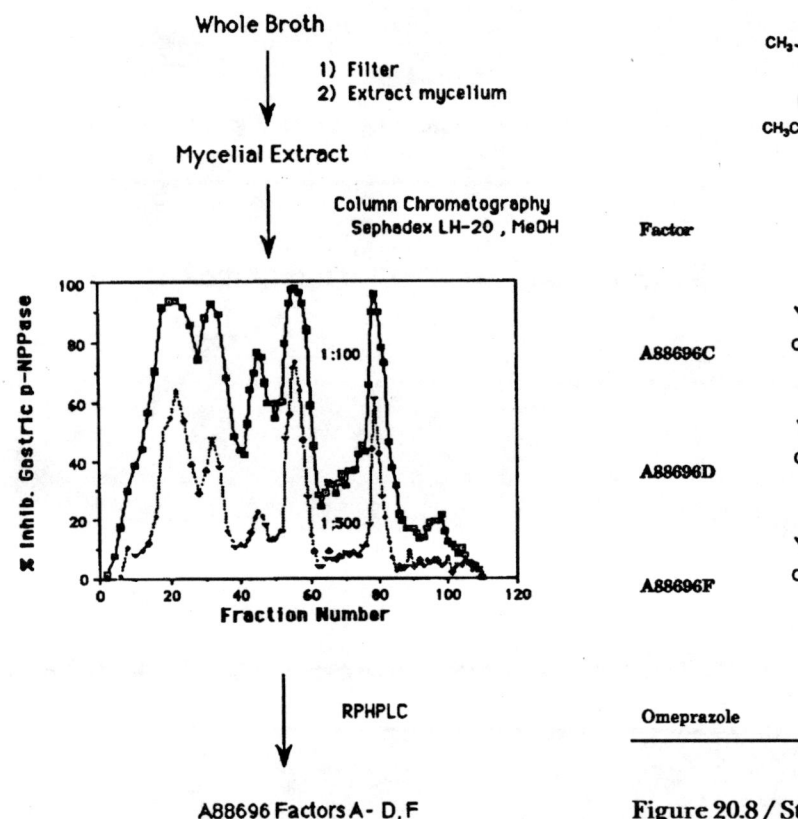

Figure 20.7 / Assay-guided isolation of H^+-K^+-ATPase inhibitors produced by culture A88696 (19).

Factor	Structure	IC_{50} (μM)	Reversal of inh. by DTT
A88696C		>116	ND
A88696D		4.9	-
A88696F		0.6	+
Omeprazole		2.8	+

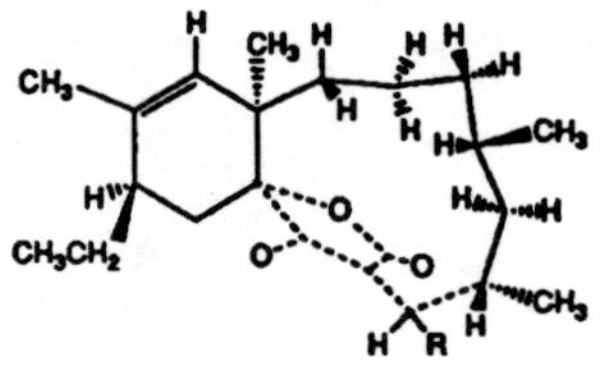

Figure 20.8 / Structures of A88696 factors and inhibition of H^+-K^+-ATPase activity. Reversal of inhibition by the presence of 5 mM DTT in the incubation mixture is denoted by a plus (+) and no significant effect of DTT is denoted by a minus (–) (19).

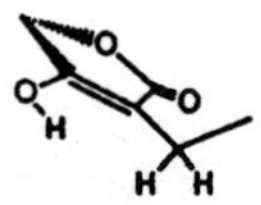

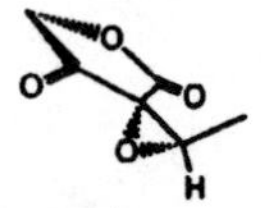

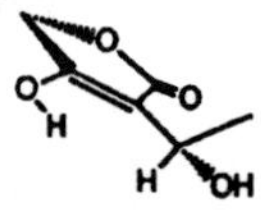

tetronolide (fig. 20.8), the aglycone of the tetrocarcins produced by *Micromonospora chalcea* (11), and kijanolide (fig. 20.9), the aglycone of kijanimycin produced by *Actinomadura kijaniata* (17). The A88696 factors do not contain the β-diketo functionality found in tetronolide and kijanolide.

A88696F is significantly more potent than omeprazole and factor D. Factor C is only slightly active, in part due to poor solubility. Interestingly the inhibition of gastric H^+-K^+-ATPase by A88696 factor F is appreciably diminished by incubation in the presence of the reducing agent dithiothreitol, indicating the involvement of enzyme

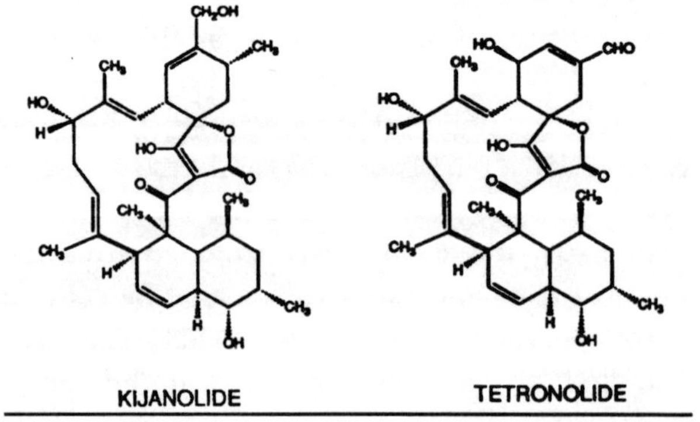

Figure 20.9 / Structures of related macrocyclic antitumor macrolides kijanolide and tetronolide (11, 17).

sulfhydryl groups. By contrast, the presence of DTT is without effect on inhibition by factor D (19).

SUMMARY

Microorganisms continue to be a rich source for the discovery of chemically novel compounds as inhibitors of enzymes. Purified new antibiotic factors produced by *Streptomyces aculeolatus* were found to be inhibitors of gastric H^+-K^+-ATPase in our *in vitro* assay, and to have potencies equal to or better than the clinically efficacious drug omeprazole in the assay. The mechanism of action of A80915A is distinct from that of omeprazole or SCH 28080. The acylpeptide pumilacidins also are potent inhibitors of H^+-K^+-ATPase, although studies on the mechanism of action have not been reported. Moreover, the direct screening of fermentation broths and cyanophyte extracts resulted in the detection and isolation of the structurally unique A88696 factors. Although inhibition by the most potent member of the A88696 series (A88696F) is prevented by the presence of reducing agent, inhibition by another factor (A88696D) is unaffected, indicating that these compounds differ in their action as inhibitors of H^+-K^+-ATPase. The discovery and the characterization of new classes of inhibitors such as these natural products may lead to a better understanding of the function of the proton pump. Ultimately the development of new clinical agents with different therapeutic profiles may be achieved.

REFERENCES

1. Berglindh T. and Sachs G. 1985. Emerging strategies in ulcer therapy: Pumps and receptors. *Scand J Gastroenterol.* 20: 7–14.
2. Bonjouklian R., Mynderse J. S., Deeter J. B. and Hunt A. H. 1991. Structures and chemistry of A88696 gastric H^+-K^+-ATPase inhibitors. American Chemical Society 1991 Joint Central-Great Lakes Regional Meeting, Indianapolis, Indiana, May 29–31, 1991. Abstract O-384.
3. Calam J. 1986. Future treatment of peptic ulcers. *Scand. J. Gastroenterol.* 20: 47–53.
4. Cuppoletti J. 1990. [^{125}I] Azidosalicylyl melittin binding domains-evidence for a polypeptide receptor on the gastric ($H^+ + K^+$)ATPase. *Arch. Biochem. Biophys.* 278: 409–415.
5. Cuppoletti J. and Abbott A. J. 1990. Interaction of melittin with the ($Na^+ + K^+$)ATPase: evidence for a melittin-induced conformational change. *Arch. Biochem. Biophys.* 283: 249–257.
6. Cuppoletti J., Blumenthal K. M. and Malinowska D. H. 1989. Melittin inhibition of the gastric ($H^+ + K^+$)ATPase and photoaffinity labeling with [^{125}I]azidosalicylyl melittin. *Arch. Biochem. Biophys.* 265: 263–270.
7. Dantzig A. H., Minor P. L., Garrigus J. L., Fukuda D. S. and Mynderse J. S. 1991. Studies on the mechanism of action of A80915A, a semi-naphthoquinone natural product, as an inhibitor of gastric (H^+-K^+)-ATPase. *Biochem. Pharmacol.* in press.
8. Forte J. G. and Wolosin J. M. 1987. "HCl secretion by the gastric oxyntic cell." Physiology of the Gastrointestinal Tract. Johnson. New York. Raven. 853–863.
9. Fukuda D. S., Baker P. J., Ott J., Counter R. T., Ensminger P. W., Allen N. E., Alborn W. E. Jr., Hobbs J. N. Jr. and Mynderse J. S. 1990. A80915, a new antibiotic complex produced by *Streptomyces aculeolatus*. Isolation, characterization, and antibacterial activity. *J. Antibiotics.* 43: 623–633.
10. Gomi S., Ohuchi S., Sasaki T., Itoh J. and Sezaki M. 1987. Studies on new antibiotics SF2415. II. The structural elucidation. *J. Antibiotics.* 40: 741–749.
11. Hirayama N., Kasai M., Shirahata K., Ohashi Y. and Sasada Y. 1980. The structure of tetronolide, the aglycone of antitumor antibiotic tetrocarcin. *Tetrahedron Lett.* 29: 6951–6954.

12. Im W. B., Sih J. C., Blakeman D. P. and McGrath J. P. 1985. Omeprazole, a specific inhibitor of gastric (H^+-K^+)-ATPase, is a H^+-activiated oxidizing agent of sulfhydryl groups. *J. Biol. Chem.* 260: 4591–4597.
13. Ito S. 1987. "Functional gastric morphology." Physiology of the Gastrointestinal Tract. Johnson. New York. Raven. 817–851. Second.
14. Keeling D. J., Fallowfield C. and Underwood A. H. 1987. The specificity of omeprazole as an ($H^+ + K^+$)-ATPase inhibitor depends upon the means of its activation. *Biochem. Pharmacol.* 36: 339–344.
15. Keeling D. J., Laing S. M. and Senn-Bilfinger J. 1988. SCH 28080 is a luminally acting, K^+-site inhibitor of the gastric ($H^+ + K^+$)-ATPase. *Biochem. Pharmacol.* 37: 2231–2236.
16. Koshino T., Suzuki K., Miyazaki S., Yamamoto H., Tsunoda S., Shikama H. and Ohta A. 1988. Daitocidin and its production. Jpn. Kokai. 255298: October 21, 1988.
17. Mallams A. K., McPhail A. T., Macfarlane R. D. and Stephens R. L. 1983. Kijanimicin. Part 3. Structure and absolute stereochemistry of kijanimicin. *J. Chem. Soc. Perkin Trans I.* 1497–1534.
18. McArthur K. E., Jensen R. T. and Gardner J. D. 1986. Treatment of acid-peptic diseases by inhibition of gastric H^+,K^+-ATPase. *Ann. Rev. Med.* 37: 97–105.
19. Mynderse J. S., Fukuda D. S., Foster R. S., Berry D. R., McKinney E. R., Godfrey O. W., Nakatsukasa W. M., Hunt A. H., Bonjouklian R., Minor P. L., Garrigus J. L. and Dantzig A. H. 1990. Microbial metabolites as gastric H^+/K^+-ATPase inhibitors. Second Interscience Conference on the Biotechnology of Microbial Products: Novel Pharmacological and Agrobiological Activities, Sarasota, Florida, October 14–17, 1990. Abstract S-24.
20. Naruse N., Tenmyo O., Kobaru S., Kamei H., Miyaki T., Konishi M. and Oki T. 1990. Pumilacidin, a complex of new antiviral antibiotics. Production, isolation, chemical properties, structure and biological activity. *J. Antibiotics.* XLIII: 267–280.
21. Robert A. 1987. "Effect of drugs on gastric secretion." Physiology of the Gastrointestinal Tract. Johnson. New York. Raven. 1071–1088. Second.
22. Ryberg B., Bishop A. E., Bloom S. R., Carlsson E., Hakanson R., Larsson H., Mattsson H., Polak J. M. and Sundler F. 1989. Omeprazole and ranitidine, antisecretagogues with different modes of action, are equally effective in causing hyperplasia of enterochromaffin-like cells in rat stomach. *Regul Peptides.* 25: 235–246.
23. Sachs G., Carlsson E., Lindberg P. and Wallmark B. 1988. Gastric H,K-ATPase as therapeutic target. *Ann. Rev. Pharmacol. Toxicol.* 28: 269–284.
24. Sachs G., Tache Y., Debas H. T., Jacobson E. D. and Freston J. W. 1987. Control of gastric secretion. *Am. J. Med.* 83: 307–328.
25. Shiomi K., Nakamura H., Iinuma H., Naganawa H., Isshiki K., Takeuchi T., Umezawa H. and Iitaka Y. 1986. Structures of new antibiotics napyradiomycins. *J. Antibiotics.* 39: 494–501.
26. Walan A. 1989. Clinical utility and safety of omeprazole. *Meth Find Exp Clin Pharmacol.* 11: 107–111.
27. Walan A., Bader J.-P., Classen M., Lamers C. R. H. W., Piper D. W., Rutgersson K. and Eriksson S. 1989. Effect of omeprazole and ranitidine on ulcer healing and relapse rates in patients with benign gastric ulcer. *New Eng. J. Med.* 320: 69–75.

21

Microbial Secondary Metabolites which Regulate Mammalian Cell Growth

Hiroyuki Osada and Kiyoshi Isono

Antibiotics Laboratory, RIKEN (The Institute of Physical and Chemical Research), Wako, Saitama 351-01, Japan

Key Words: Oncogene, Growth factor, Mitogen inhibitor, Protein kinase C inhibitor, Protein phosphatase inhibitor

INTRODUCTION

Recently, knowledge regarding the molecular basis of mammalian cell growth has been accumulated. It is now considered that most oncogene products (e.g., growth factors, growth factor receptors, and protein kinases) play essential roles in cell differentiation and proliferation (8, 10, 26). Therefore, regulators of cell growth factors and protein kinases are useful tools for cell biology as well as possible candidates of antitumor antibiotics (2, 30). For this reason, screening programs for the inhibitors of cell growth factors and for the regulators of protein phosphorylation from microbial origin were initiated. Described in this chapter are the screening methods for active compounds from microbial secondary metabolites, isolation of several new compounds and their activity against mammalian cells.

SCREENING FOR INHIBITORS OF CELL GROWTH

Epidermal growth factor (EGF) is one of the best characterized growth factors (27). EGF is a homologous protein to transforming growth factor alpha (TGF-α) and both growth factors share the same receptor which is encoded on proto-oncogene c-

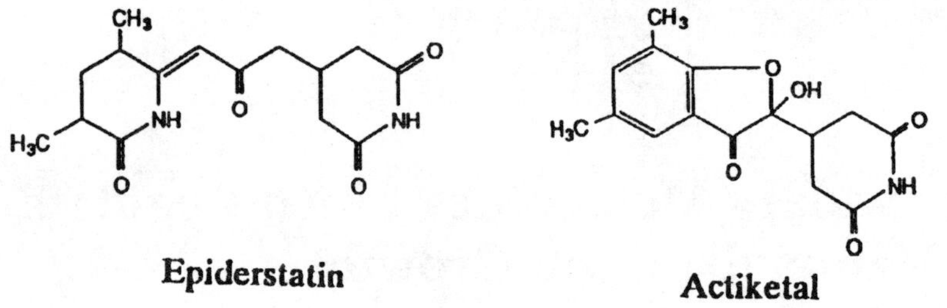

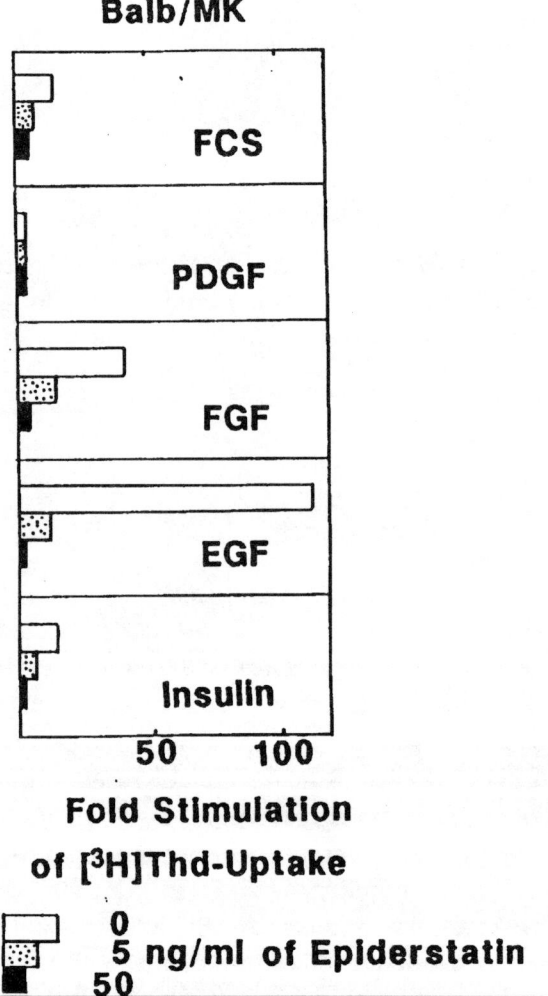

Figure 21.1

-*erb* B (6, 14). Therefore, the inhibitor of EGF-induced cell growth is expected to be a growth inhibitor of some tumor cells.

We used a mouse epidermal keratinocyte, Balb/MK cell (31) for the assay as a sensitive indicator cell to detect the inhibitors of mitogenic activity of EGF. The cell lacks the receptor of platelet derived growth factor (PDGF) and its growth depends on EGF.

Mitogenic activity of EGF was measured by the incorporation of [^{3}H]thymidine into quiescent Balb/MK cells (21). Ninety-six-multiwell plates were precoated with human fibronectin prior to seeding with Balb/MK cells. The cells were cultured in Eagle's low-Ca^{2+} minimal essential medium supplemented with 10% fetal bovine serum (FBS) and 5 ng/ml EGF. At the confluent stage, the medium was changed to Eagle's low-Ca^{2+} minimal essential medium supplemented with 5 µg/ml transferrin and Na_2SeO_3. Cells were incubated in the FBS/EGF free medium for 48 hours to arrest in G_0 phase of the cell cycle. Incorporation of [^{3}H]thymidine was monitored during a 5-hour period beginning 17-hour after addition of microbial broth samples with 10 ng/ml EGF. By use of this assay, three novel antibiotics which inhibit the EGF dependent incorporation of [^{3}H]thymidine were isolated.

Figure 21.2

EPIDERSTATIN AND ACTIKETAL Novel glutarimide antibiotics, epiderstatin (18, 23) and actiketal (25) (Fig. 21.1) were isolated from *Streptomyces pulveraceus* subsp. *epiderstagenes*. Epider-

Figure 21.3

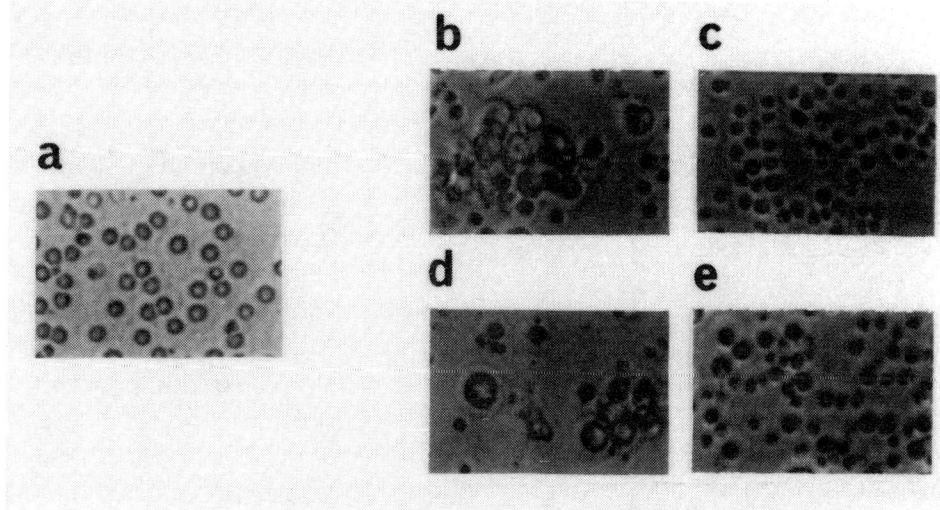

statin inhibited EGF and other peptide growth factors, such as basic fibroblast growth factor (FGF), PDGF, and insulin (Fig. 21.2). However, the antibiotic did not inhibit the tyrosine kinase activity of EGF receptor. As shown in Fig. 21.3, epiderstatin also inhibited the mitogenic activity of concanavalin A and lipopolysaccharide in mouse spleen cells (24). In addition, this antibiotic inhibited the transcription of c-myc mRNA in EGF stimulated Balb/MK cells and a human leukemia HL60 cell (Fig. 21.4). The transient expression of mRNA of nuclear oncogenes occurs as the early event of mitogen response of quiescent cells (9, 11).

In general, growth of virus transformed cells is faster than that of the normal cells. Most of the transformed cells are at S phase in the cell cycle, because the cells are actively growing and do not stay at G_0 phase. To see if epiderstatin arrests the cell cycle in G_0/G_1 phase, temperature sensitive Rouse sarcoma virus transformed rat kidney cells (src^{ts}-NRK) were used. Under permissive temperature (32°C), the cells grew rapidly and exhibited a round morphology. Cell population at S phase was about 60% in this condition. When the temperature was increased to 39°C, the cell growth decreased and the transformed morphology was reversed to the normal flat one. Cell population at S phase was only 28% at 39°C. When 10 ng/ml of

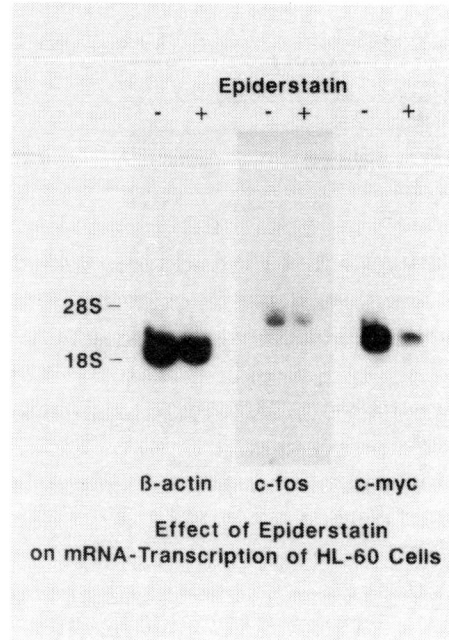

Figure 21.4

epiderstatin was added to the cell culture at 32°C, the growth rate decreased to the similar level observed at 39°C and cell population at S phase was about 32%. In this condition, the morphology of the cell is completely flat (Fig. 21.5). Biosynthesis of $p60^{v-src}$ was suppressed by the antibiotic without inhibition of $p60^{v-src}$ kinase activity.

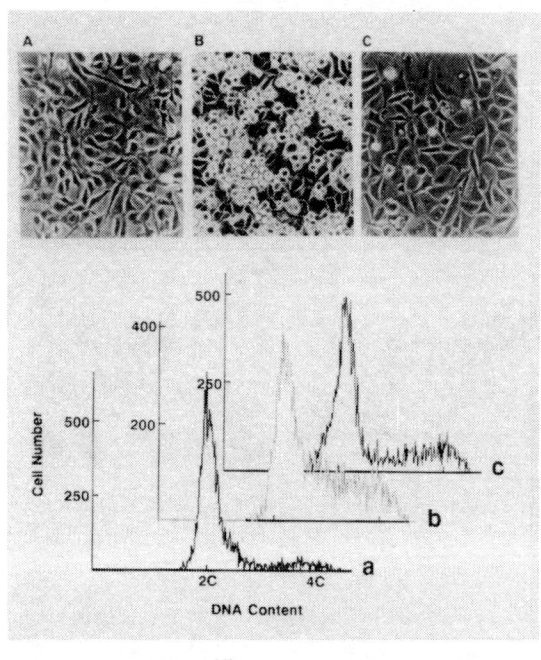

Figure 21.5

K562 cells were cultured in RPMI 1640 medium supplemented with 10% FBS and seeded at a density of 1×10^4 cells per well in 96-multiwell plates. Cell morphology was observed under a microscope after 30-minute treatment with microbial broth samples. Tumor promoting phorbol esters and teleocidins are known to activate protein kinase C (PKC) directly (3, 7). When phorbol dibutyrate (PDBu) is added to K562 cells, many blebs appeared on the cell surface within 10 min. This bleb forming activity correlated with increase of phosphorylated proteins.

It appeared that the bleb formation was induced not only by the activators of PKC but also by the inhibitors of protein phosphatases (13). Okadaic acid is a non-TPA type tumor promoter and is a strong inhibitor of protein phosphatase (29). Okadaic acid also induced the bleb formation (ED_{50} = 10 µg/ml). The inhibitors of PKC including staurosporine suppressed the bleb formation induced by PDBu (20).

REVEROMYCIN A Reveromycin A (Fig. 21.6) is a newly isolated polyketide type antibiotic (20), which inhibits the mitogenic activity of EGF (IC_{50} = 1 µg/ml). The biological activity is similar to that of epiderstatin with regard to inhibition of the cell cycle progression at G_0/G_1 phase as well as the reversion of transformed morphology of src^{ts}-NRK cells.

SCREENING FOR INHIBITORS OF PROTEIN PHOSPHORYLATION

Organization of cytoskeleton of mammalian cells is regulated by phosphorylation and dephosphorylation (15). A unique assay system, the bleb forming assay (16), was developed using a human leukemia cell K562 to detect regulators of protein phosphorylation.

TAUTOMYCIN AND TAUTOMYCETIN Two novel compounds, tautomycin (12) and tautomycetin (5) (Fig. 21.7) were found to induce bleb formation at the concentration of 10–30 µg/ml (ED_{50}). Both compounds were originally isolated as antifungal antibiotics in our laboratory (4, 5). Tautomycin apparently activated protein phosphorylation at cell level, however, it did not compete for binding of [^{3}H]-PDBu. It was reported that okadaic acid apparently activates the cellular protein phosphorylation without activation of PKC (22). Because of the structural similarity between tautomycin and okadaic acid, we compared the biological activity of the two compounds. With regard to bleb formation and phosphatase inhibition, the two compounds exhibited close similarities as illustrated in Fig. 21.8. Tautomycin inhibited the specific binding of [^{3}H]okadaic acid to protein phosphatase and also the phosphatase activity in vitro (IC_{50} = 2 ng/ml).

Figure 21.6

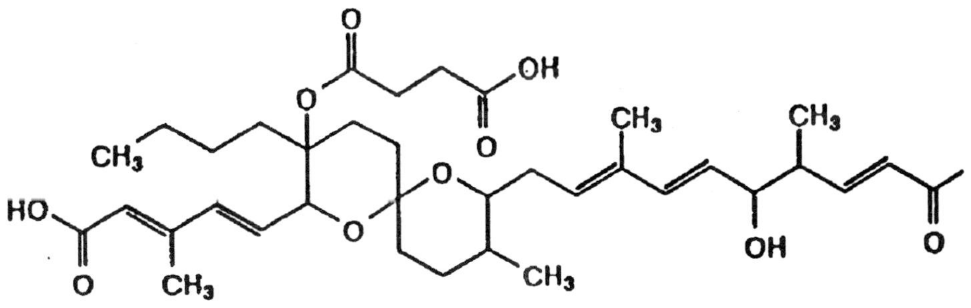

Reveromycin A

SANGIVAMYCIN AND RK-286C Sangivamycin (17) and RK-286C (19, 28) (Fig. 21.9) were identified as inhibitors of the bleb formation induced by PDBu at the IC_{50} 3.0 µg/ml and 1.0 µg/ml, respectively.

Among the three pyrrolopyrimidine antibiotics, sangivamycin, tubercidin, and toyocamycin, the inhibitory activity of sangivamycin against PKC is the strongest (Fig. 21.10). The carboxyamide group of sangivamycin is thought to be important for the activity.

RK-286C, a new member of the indolocarbazole antibiotic, was discovered as a new inhibitor of PKC. The chemical structure was established as 4'-demethylamino-4'-hydroxystaurosporine. However, the inhibition of the cell cycle progression of 3Y1 (a rat fibroblast) cells was different from that of staurosporine. Staurosporine arrested 3Y1 cells at G_1 phase of the cell cycle at a lower concentration (10 ng/ml) and at G_2 phase at a higher concentration (100 ng/ml) (1). The DNA content of the G_2

Figure 21.7

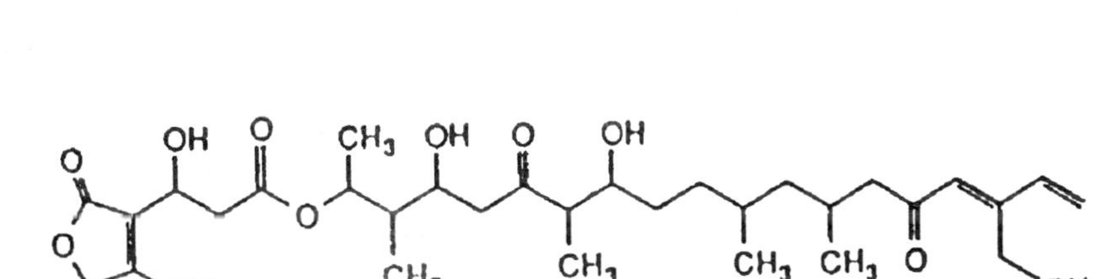

Tautomycin

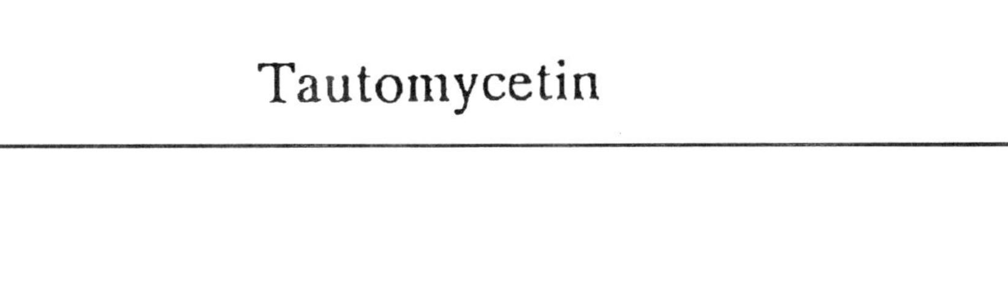

Tautomycetin

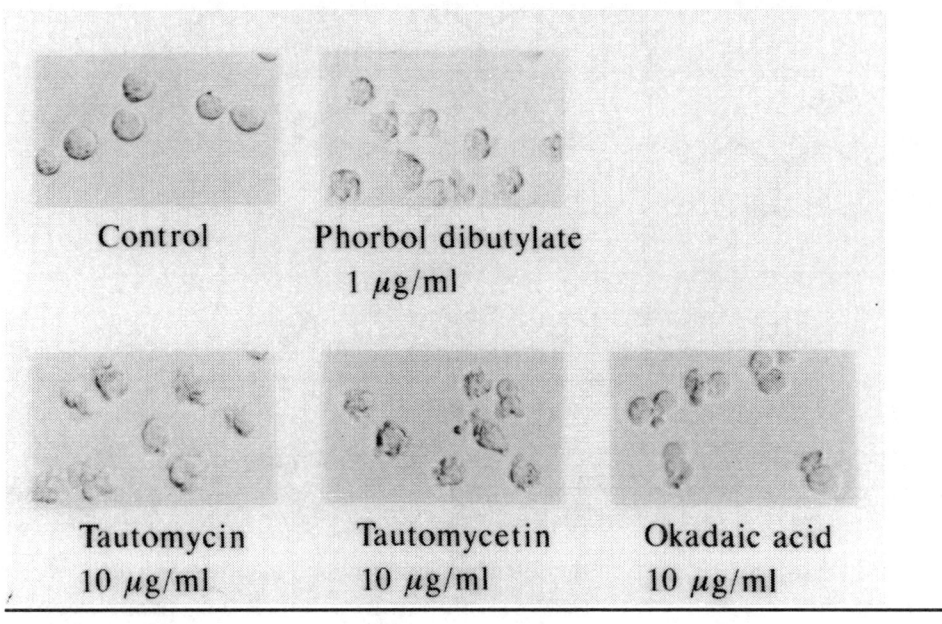

Figure 21.8

Figure 21.9

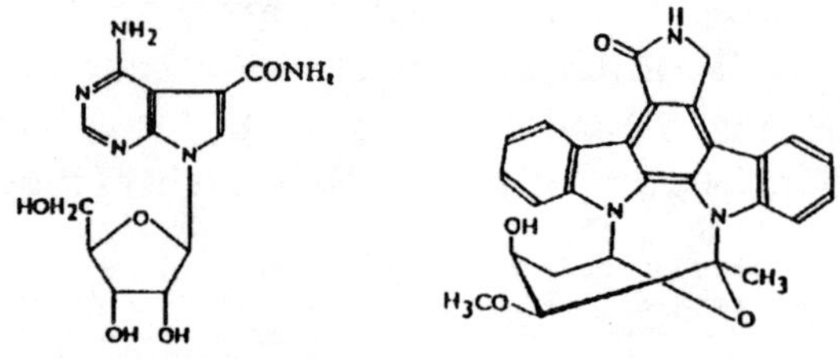

Figure 21.10

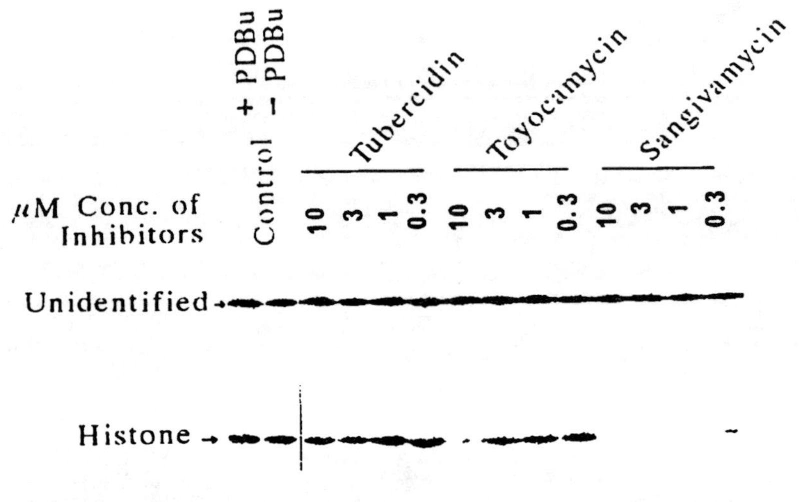

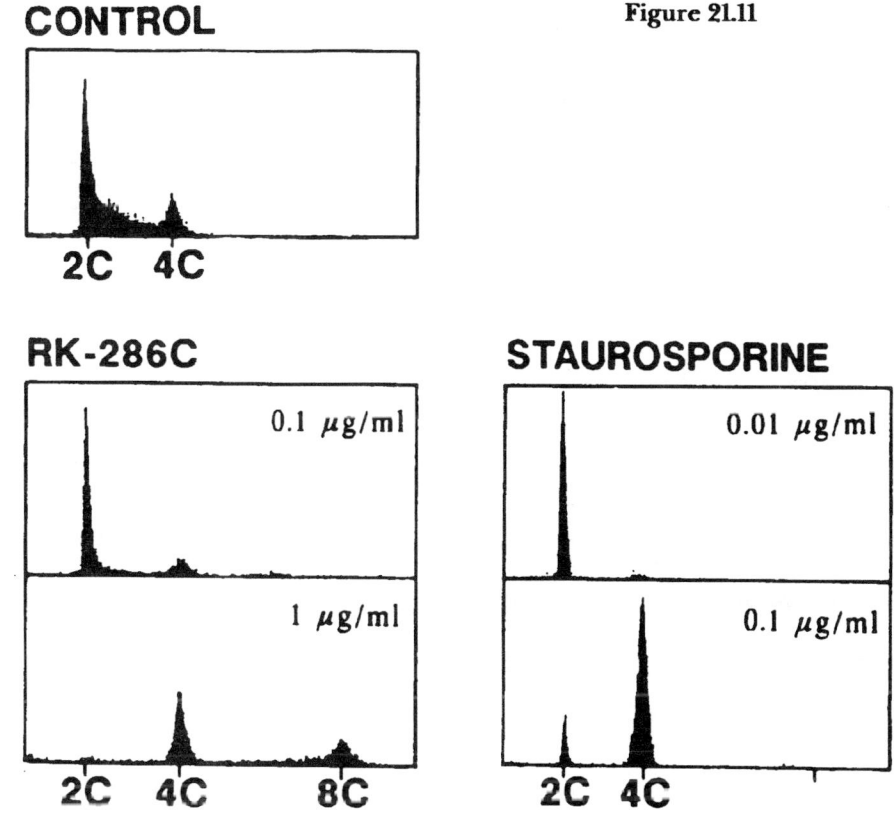

Figure 21.11

arrested cells is 2-fold that of G_1 cells in this condition. On the other hand, RK-286C treated cells were arrested at G_2 phase of the cell cycle and the DNA content of the cell is 4-fold that of the G_1 cells (Fig. 21.11). This observation suggests that a higher concentration of staurosporine inhibits the cell cycle progression at G_2 phase completely and that RK-286C inhibited the cell cycle at the same G_2 but allowed cell cycle progression without cell division.

SUMMARY

Two new screening systems for regulators of animal cell growth from microbial products were developed, one for the inhibitors of epidermal growth factor (EGF) signal transduction and the other for the regulators of protein phosphorylation. Epiderstatin, actiketal, and reveromycin A were isolated as novel inhibitors for mitogenic activity induced by EGF. Epiderstatin and reveromycin A arrested the cell cycle of temperature sensitive Rous sarcoma virus transformed rat kidney cell (src^{ts}-NRK cell) at G_0/G_1 phase, and induced the morphological reversion of transformed cells to flat normal cells at the permissive temperature.

During the screening for the regulators of protein phosphorylation, two structurally related antifungal antibiotics, tautomycin and tautomycetin were found to show interesting activity. The antibiotics induced the rapid morphological change of K562 human leukemia cells, which is termed "bleb" formation. A tumor promoter, okadaic acid, which is a strong inhibitor of protein phosphatase, induced blebs on the cell surface of K562. It was demonstrated that tautomycin is an inhibitor of protein phosphatase of mouse brain and shares the same binding site with okadaic acid.

Sangivamycin and a new indolocarbazole antibiotic, RK-286C, were isolated as inhibitors of protein kinase C. RK-286C inhibited the progression of the cell cycle at G_2 phase and caused the accumulation of DNA. These inhibitors are required as potential tools to study molecular mechanism of mammalian cell growth.

REFERENCES

1. Abe, K., M. Yoshida, T. Usui, S. Horinouchi and T. Beppu. 1991. Highly synchronous culture of fibroblasts from G2 block caused by staurosporine, a potent inhibitor of protein kinases. *Exp. Cell Res.* 192: 122–127.
2. Beppu, T. and M. Yoshida. 1989. Trichostatin and leptomycin: specific inhibitors of the G1 and G2 phases of the eukaryotic cell cycle. In *Novel Microbial Products for Medicine and Agriculture.* (Demain, A. L., G. A. Somkuti, J. C. Hunter-Cevera and H. W. Rossmoore, eds), pp. 73–78.
3. Castagna, M., Y. Takai, K. Kaibuchi, K. Sano, U. Kikkawa and Y. Nishizuka. 1982. Direct activation of calcium-activated, phospholipid-dependent protein kinase by tumor-promoting phorbol esters. *J. Biol. Chem.* 257: 7847–7851.
4. Cheng, X.-C., T. Kihara, H. Kusakabe, J. Magae, Y. Kobayashi, R.-P. Fang, Z.-F. Ni, Y.-C. Shen, K. Ko, I. Yamaguchi and K. Isono. 1987. A new antibiotic, tautomycin. *J. Antibiot.* 39: 606–608.
5. Cheng, X.-C., T. Kihara, X. Ying, M. Uramoto, H. Osada, H. Kusakabe, B.-N. Wang, Y. Kobayashi, K. Ko, I. Yamaguchi, and K. Isono. 1989. A new antibiotic, tautomycetin. *J. Antibiot.* 42: 141–144.
6. Downward, J., Y. Yarden, E. Mayes, G. Scrace, N. Totty, P. Stockwell, A. Ullrich, J. Schlessinger and M. D. Waterfield. 1984. Close similarity of epidermal growth factor receptor and v-erb-B oncogene protein sequences. *Nature* 307: 521–525.
7. Fukiki, H., M. Mori, M. Nakayasu, M. Terada, T. Sugimura and R. E. Moore. 1981. Indole alkaloids: Dihydroteleocidin B, teleocidin, and lyngbyatoxin A as members of new class of tumor promoters. *Proc. Natl. Acad. Sci. U.S.A.* 78: 3872–3876.
8. Goustin, A. C., E. D. Leof, G. D. Shipley and H. L. Moses. 1986. Growth factors and cancer. *Cancer Res.* 46: 1015–1029.
9. Greenberg, M. E. and E. B. Ziff. 1984. Stimulation of 3T3 cells induces transcription of the c-fos proto-oncogene. *Nature* 311: 433–442.
10. Heldin, C.-H. and B. Westermark. 1984. Growth factors: Mechanism of action and relation to oncogenes. *Cell* 37: 9–20.
11. Kelly, K., B. H. Cochran, C. D. Stiles and P. Leder. 1983. Cell-specific regulation of the c-myc gene by lymphocyte mitogens and platelet-derived growth factor. *Cell* 35: 603–610.
12. Magae, J., C. Watanabe, H. Osada and K. Isono. 1988. Induction of morphological change of human myeloid leukemia and activation of protein kinase C by a novel antibiotic, tautomycin. *J. Antibiot.* 41: 932–937.
13. Magae, J., H. Osada, H. Jujiki, T. C. Saido, K. Suzuki, K. Nagai, M. Yamasaki and K. Isono. 1990. Induction of morphological change of human myeloid leukemia K562 cells by tautomycin through the inhibition of protein phosphatases. *Proc. Jpn. Acad.* 66: 209–212.
14. Marquardt, H., M. W. Hankapiller, L. E. Hood, D. W. Twardzik, J. E. De Larco, J. R. Stephenson and G. J. Todaro. 1983. Transforming growth factors produced by retrovirus-transformed rodent fibroblasts and human melanoma cells: Amino acid sequence hololology with epidermal growth factor. *Proc. Natl. Acad. Sci. U.S.A.* 80: 4684–4688.
15. Miyata, Y., E. Nishida, S. Koyasu, I. Yahara and H. Sakai. 1989. Protein kinase C-dependent and -independent pathways in the growth factor-induced cytoskeletal reorganization. *J. Biol. Chem.* 264: 15565–15568.
16. Osada, H., J. Magae, C. Watanabe and K. Isono. 1988. Rapid screening method for inhibitors of protein kinase C. *J. Antibiot.* 41: 925–931.

17. Osada, H., T. Sonoda, K. Tsunoda and K. Isono. 1989. A new biological role of sangivamycin: inhibition of protein kinases. *J. Antibiiot.* 42: 102–106.
18. Osada, H., T. Sonoda, H. Kusakabe and K. Isono. 1989. Epiderstatin, a new inhibitor of the mitogenic activity of epidermal growth factor. I. Taxonomy, fermentation, isolation and characterization. *J. Antibiot.* 42: 1599–1606.
19. Osada, H., H. Takahashi, K. Tsunoda, H. Kusakabe and K. Isono. 1990. A new inhibitor of protein kinase C, RK-286C (4'-demethylamino-4'-hydroxystaurosporine). I. Screening, taxonomy, fermentation and biological activity. *J. Antibiot.* 43: 163–167.
20. Osada, H., H. Koshino, K. Isono, H. Takahashi and G. Kawanishi, 1991. Reveromycin A, a new antibiotic which inhibits the mitogenic activity of epidermal growth factor. *J. Antibiot.* 44: 259–261.
21. Rubin, J. F., H. Osada, P. W. Finch, W. G. Taylor, S. Rudikoff and S. A. Aaronson. 1989. Identification and characterization of a novel growth factor specific for epithelial cells. *Proc. Natl. Acad. Sci. U.S.A.* 86: 802–806.
22. Sassa, T., W. W. Richter, N. Uda, M. Suganuma, H. Suguri, S. Yoshizawa, M. Hirota and H. Fujiki. 1989. *Biochem. Biophys. Res. Commun.* 159: 939–944.
23. Sonoda, T., H. Osada, M. Uramoto, J. Uzawa and K. Isono. 1989. Epiderstatin, a new inhibitor of the mitogenic activity of epidermal growth factor. II. Structure elucidation. *J. Antibiot.* 42: 1607–1609.
24. Sonoda, T., H. Osada, J. Magae and K. Isono. 1990. Epiderstatin and its related glutarimide antibiotics inhibits the cell growth induced by mitogen stimulation. *Agric. Biol. Chem.* 54: 1259–1263.
25. Sonoda, T., H. Osada and K. Isono. 1991. Actiketal, a new member of the glutarimide antibiotics. *J. Antibiot.* 44: 1760–163.
26. Sporn, M. B. and A. Roberts. 1985. Autocrine growth factors and cancer. *Nature* 313: 745–747.
27. Stoscheck, C. M. and L. E. King, Jr. 1986. Role of epidermal growth factor in carcinogenesis. *Cancer Res.* 46: 1030–1037.
28. Takahashi, H., H. Osada, M. Uramoto and K. Isono. 1990. A new inhibitor of protein kinase C, RK-286C (4'-demethylamino-4' hydroxystaurosporine). II. Isolation, physico-chemical properties and structure. *J. Antibiot.* 43: 163–173.
29. Takai, A., C. Bialojan, M. Troschka and J. C. Ruegg. 1987. Smooth muscle myosin phosphatase inhibition and force enhancement by black sponge toxin. *FEBS Lett.* 217: 81–84.
30. Umezawa, K., M. Hori and T. Takeuchi. 1989. Microbial secondary metabolites inhibiting oncogene functions. In *Novel Microbial Products for Medicine and Agriculture.* (Demain, A. L., G. A. Somkuti, J. C. Hunter-Cevera and H. W. Rossmoore, eds), pp. 57–62.
31. Weissman, B. E. and S. A. Aaronson. 1983. Balb and Kirsten murine sarcoma viruses alter growth and differentiation of EGF-dependent Balb/c mouse epidermal keratinocyte lines. *Cell* 32: 599–606.

22

The Starfish Embryo Assay Useful for Screening of New Inhibitors of RNA Synthesis

Susumu Ikegami

Faculty of Applied Biological Sciences, Hiroshima University, 1-chome, Kagamiyama, Saijo-cho, Higashihiroshima-shi, Hiroshima 724, Japan

A selective assay system has been developed using oocytes and embryos of the starfish, Asterina pectinifera. This is a useful probe for detecting bioactive microbial metabolites that affect biochemical events taking place in eukaryotes; e.g., protein synthesis, nucleic acid synthesis, microtubule formation, actin polymerization, and others.

Fertilized starfish eggs were examined for their ability to undergo the early events of embryonic development in the presence of the culture broth conditioned by many bacterial strains. A strain of Streptomyces albus *produced a substance which halts embryonic development specifically just after completion of blastulation. The causative substance was purified to homogeneity and was found to be a nucleoside analogue which inhibited RNA synthesis in starfish embryos at blastulation. These results exemplify the usefulness of the starfish embryo assay for the detection of specific inhibitors of RNA synthesis.*

INTRODUCTION

There are many microbial metabolites that inhibit cell division of eukaryotes. However, most of them show a broad spectrum of activity, most often accompanied by non-specific toxicity toward mammalian cells. It was therefore desirable to adopt an assay method which was specific or selective to a particular activity as well as simple. Recently, we devised an assay system using starfish oocytes and fertilized sea urchin eggs to discover potential antitumor drugs from microbial metabolites [4]. This assay method can detect such selective agents as DNA synthesis inhibitors, RNA synthesis inhibitors, and microtubule assembly inhibitors, which may lead to development of anticancer drugs. This paper describes the starfish embryo assay method to detect RNA synthesis inhibitors and isolation of an active microbial metabolite found by the starfish embryo assay.

STARFISH OOCYTE MATURATION AND EARLY EMBRYONIC DEVELOPMENT

The eggs of the starfish *Asterina pectinifera* are quite desirable for studying cell division, since they can be obtained in great numbers throughout the year in Japan and are larger than sea urchin eggs. Oocytes in a ripe ovary are arrested at the germinal vesicle stage of first meiotic division (Fig. 22.1A). Addition of 1-methyladenine at a final concentration of 150 ng/ml to the oocytes results in synchronous oocyte maturation [7]; germinal vesicle disintegrated at 40 minutes at 20°C after the start of 1-methyladenine treatment (Fig. 22.1B). At this stage, the oocytes are fertilizable and are called eggs. The first polar body is expelled 1 hour after the start of 1-methyladenine treatment and the second one 70 minutes later. The optimum period for insemination is between germinal vesicle breakdown and the formation of the first polar body; insemination after the formation of the first polar body results in high frequency of polyspermy which leads to abnormal embryonic development. The formation of the fertilization membrane that occurs within 5 minutes after fertilization (Fig. 22.1C) and two polar bodies are formed in precise time (Fig. 22.1D). The first cleavage occurs at 175 minutes after the start of 1-methyladenine treatment (Fig. 22.1E). Blastomeres of the embryo divide rapidly and synchronously without growth for a total of eight to nine cleavages (Fig. 22.1F). Completion of these rapid cleavage periods is followed by the immediate activation of a new developmental program, blastulation, which occurs at 4 hours after the first cleavage. Blastulation involves active morphological movement of embryonic cells, during which cells acquire the highly cooperative nature of epithelial cells and become packed into a sheet, which forms a sphere (Fig. 22.1G). In the midblastula stage prior to hatching, the synchrony of cell division vanishes simultaneously with the formation of cilia, then the division proceeds asynchronously. Blastulae hatch at 11.5 hours after the first cleavage and swimming embryos perform a complex series of gastrulation movement, which begins at 2 hours after hatching. This dramatic process transforms the simple hollow ball of cells into a multilayered structure with a central axis and bilateral symmetry: by a complicated invagination, a large area of cells on the outside of the embryo is brought to lie inside it (Fig. 22.1H).

INHIBITORS OF STARFISH OOCYTE MATURATION AND EMBRYONIC DEVELOPMENT

Oocyte maturation and early embryonic development were examined in the presence of various microbial metabolites that affect cellular functions in eukaryotes.

Amphotericin B, a polyene antibiotic produced by *Streptomyces nodosus*, has been shown to form an adduct with sterols in the plasma membrane resulting in cytolysis in animal cells [4]. Germinal vesicle breakdown did not take place in oocytes in the presence of 1-methyladenine if the culture medium contained amphotericin B at concentrations greater than 1 µg/ml. Germinal vesicle breakdown was also prevented by inhibitors of protein synthesis such as cycloheximide, produced by *Streptomyces griseus* (100 µg/ml), and those of the electron transport system such as antimycin A_1, produced by *Streptomyces* sp. (0.1 µg/ml).

Although meiosis was reinitiated by the addition of 1-methyladenine as revealed by germinal

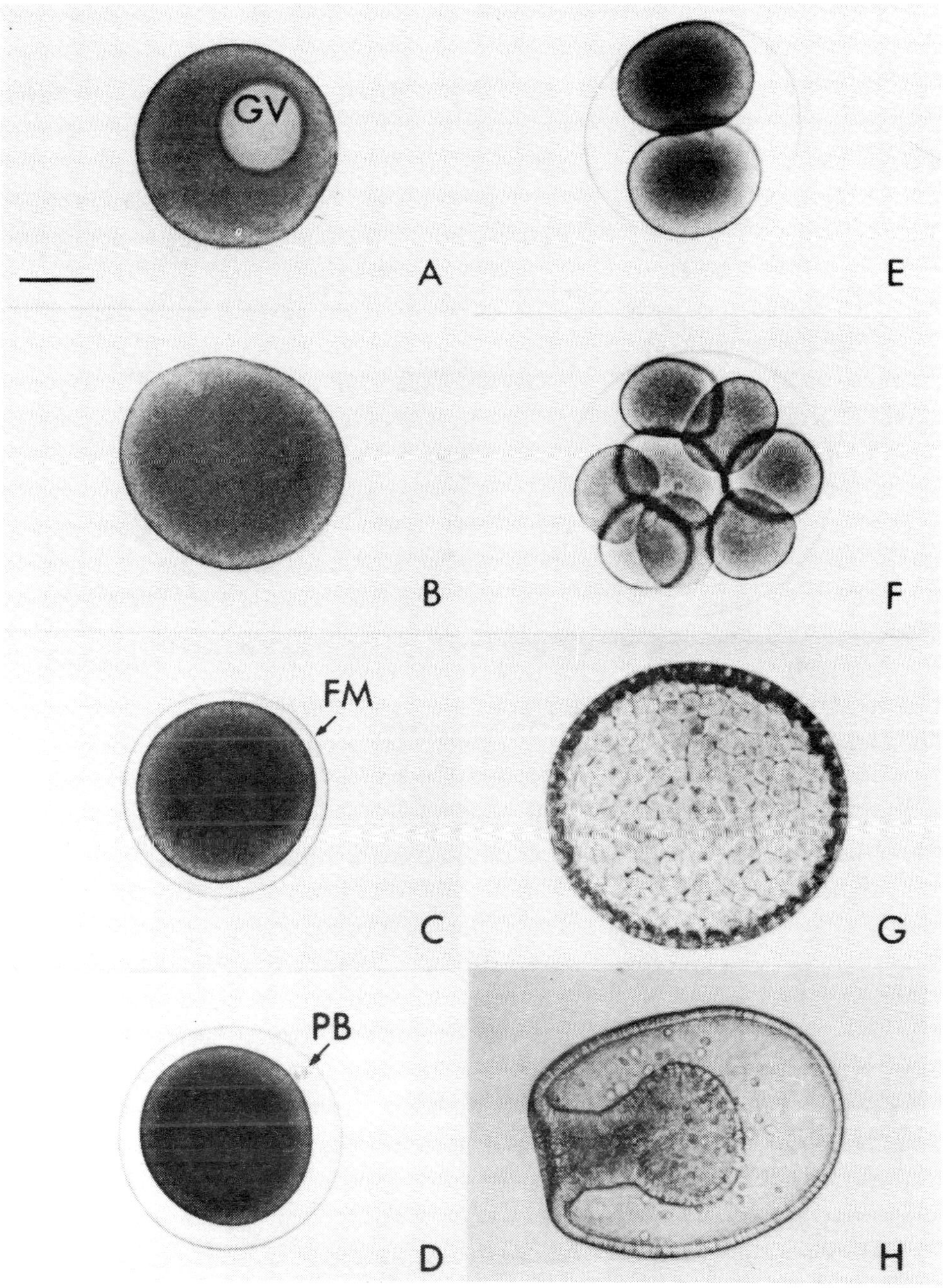

Figure 22.1

vesicle breakdown, the formation of polar bodies was prevented in oocytes which were exposed to inhibitors of microtubule assembly such as ansamytosin produced by *Actinosynnema brctiosum* (0.025 µg/ml) or those of microfilament formation such as cytochalasin B produced by *Phoma* sp. (100 µg/ml) [10].

Inhibitors of nucleic acid biosynthesis did not affect the processes of oocyte maturation since chromosomal replication does not occur during meiotic maturational divisions and the RNAs required for protein synthesis in oocytes and eggs are synthesized during oogenesis and stored in the cytoplasm [5].

Fertilized eggs cultured in the presence of 10 µg/ml aphidicolin, a selective inhibitor of DNA synthesis produced by *Harziella entomophila*, underwent cell division, despite an almost total absence of DNA synthesis, up to the 512-cell stage [8]. The treated embryos never blastulated and died when control embryos reached the early blastula stage. In the presence of aphidicolin, association of chromosomes with the mitotic apparatus was interrupted. Therefore, chromosomes were not distributed in newly formed blastomeres at the first cleavage, resulting in formation of blastomeres that were devoid of nuclei and chromosomes [11]. However, they were still capable of dividing through the formation of spindles and asters.

Blastomeres of embryos which were treated with 10µg/ml methotrexate (amethopterin), a synthetic analogue of folic acid and an inhibitor of thymidine 5'-monophosphate biosynthesis, divided achromosomally up to the 512-cell stage starting from the 64-cell stage, when the intracellular pool of thymidine 5'-triphosphate became exhausted [2, 9].

The rate of RNA synthesis in starfish embryos is quite low during the cleavage period and becomes high at blastulation. Concomitant with this increase in the rate of RNA synthesis, the pool of nucleoside 5'-triphosphates becomes larger. Addition of 25 µg/nl formycin, an inhibitor of RNA synthesis obtained from *Nocardia interforma*, to a culture of starfish embryos interfered with the formation of uridine 5'-triphosphate and cytidine 5'-triphosphate, thereby inhibiting RNA synthesis and halting embryonic development at the early blastula stage [6]. However, nuclei were present in blastomeres of embryos whose development had been halted by formycin at the early blastula stage. Therefore, it is easily possible to distinguish between inhibitors of RNA synthesis and those of DNA synthesis by observing the presence of nuclei in blastomeres of treated embryos in which development has been arrested just before or after completion of blastulation.

Embryos reared in the presence of 10 µg/ml tunicamycin, a product of *Streptomyces lysosuperificus* and an inhibitor of the glycosylation of many membrane proteins, developed normally passing through the blastula stage but not beyond the early gastrula stage [1].

EFFECTS OF ACTINOMYCIN D ON RNA SYNTHEISIS IN STARFISH EMBRYOS

The effect of actinomycin D, a product of *Streptomyces antibioticus* and a widely used inhibitor of RNA synthesis, was examined on RNA synthesis occurring in starfish embryos and the developmental consequences of the resulting inhibition [5]. The finding that actinomycin D at the minimum effective concentration necessary to block RNA synthesis at blastulation in early starfish embryos, 25µg/ml, prevented embryonic development just before the onset of blastulation indicated that embryonic transcription is essential for progression of development up to the blastula stage in this species. Since this is in striking contrast to the case of formycin as described, it seems likely that starfish embryos are very susceptible to some action of actinomycin D other than inhibition of RNA synthesis. When the chromatin of actinomycin D-treated embryos was analyzed, it disintegrated specifically at the 32-cell stage, although blastomeres kept dividing achromosomally until just before the beginning of blastulation. At the 8- to 16-cell stages, actinomycin D inhibited DNA synthesis severely but did not affect RNA synthesis at all. Generally, the main drawback of experiments involving the introduction of actinomycin D to cell or tissue cultures to block RNA synthesis is that the effect is usually irreversible, therefore leaving ambiguity in concluding that the observed effect is due to inhibition of RNA synthesis. The starfish embryo assay demonstrates clearly that the undesirable side effect of actinomycin D is quite large.

SCREENING OF MICROBIAL EXTRACTS BY THE STARFISH EMBRYO ASSAY

A search for microbial products capable of arresting the embryonic development of fertilized starfish eggs at the early blastula stage was implemented as follows.

Conditioned culture broth specimens of 3,000 different microbial strains were prepared. Each broth filtrate was diluted 640 times with artificial seawater to produce a dilute sample solution. A small number of fertilized starfish eggs were placed in the diluted sample solution immediately after fertilization. A total of 232 different diluted sample solutions inhibited the fertilized eggs to reach the gastrula stage. The other 2,726 diluted sample solutions inhibited the fertilized eggs to reach the gastrula stage. Filtrates of culture broths that produced inhibition of embryonic development after a 640-fold dilution were diluted ten times with artificial seawater, and applied to suspensions of starfish oocytes. Two hours later, 1-methyladenine was added to each solution to give a final concentration of 150 ng/ml. Initiation of oocyte maturation was monitored by the occurrence of germinal vesicle breakdown in the oocyte. Forty minutes after application of 1-methyladenine, the maturing oocytes or eggs were inseminated. Those filtrates which caused oocyte lysis or prevented germinal vesicle breakdown in 1-methyladenine-treated oocytes were considered to contain cytotoxic substances or substances having actions other than inhibition of RNA synthesis, since oocytes are quiescent cells and germinal vesicle breakdown is independent of RNA synthesis. These culture broths were not investigated further. Culture filtrates that prevented fertilization or early cleavages were also discarded, since RNA synthesis is not required for these cytological events. Only 5 filtrates produced inhibition of embryonic development over a range of 10- to 640-fold dilution at or shortly after the completion of blastulation (256- to 1,024-cell stage). However the process of oocyte maturation, fertilization and early cleavages were not affected. Of these 5 filtrates, two showed reproducible inhibition of embryonic development. One filtrate was the product of a soil isolate identified as a subspecies of *Streptomyces albus* (A282).

ISOLATION OF A NEW INHIBITOR OF RNA SYNTHESIS

The active component present in culture broth produced by *Streptomyces albus* A282 was purified to homogeneity by successive chromatographies using Diaion HP-20 (eluent: methanol), silanized silica gel (5% aqueous methanol), alumina (70% methanol) and Robar RP-8 (10% methanol). The molecule (1) had not been found in nature and its structure was identified by Otter *et al.* as 7-ß-D-ribofuranosyl-4-amino-[3,2-*d*]pyrimidine [3].

Fertilized starfish eggs were placed in artificial seawater containing 1 at a concentration of 0.8 µg/ml or greater immediately after fertilization. Cleavages were normal and showed the same timing in both 1-treated and control embryos up to the 32-cell stage. The sixth cleavage was delayed, however. The ninth cleavage occurred 7.6 hours after fertilization; the embryo failed to cleave further and remained at the beginning of the early blastula stage. It was notable that the nucleus was present in each blastomere of the treated, development-arrested embryos, suggesting that chromosomal DNA synthesis was not affected by the treatment. 1 inhibited RNA synthesis of the blastulating embryo. Furthermore, microinjection of 1 into the cytoplasm of *Xenopus laevis* oocytes, which contain huge amounts of ribonucleoside 5'-triphosphates, suppressed RNA synthesis, suggesting that the RNA polymerase reaction was inhibited by 1 or its active metabolite. 1 markedly inhibited the proliferation of mouse carcinoma FM3A cells, suggesting that 1 arrests the mitotic division of animal cells through blockage of RNA synthesis.

CONCLUSION

A selective assay system using fertilized starfish eggs appears to be a useful probe for detecting microbial metabolites that inhibit RNA synthesis but not DNA synthesis. This has been shown by our isolation of a nucleoside analogue from a culture of *Streptomyces albus*. Further studies, in which more sophisticated systems are involved, are required to establish the site and mode of action. It is desirable to use the starfish embryo assay in combination with cell growth inhibition tests using mammalian cell lines to discover substances which may have activity against tumor cells without nonspecific toxicity.

ACKNOWLEDGEMENTS

This study was carried out in collaboration with Drs. H. Ohkishi, T. Yugami, T. Matsuzaki, T. Hayase, Y. T. Osano, T. Mikawa, H. Isomura and N. Tsuchimori. This study was supported in part by a Grant-in-Aid for Cancer Research from the Ministry of Education, Science and Culture, Japan (02151037).

REFERENCES

1. Dan-Sohkawa, M., G. Tamura and H. Mitsui. 1980. Mesenchyme cells in starfish development: effect of tunicamycin of their differentiation, migration and function. *Dev. Growth Differ.* 22: 495–502.
2. Ikegami, S., J. Imayoshi, N. Taahashi and H. Nagano. 1985. Dihydrofolate reductase in starfish oocytes and embryos: developmental consequences of its inhibition by methotrexate. *Dev. Growth Differ.* 27: 393–403.
3. Otter, B. A., S. A. Patil, R. S. Klien and S. E. Ealickt. 1991. A corrected structure for "pyrrolosine." *J. Am. Chem. Soc.* in press.
4. Ikegami, S., K. Kawada, Y. Kimura and A. Suzuki. 1979. A rapid and convenient procedure for the detection of inhibitors of DNA synthesis using starfish oocytes and sea urchin embryos. *Agric. Biol. Chem.* 43: 161–166.
5. Ikegami, S., Y. Ozaki, Y. Ooe and N. Itoh. 1991. Achromosomal cleavage of early starfish embryos cultured in the presence of actinomycin D. *Dev. Growth Differ.* 33: 193–200.
6. Isomura, H., N. Itoh and S. Ikegami, 1989. RNA synthesis in starfish embryos: developmental consequences of its inhibition by formycin. *Biochim. Biophys. Acta* 1007: 343–349.
7. Kanatani, H. 1973. Maturation-inducing substance in starfishes. *Int. Rev. Cytol.* 35: 253–298.
8. Nagano, H., S. Hirai, K. Okano and S. Ikegami. 1981. Achromosomal cleavage of fertilized starfish eggs in the presence of aphidicolin. *Develop. Biol.* 85: 40–415.
9. Nagano, H., K. Okano and S. Ikegami. 1983. Changes in deoxyribonuecleoside triphosphate pools in the starfish oocyte during maturation and early embryogenesis. *Exp. Cell Res.* 145: 219–222.
10. Tsuchimori, N., S. Miyashiro, T. Tsuji, T. Kida, H. Shibai and S. Ikgami. 1986. Development of fertilized starfish eggs in which cytokinesis is prevented by iturin A-2. *Dev. Growth Differ.* 28: 619–627.
11. Yamada, H., S. Hirai, S. Ikegami, Y. Kawarada, E. Okuhara and H. Nagano. 1985. The fate of DNA originally existing in the zygote nucleus during achromosomal cleavage of fertilized echinoderm eggs in the presence of aphidicolin: microscopic studies with anti-DNA antibody. *J. Cell. Biol.* 124: 9–12.

23

New Anthracyclines Effective on Adriamycin-Resistant Cell Lines

Tskeshi Uchida[1], Noboru Otake[2] and Tomio Takeuchi[3]

Authors' affiliations: [1]Pharmaceutical Research Laboratory, Kirin Brewery Co., Ltd., Maebashi-shi, Gunma 371, Japan; [2]Teikyo University, Utsunomiya-shi, Tochigi 320, Japan; [3]Institute of Microbial Chemistry, Shinagawa-ku, Tokyo 141, Japan

Multidrug resistance is one of the major problems in cancer chemotherapy. In screening for antitumor substances effective on multidrug-resistant cells which expressed P-glycoprotein, new anthracyclines, ditrisarubicins, barminomycins and their degradation products were found. The relationship between their structures and activities suggested that by modifing the amino residue at position C-3' in anthracycline compounds, the multidrug resistance may be eliminated. Therefore, derivatives of 13-deoxocarminomycin and 13-deoxo-10-hydroxycarminomycin at their 3'-amino were synthesized. One compound, 3'-deamino-3'-morpholino-13-deoxo-10-hydroxycarminomycin (MX2) was effective on multidrug-resistant cells in vitro and in vivo. Moreover, this compound was active against experimental brain tumors and exhibited less cardiotoxicity.

Key words: multidrug resistance, anthracycline, morpholino, P-glycoprotein, MX2

INTRODUCTION

Anthracycline antibiotics, adriamycin and daunomycin, are very useful drugs for cancer chemotherapy. Recently their use is restricted by the frequent emergence of drug resistance to cancer chemotherapy. One reason for the clinical resistance is the metabolic inactivation or excretion of the drugs by internal organs [1][2]. The other is the drug resistance at cellular level [3]. Despite the multitude of chemotherapeutic agents available, tumor cells often become resistant to drugs that exhibit different mechanicsms of cytotoxic action. This type of resistance is widely observed in various experimental tumors and is referred to as multidrug resistance (MDR) or pleiotropic-drug resistance [4]. The cellular resistance is classified into two types. One is natural (*de novo*) drug resistance, evidenced by colon cancer, renal cancer, adenocarcinoma of the lung and malignant melanoma. These tumor types are refractory to chemotherapy at initial diagnosis. The other is acquired drug resistance, and in this type tumor cells can acquire broad resistance to chemotherapy following initial successful therapy. Recently, it has been suggested that there is a common mechanism of drug resistance in these two types [5]. In the study of MDR phenomenon, 170 kilodalton membrane glycoproteins were shown to be hyperexpressed in various MDR cells [6] and believed to serve as an energy-dependent efflux pump [7]. This protein is called P-glycoprotein, and the increased expression of P-glycoprotein in tumors was widely correlated with clinical treatment [3]. Moreover, recent study revealed the P-glycoprotein was also expressed by endothelial cells at blood-brain barrier sites [8].

In terms of therapeutic approaches for overcoming the MDR phenomena, there are three directions of studies to pursue. (1) The development of new antitumor agents effective on MDR cells; (2) the development of calcium channel blockers and calmodulin inhibitors, which can inhibit the efflux of drugs from MDR cells [9]; (3) the development of therapy using monoclonal antibody specific to P-glycoprotein [10][11]. In this article, the isolation and chemical modification of new anthracyclines active against adriamycin-resistant cells are described. One compound, MX2, warrants consideration for futher development as a new chemotherapeutic agent.

RESULTS AND DISCUSSION

DITRISARUBICINS AND BAUMYCINS

In screening for new antitumor antibiotics, which are effective on both drug-sensitive (P388/S) and adriamycin-resistant (P388/ADM) sublines of P388 mouse leukemia, new anthracyclines, ditrisarubicins [12] and serirubicins [13], were isolated. Each ditrisarubicin consisted of β-rhodomycinone as its aglycone moiety, and two sugar-chains composed of three hexoses at positions C-7 and C-10 in its aglycone structure. Serirubicin and 1-hydroxyserirubicin, which were present in the same culture broths as ditrisarubicins, contained the same sugar-chain as ditrisarubicin B, but their aglycone moieties were α-cytromycinone and $α_2$-rhodomycinone, respectively. In our screen, all ditrisarubicin-type anthracyclines were active against both sensitive and resistant cell lines, as shown in table 23.1. P388/ADM cells are well known to show the decreased uptake of adriamycin as compared with P388/S cells. However, the uptake of ditrisarubicin B by P388/S and P388/ADM was very rapid, and the intracellular concentration of the drug in P388/ADM was almost the same as that in P388/S after 10-minutes of incubation (unpublished data). Further studies with ditrisarubicins were not performed since they did not exhibit good antitumor activity *in vivo*.

In a previous study, baumycins A1, A2, B1 and B2, daunomycin analogues containing unique acetal moieties at their 4'-position, were described [14]. In the continuing search for new antitumor

Table 23.1 / Cytotoxicity of anthracycline compounds against P388/S and P388/ADM leukemia cells.

Compound	IC$_{50}$(ng/ml)		Ratio of IC$_{50}$ (R/S)
	P388/S	P388/ADM	
Adriamycin	12	310	25.8
Daunomycin	18	390	21.7
Ditrisarubicin A	4.3	5.3	1.5
Ditrisarubicin B	4.0	5.1	1.2
Ditrisarubicin C	3.8	4.4	1.3
Serirubicin	9.8	14.7	1.5
Carminomycin I	2.5	50	20
Carminomycin III	11	70	6.4
Barminomycin I	0.013	0.075	5.8
Barminomycin Ir	1.5	6.2	4.2

The IC$_{50}$ value was determined by 48-hour exposure with drug. The R/S value indicates the degree of resistance compared with sensitive cells.

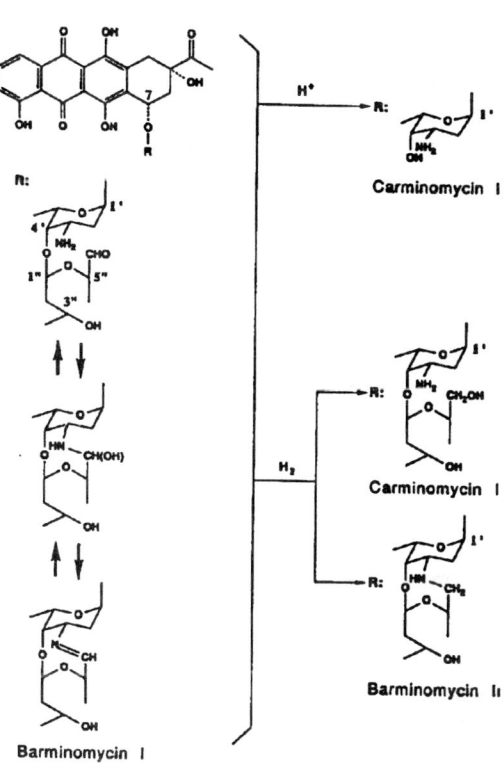

Figure 23.1 / Structures of barminomycin I and its degradation products.

natural products, extremely potent anthracyclines, barminomycins I and II were isolated from a carminomycin-producing strain [15]. The structures of 4'-O-substituents were difficult to elucidate by ^{1}H-NMR spectra. The reduction of barminomycin I with NaBH$_3$CN produced carminomycin III and barminomycin Ir. Based on these compounds, the structure of barminomycin I is proposed(Fig. 23.1). The residues at position C-6" of baumycins A1 and A2 are -CH$_2$OH, and those of baumycins B1 and B2 are -COOH. On the other hand, barminomycin I is the tautomer of aldehyde-type, carbinol amine-type and imine-type at the 6"-position and/or 3'-position in its structure. The basic molecule of 4'-O-substituent of barminomycin I may be produced by the same biosynthetic pathway as baumycins. As shown in table 23.1, barminomycin I and its degradation products exhibited a partial cross-resistance. However, the comparison of the R/S values and the structures for sugar moieties suggested that the alkylation of the amino residue at position C-3' of daunosamine may be very important in overcoming multidrug resistance.

In order to verify this hypothesis, some degradation products of ditrisarubicin B(DTR-B) were prepared by partial hydrolysis (Fig. 23.2). Oxaunomycin (R20X2, 13-deoxo-10-hydroxycarminomycin) [16] (Fig. 23.2) was found in the culture broth from a mutant strain of a daunomycin-producing microorganism. All these compounds in Fig. 23.2 contain β-rhodomycinone as their aglycone moietes, and the degree of cross-resistance for these anthracyclines was compared. The results indicate that the modification of the amino residue at position C-3' of anthracycline structure and the molecular size of the sugar moiety at position C-7 are important to circumvent MDR.

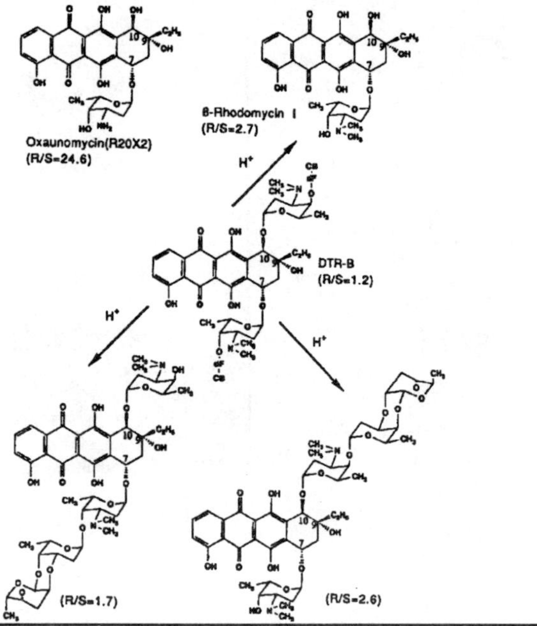

Figure 23.2 / Degree of cross-resistance for ditrisarubicin B(DTR-B) derivatives. The R/S values indicate the degree of cross-resistance in P388/ADM leukemia cells *in vitro*. dF: 2-deoxyfucose, CB: cinerulose B

3'-Amine	R/S[a]	ILS(%)[b]	T.I.[c]
1. NH$_2$ (R20X)	9.8	177	3.5
2. NHEt	6.5	48	1.4
3. NHCH$_2$CH=CH$_2$	6.0	72	1.6
4. NHCH$_2$CH(OH)CH$_3$	20.6	110	3.4
5. NHCH$_2$ (dioxolane)	4.4	122	2.5
6. NHCH$_2$ (phenyl)	1.9	90	1.5
7. NH (cyclopentyl)	2.4	4	-
8. NH (NCOCF$_3$)	4.4	110	3.5
9. NH (O)	3.0	4	-
10. N(Et)$_2$	3.7	8	-
11. N(CH$_2$CH=CH$_2$)	2.0	55	-
12. N (pyrrolidine)	2.7	102	2.9
13. N-NCOCH$_3$	3.0	67	5.1
14. N-O (MX)	1.2	111	3.9
15. Adriamycin	37.4	115	6.5

Figure 23.3 / Degree of cross-resistance in P388/ADM and antitumor activity for R20X (13-deoxocarminomycin) derivatives against P388 bearing mice.
[a]The R/S value indicates the degree of cross-resistance in P388/ADM leukemia cells.
[b]10^6 P388 leukemia cells were inoculated i.p. on day 0, and drug was administered i.v. on days 1 and 5. ILS (increase of life span) was caluculated from the mean survival times of the treated and the control mice.
[c]Therapeutic index (T.I.) is the ratio of dose at ILSmax to dose at ILS30.

MODIFICATIONS OF 3'-AMINO IN ANTHRACYCLINES

In order to find new anthracyclines which have activity against multidrug-resistant tumors, 13-deoxocarminomycin (R20X) and 13-deoxo-10-hydroxycarminomycin(R20X2) were selected as parent compounds for modification, as many chemical modifications of anthracycline had been studied using adriamycin and daumomycin [17]. More than fifty derivatives were synthesized by chemical modification, and their R/S values in P388/ADM and their antitumor activity against P388 leukemia in CD2F1 mice were tested. Fig. 23.3 lists the results for one part of derivatives at the 3'-position of R20X. In general, the R/S values for R20X derivatives containing tertiary amines at their 3'-position were lower than those for secondary amines and the *in vivo* activity for the derivatives attaching large moieties were superior to those for small moieties. Among these derivatives, MX (3'-deamino-3'-morpholino-13-deoxocarminomycin), showed the lowest R/S value and its antitumor activity in P388-bearing mice was classified into the most active group.

Regarding morhpolino anthracyclines, Acton and coworkers [17,18] reported that the antitumor activity of anthracycline derivatives was enhanced dramatically by morpholino derivation at their 3'-position, using adriamycin and daunomycin. Sikic et al [19] reported that cyanomorpholinyl adriamycin was non-cross-resistant in MDR cells.

Table 23.2 / Antitumor acitvity of morpholino anthracyclines and their parent drugs against P388 leukemia cells.

Parent				Morpholino			
Drug	ILSmax	(mg/kg)	T.I.	Drug	ILSmax	(mg/kg)	T.I.
R20X	177%	(4.1)	3.5	MX	111%	(6.7)	3.9
R20X2	152%	(1.0)	4.2	MX2	140%	(3.3)	8.9
R20Y5	58%	(16.0)	2.5	MY5	142%	(60.0)	7.0
CM	169%	(4.0)	5.3	MCM	91%	(2.0)	3.7
ADM	140%	(13.3)	6.5	MADM	54%	(0.05)	1.6

CM: carminomycin I, ADM: adriamycin

10^6 P388 cells were inoculated i.p. in CD2F1 mice on day 0. Drug was administered i.v. on days 1 and 5. Numbers in parentheses are the doses for ILSmax. Therapeutic index (T.I.) is the ratio of the dose for ILSmax to the dose for ILS_{30}.

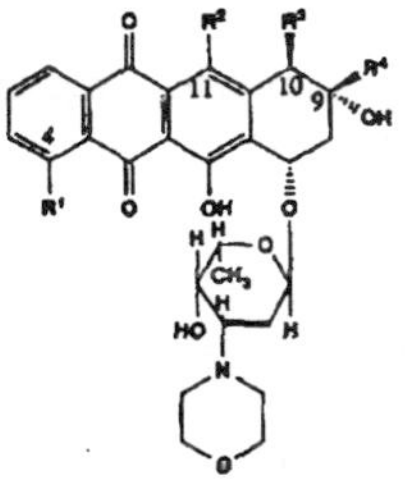

Drug	R^1	R^2	R^3	R^4
MX	OH	OH	H	C_2H_5
MX2	OH	OH	OH	C_2H_5
MY5	OH	H	H	C_2H_5
MCM	OH	OH	H	$COCH_3$
MADM	OCH_3	OH	H	$COCH_2OH$

Figure 23.4 / Structures of morphorino anthracyclines.

Therefore, five of morpholino anthracyclines, i.e. MX, MX2(3'-deamino-3'-morpholino-13-deoxo-10-hydroxycarminomycin), MY5(3'-deamino-3'-morpholino-13-deoxo-11-deoxycarminomycin), MCM(3'-deamino-3'-morpholino-carminomycin) and MADM(3'-deamino-3'-morpholino-adriamycin) were synthesized (Fig. 23.4), and compared their antitumor activity with those of their parental drugs (Table 23.2) [20]. In the case of carminomycin(CM) and adriamycin(ADM), morpholino derivation decreased thier ILSmax and therapeutic index (T.I.), when compared with parent drugs. In contrast, the activity of morpholino compounds attaching ethyl residue at their 9-position was similar or superior to those of the parent compounds. These data indicate that the substituent at position C-9 in morpholino anthracycline is very important to give higher efficacy *in vivo*. As judged by ILSmax and T.I. values, MX2 was the most effective morpholino compound we examined.

MX2

The results of the *in vitro* cytotoxicity of MX2 and adriamycin against several multidrug-resistant cell lines of P388 mouse leukemia and K562 human lymphoblastic leukemia are summarized in Table 23.3. These data suggested MX2 had an ability to inhibit MDR in P-glycoprotein positive tumor. The time course of MX2 accumulation in resistant cells was almost the same as that in sensitive cells [21]. Moreover, the *in vivo* evaluation gave the potential advantage to MX2. Adriamycin was not effective on P388/ADM, but MX2 exhibited an apparent effect on the increase of life span of P388/ADM-bearing mice [22]. In addition, MX2 had broad antitumor spectra against murine and human tumors. Its efficacy against L1210 leukemia, P388 leukemia, Lewis lung carcinoma and colon 26 adenocarcinoma was superior to that of adriamycin [22], and MX2 was also effective on several kinds of human tumor xenografts. In addition, the drug showed a marked therapeutic effect against

Table 23.3 / Cytotoxicity of MX2 and adriamycin against drug-sensitive and drug-resistant cell lines.

Cell line	IC_{50}(nM)	
	MX2	Adriamycin
P388 (parent)	8.5	10.5
P388/ADM	12.9 (1.5)	1420 (135)
P388/ACR	12.7 (1.5)	769 (73)
P388/MMC	10.1 (1.2)	22.4 (2.1)
K562 (parent)	31.6	13.8
K562/ADM	42.5 (1.3)	2820 (204)
K562/VCR	40.3 (1.3)	347 (25)

P388/ADM: adriamycin-resistant cells, P388/ACR: aclacinomycin-resistant cells, P388/MMC: mitomycin C-resistant cells, K562L human myelocytic leukemia cells, K562/ADM: adriamycin-resistant cells, K562/VCR: vincristin-resistant cells. Cytotoxicity was determined by 72-hour exposure with drug. Numbers in parentheses indicate the degree of resistance compared with parent cells.

MX 1 human mammary adenocarcinoma [22]. Anthracycline antibiotics have never been reported to have an efficacy on brain tumor, but MX2 administered i.v. showed antitumor effect against intracerebrally-inoculated L1210 leukemia [20] and intracisternal-inoculated Walker 256 adenocarcinoma [23].

One of the clinical problems of adriamycin is the irreversible cardiotoxicity which is produced at total dosage levels of adriamycin treatment, exceeding 500–550 mg/m^2. The study of cardiotoxocity evaluation using New Zealand white rabbits indicated that MX2 had no significant toxicity (unpublished data). These results suggest that MX2 warrents consideration for futher development as a therapeutic agent.

ACKNOWLEDGEMENT

We are grateful to Prof. T. Tsuruo and Dr. M. Inaba for kindly providing P388 sublines and helpful suggestions throughout the study.

REFERENCES

1. Goldie, J. H. and Coldman, A. J. 1983. Quantitative model for multiple levels of drug resistance in clinical tumors. *Cancer Treat. Rep.* 67:923–931.
2. Morrow, C. S. and Cowan, K. H. 1988. Mechanisms and clinical significance of multidrug resistance. *Oncology* 2:55–63.
3. Schneider, M. B., Kaufmann, T. E. M., Mattern, J. and Volm, M. 1989. P-Glycoprotein expression in treated and untreated human breast cancer. *Br. J. Cancer* 60:815–818.
4. Chabner, B. A., Clendeninn, N. J. and Curt, G. A. 1983. Symposium on cellular resistance to anticancer drugs. *Cancer Treat. Rep.* 67:855–932.
5. Tsuruo, T. 1988. Mechanisms of multidrug resistance and implications for therapy. *Jpn. J. Cancer Res.* 79:285–296.
6. Juliano, R. L. and Ling, V. 1976. A surface glycoprotein modulating drug permeability in Chinese hamster ovary cell mutants. *Biochim. Biophys. Acta.* 455:152–162.
7. Chen, C. J., Chin, J. E., Ueda, K., Clark, D. P., Pastan, I., Gottesman, M. M. and Roninson, I. B. 1986. Internal duplication and homology with bacterial transport proteins in the *mdr*1 (P-glycoprotein) gene from multidrug-resistant human cells. *Cell* 47:381–389.
8. Cordon-Cardo, C., O'Brien, J. P., Casals, D., Rittman-Grauer, L., Biedler, J. L., Melamed, M. R. and Bertino, J. R. 1989. Multidrug-resistance gene (P-glycoprotein) is expressed by endothelial cells at blood-brain barrier sites. *Proc. Natl. Acad. Sci. USA* 86:695–698.
9. Tsuruo, T., Iida, H., Tsukagoshi, S. and Sakurai, Y. 1981. Overcoming of vincristine resistance in P388 leukemia *in vivo* and *in vitro* through enhanced cytotoxicity of vicristine and vinblastine by verapamil. *Cancer Res.* 41:1967–1972.

10. Fitzgerald, D. J., Willingham, M. C., Cardarelli, C. O. et al. 1987. A monoclonal antibody-*Pseudomonas* toxin conjugate that specifically kills multidrug-resistant cells. *Proc. Natl. Acad. Sci. USA.* 84: 4288–4292.

11. Tsuruo, T., Hamada, H., Sato, S. and Heike, Y. 1989. Inhibition of multidrug resistant human tumor growth in athymic mice by anti-P-glycoprotein monoclonal antibodies. *Jpn. J. Cancer Res.* 80: 627–631.

12. Uchida, T., Imoto, M., Masuda, T., Imamura, K., Hatori, Y., Sawa, T., Naganawa, H., Hamada, M., Takeuchi, T. and Umezawa, H. 1983. New antitumor antibiotics, ditrisarubicins A, B and C. *J. Antibiot.* 36: 1080–1083.

13. Uchida, T., Imoto, M., Masuda, T., Yoshimoto, H., Imamura, K., Sawa, T., Naganawa, H., Takeuchi, T. and Umezawa, H. 1985. New anthracycline antibiotics: serirubicin and 1-hydroxyserirubicin. *J. Antibiot.* 38: 795–798.

14. Takahashi, Y., Naganawa, H., Takeuchi, T. and Umezawa, H. 1977. The structure of baumycin A1, A2, B1, B2, C1 and C2. *J. Antibiot.* 30: 622–624.

15. Uchida, T., Imoto, M., Takahashi, Y., Odagawa, A., Sawa, T., Tatsuta, K., Naganawa, N., Hamada, M., Takeuchi, T. and Umezawa, H. 1988. New potent anthracyclines, barminomycins I and II. *J. Antibiot.* 41: 404–408.

16. Yoshimoto, A., Fujii, S., Johdo, O., Kubo, K., Ishikura, T., Naganawa, H., Sawa, H., Takeuchi, T. and Umezawa, H. 1986. Intensely potent anthracycline antibiotic oxaunomycin produced by a blocked mutant of a daunorubicin-producing microorganism. *J. Antibiot.* 39: 902–909.

17. Morsher, C. W., Wu, H. Y., Fujiwara, A. N. and Acton, E. M. 1982. Enhanced antitumor properties of 3′-(4-morpholinyl) and 3′-(4-methoxy-1-piperidinyl) derivatives of 3′-deaminodaunorubicin. *J. Med. Chem.* 25: 18–24.

18. Acton, E. M., Tong, G. L., Mosher, C. W. and Wolgemuth, R. L. 1984. Intensely potent morpholinyl anthracyclines. *J. Med. Chem.* 27: 638–645.

19. Sikic, B. I., Ehsan, M. N., Harker, W. G., Friend, N. F., Brown, B. W., Newman, R. A., Hacker, M. P. and Acton, E. M. 1985. Dissociation of antitumor potency from anthracycline cardiotoxicity in a doxorubicin analog. *Science* 228: 1544–1546.

20. Komeshima, N., Tsuruo, T. and Umezawa, H. 1988. Antitumor activity of new morpholino anthracyclines. *J. Antibiot.* 41: 548–553.

21. Watanabe, M., Komeshima, N., Naito, N., Isoe, T., Otake, N. and Tsuruo, T. 1990. Cellular pharmacology of MX2, new morpholino anthracycline, in human pleiotropic drug-resistant cells.

22. Watanabe, M., Komeshima, N., Nakajima, S. and Tsuruo, T. 1988. MX2, a morpholino anthracycline, as a new antitumor agent against drug-sensitive and multidrug-resistant human and murine tumor cells. *Cancer Res.* 48: 6653–6657.

23. Izumoto, S., Arita, N., Hayakawa, T., Ohnishi, T., Taki, T., Yamamoto, H. and Ushio, Y. 1990. Effect of MX2, a new morpholino anthracycline, against experimental brain tumors. *Anticancer Res.* 10: 735–740.

24

Progress in Esperamicin Research

Kin Sing Lam, Salvatore Forenza, Judith A. Veitch, Donald R. Gustavson, Jerzy Golik, and Terrence W. Doyle

Bristol-Myers Squibb Company, Pharmaceutical Research Institute, P.O. Box 5100, 5 Research Parkway
Wallingford, CT 06492 U.S.A.

Esperamicin A_1, an extremely potent antitumor antibiotic, was isolated from cultures of Actinomadura verrucosospora ATCC 39334. The estimated titer of esperamicin A_1 in the initial fermentation was about 0.05 µg/ml. Through extensive media development and strain improvement studies, the titer of esperamicin A_1 was increased to 25–30 µg/ml in the new improved medium (H946) by the hyperproducer mutant (MU-5019). Adequate quantities of esperamicin A_1 became available from fermentation for structure elucidation and expanded biological evaluation. In addition, isolation of blocked mutants enabled us to isolate and determine the structures of most of the minor components of the esperamicin complex. Biosynthetic studies demonstrated that the diyne-ene portion of esperamicin A_1, the moiety responsible for the potent activity of esperamicin A_1, is derived from acetate. Phase I clinical trials of esperamicin A_1 have just been completed.

INTRODUCTION

In 1985 we reported the isolation of several members of a family of extremely potent antitumor antibiotics from cultures of *Actinomadura verrucosospora* ATCC 39334 (16). The producing organism was isolated from a soil sample collected at Pto Esperanza, Misiones, Argentina and the novel class named esperamicins. Only the partial structures of the major components, esperamicin A_1 (esp A_1) and esperamicin A_2 (esp A_2), were reported at that time. The estimated titer of the target compound, esp A_1, in the initial fermentation was about 0.05 µg/ml. Because of such a low titer and the complexity of the mixture of antibiotics produced in the fermentation, it was very difficult to isolate enough pure compound for structure elucidation and biological evaluation. The isolation of veractamycins (3), FR-900405 and FR-900406 (12, 14), diyne-ene antibiotics closely related to esperamicins, was also reported at about the same time. These compounds were also produced by different strains of *Actinomadura* at very low concentrations in the fermentation.

In this review paper, we describe some of our progress in the microbiological aspects of esperamicin research since 1985. In particular, the development of an improved production medium and the isolation of hyperproducer mutants has led to a 500–600 fold increased in the titer of esp A_1 in the fermentation. Minor congeners in the fermentation have been identified using blocked mutant studies. In addition, the elucidation of the biosynthetic origin of the moiety responsible for the potent activity of esp A_1 will be discussed.

ISOLATION OF ESPERAMICIN-PRODUCING STRAIN

In our continuing search for microorganisms which produce novel antitumor chemotypes, the supernatant from an actinomycete culture SA-24868 showed extremely potent antitumor activity against P388 leukemia implanted in mice (Table 24.1). At 160 fold dilution the culture supernatant still showed significant antitumor activity with % T/C of 170. The extreme potency of the antibiotics produced by strain SA-24868 was further illustrated by the fact that, extracting and processing the fermentation broth down to very low mass (<1 mg, at 5–10% antibiotic content), the extract still demonstrated good *in vivo* antitumor activity. The potency of the antibiotics extracted from strain SA-24868 against P388 leukemia in murine models was estimated to be about 5–10,000 fold more active than the extract from other soil isolate cultures in our screening program.

Table 24.1 / Effect of supernatant from strain SA-24868 on P388 leukemia

Dilution	Median Survival Time (Day)	% T/C
1–10	16	160
1–40	18	180
1–160	17	170
Control	10	100

Tumor inoculum: 10^6 ascites cells, ip
Host: CDF_1 mice
Evaluation: MST = median survival time
Effect: % T/C = (MST treated/MST control) × 100
Criteria: % T/C ≥125 considered significant antitumor activity

The actinomycete strain SA-24868 was isolated from a soil sample collected at Pto Esperanza, Misiones, Argentina. The class of novel antibiotics produced by this organism was named esperamicins. The morphological, cultural and physiological characteristics and cell wall chemistry of strain SA-24868 enabled us to classify the strain as *Actinomadura verrucosopora* (4). Strain SA-24868 produced extremely small amounts of esperamicin complex. Before the development of an accurate assay for esperamicin production, we could not quantitate the production of the antitumor antibiotics in the early fermentation. Based on a retrospective analysis of the biological data, we have estimated that the original titer of the esperamicin complex produced by strain SA-24868 was about 0.2 µg/ml, with the target compound, esp A_1, accounting for approximately 20–30% of the antibiotic complex. In order to isolate enough compound from the fermentation for further chemical and biological characterization, extensive medium development and strain improvement studies were carried out in our laboratories.

Table 24.2 / Effects of media components on the production of esperamicin A_1

Carbon Sources					
Acetate	-	Fructose	-	Oat Flour	-
Beet Molasses	-	Galactose	+	Potato Dextrin	0
Cane Molasses	++	Glycerol	+/0	Ribose	0
Cellobiose	0	Lactose	-	SJ (Corn) Starch	++
Corn Dextrin	-	Maltose	-	Stadex 92	-
Corn Syrup	+/0	Mannitol	+	Sucrose	+

Nitrogen Sources					
Asparagine	-	Dried Yeast	-	Proline	-
Corn Gluten	+	Fishmeal	++	Soybean Flour	-
Corn Steep Liquor	+/0	Uric Acid	+/0	Tryptophan	-
Distillers Solubles	+	NZ - Amine	+/0	Tyrosine	-
$NaNO_3$	+/0	Peptone	-	$(NH_4)_2SO_4$	-
		Pharmamedia	-		

Trace Elements					
Co	-	Mg	+/0	NaI	+++
Cu	++	Mo	+/0	Ni	+/0
Fe	+/0			Zn	+/0

Oils					
Corn Oil	+/0	Lard Oil	-	Soy Oil	-
		Proflo Oil	-		

Others					
EDTA	0	Glucuronic Acid	+/0	Oleic Acid	0
Gluconic Acid	+/0	Glucosamine	0	Propionic Acid	-
KH_2PO_4	-	$Mg_3(PO_4)_2$	-	Vitamins	+/0

+: increase, +/0: slight increase, 0: no change, -: decrease

STRAIN IMPROVEMENT PROGRAM

A mutation program was carried out to select hyperproducer mutants from strain SA-24868 (4, 19). N-Methyl-N'-nitro-N-nitrosoguanidine (NTG) and UV irradiation were used as the mutagens. A mutant strain, SA-25262, was isolated from the NTG mutation program (4). The production of esp A_1 by strain SA-25262 in production medium H55* was 1–1.5 µg/ml. Strain SA-25262 was further treated with UV-irradiation. The survivor single colony isolates were screened for better esp A_1 production titers and hyperproducer mutant strain MU-5019 was obtained (19). The titer of esp A_1 by strain MU-5019, in medium H55, was 4–5 µg/ml.

*Medium H55 consists of cane molasses 3%, corn starch 1%, fishmeal 1%, $CuSO_4 \cdot 5H_2O$ 0.005% and $CaCO_3$ 0.1%

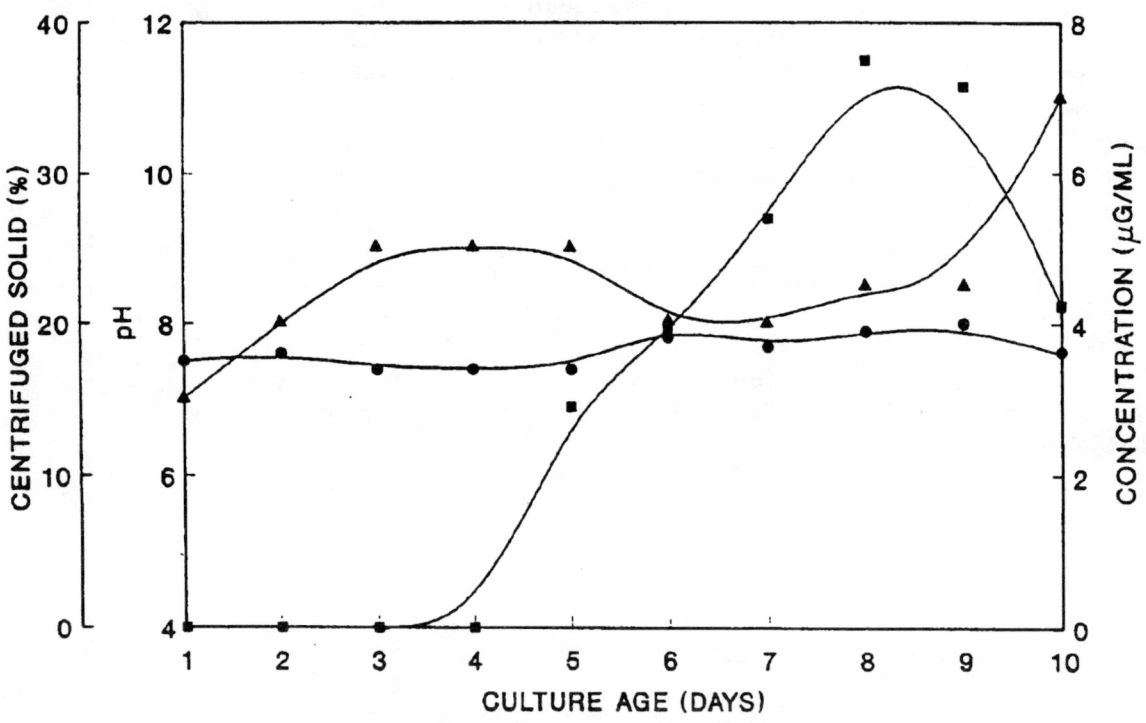

Figure 24.1 / Fermentation profiles of *Actinomadura verrucospora* SA-25262 grown in medium H946. ■, esperamicin A_1; pH; % centrifuged solid.

MEDIUM DEVELOPMENT PROGRAM

Extensive media development was carried out to improve the production of esp A_1 in the fermentation. Two to three thousands media have been tested. Table 24.2 summarizes the effect of some of the medium components we have tested on the production of esp A_1. The best carbon and nitrogen sources for esp A_1 production were cane molasses, corn starch and fishmeal. Trace element studies showed that $CuSO_4$ and NaI were important components for the production of esp A_1. NaI (0.5 mg/l) enhanced the production of esp A_1 by about 2 fold. Gathering all this information, the medium composition of production medium H946 used for large scale fermentation of esp A_1 was formulated: cane molasses 6%, corn starch 2%, fishmeal 2%, $CuSO_4 \cdot 5H_2O$ 0.01%, $CaCO_3$ 0.2% and NaI 0.5 mg/l.

The time course of esp A_1 production in medium H946 by strains SA-25262 and Mu-5019 is shown in Figs. 24.1 and 24.2. Esp A_1 production reached a maximum titer of 6.7 µg/ml at day 9 of the fermentation of strain SA-25262. The production cycle of esp A_1 by strain MU-5019 was longer, reaching a maximum titer of 30 µg/ml at day 13. Esp A_1 production in the fermentation was quantitated by HPLC analysis. Fig. 24.3 shows a typical HPLC chromatogram of the extract from cultures of strain SA-25262. The major component of the esperamicin complex produced by strain SA-25262 is esp A_1, with retention time of 8.3 minute, while the second major component of the esperamicin complex in the fermentation is esp A_2, with retention time of 16.3 minute. Using production medi-

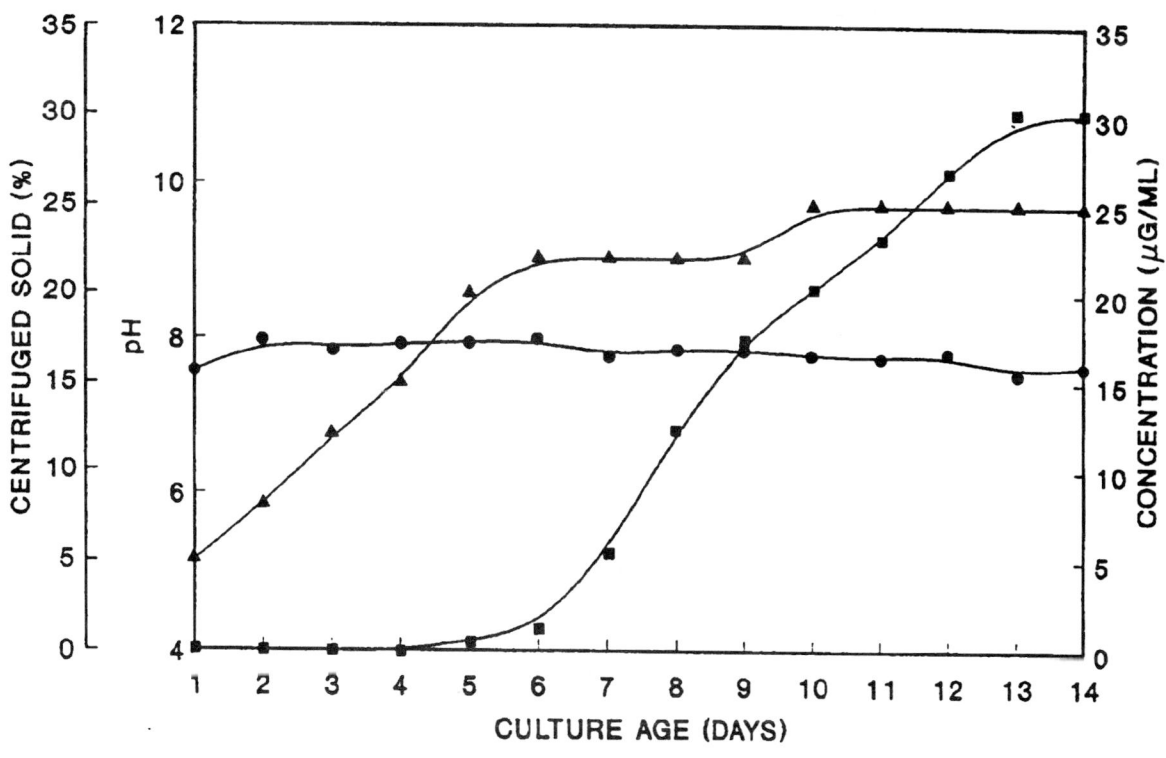

Figure 24.2 / Fermentation profiles of *Actinomadura verrucosopora* MU-5019 grown in medium H946. ■, esperamicin A_1; pH; % centrifuged solid.

um H946 and the two mutant strains, enough esp A_1 was isolated for structure determination.

STRUCTURE DETERMINATION

Fig. 24.4 shows the structure of esp A_1. The isolation and the elucidation of the structure of esp A_1 has been reported (6, 7, 16). The absolute configuration of each of the sugars and the bicyclic core have also been established (8–10, 30). Esp A_1 consists of a bicyclic core to which are attached a trisaccharide and a substituted 2-deoxy-L-fucose with an aromatic chromophore attached to the 3 position. The individual sugars of the trisaccharide are novel sugars and contain an unusual hydroxylamino sugar linked to a thiomethyl sugar via an o-glycosidic linkage at the 4 position. The hydroxylamino sugar is further attached to an isopropylamino sugar at the 2 position. The bicyclic core contains the very unusual diyne-ene, an allylic trisulfide and a bridgehead enone. Detailed SAR studies (20) have established that the interaction of these three functionalities results in a bioreductively activated, highly efficient, DNA strand scission.

As indicated earlier, the production of esperamicins in the fermentation is in a form of a complex with esp A_1 and esp A_2 being the major components of the complex. The structures of the naturally occurring esperamicins isolated to date from strain SA-25262 or blocked mutants are shown in Fig. 24.5. Esp A_1, Esp A_{1b}, Esp A_2 and esp P were isolated from large scale fermentation of strain SA-25262 (1, 5, 6, 7, 16). Recently, we were

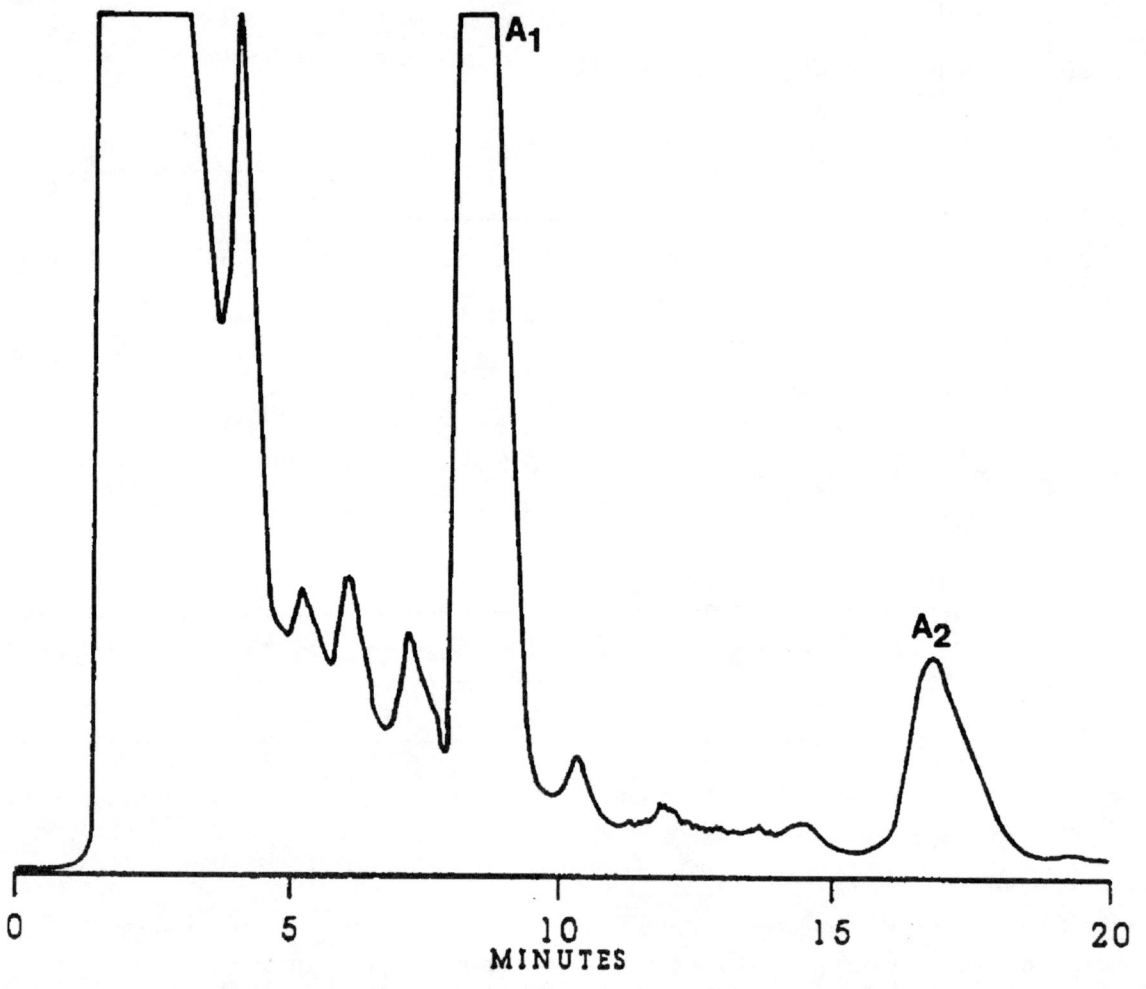

Figure 24.3 / HPLC of whole broth extract from culture of *Actinomadura verrucosospora* SA-25262 grown in medium H946. The fermentation extract was processed by extracting the culture broth with equal volume of ethyl acetate. The ethyl acetate extract was concentrated 10 fold and 25–50 µl of the concentrated extracts were used for HPLC analysis using a C-18 reversed phase column (Novapak, 3.9 × 150 mm, Waters Associates). The solvent system was 0.05M ammonium acetate (pH 4.5)-methanol-acetonitrile (1:1:1) and detector wavelength was set at 254 nm.

able to isolate esp A_{1c}, esp A_{2b} and esp A_{2c} from blocked mutants (18, K. S. Lam, unpublished observations) isolated in the course of the mutation program.

ISOLATION OF BLOCKED MUTANTS

Two blocked mutant strains, DG-108-9-3 and DG-111-10-6, were isolated from strain Mu-5019 using UV-irradiation as the mutagen (18). Strains DG-108-9-3 and DG-111-10-6 do not produce any esp A_1 and esp A_2 in the fermentation by HPLC analysis

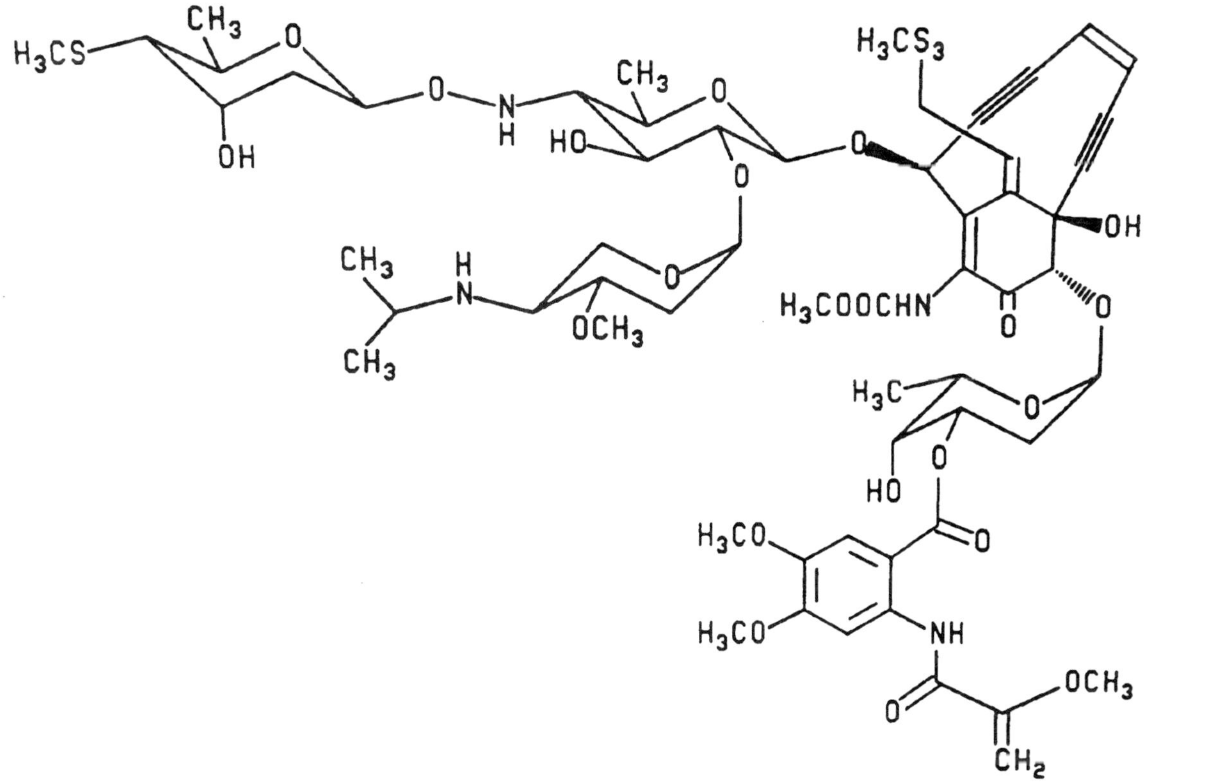

Figure 24.4 / Structure of esperamicin A1.

(Figs. 24.6 and 24.7). However, the supernatant from the culture broths of these two blocked mutants exhibited extremely potent antitumor activity against P388 leukemia in mice (Table 24.3). HPLC analysis (Fig. 24.6) of the broth extracts showed that strain DG-108-9-3 produced two new esperamicin analogs, esp A_{1c} (retention time: 6.1 min) and esp A_{2c} (retention time: 10.5 min). Strain DG-111-10-6 produced a major esperamicin metabolite which was identified as esp A_{1b} (retention time: 7.3 min, Fig. 24.7). This mutant also produced esp A_{1c}, esp A_{2c} and another new esperamicin, esp A_{2b} with retention time of 13.5 min (Fig. 24.7). The isolation of blocked mutants has proven to be a very productive way to isolate new congeners of esperamicin (esp A_{1c}, esp A_{2b} and esp A_{2c}) from the fermentation.

BIOLOGICAL ACTIVITIES AND CLINICAL EVALUATION

The *in vivo* antitumor activity of esp A_1 against P388 leukemia in mice is summarized in Table 24.4. Esp A_1 is active against P388 leukemia at a concentration as low as 0.05 µg/kg with a % T/C of 130. This kind of potency is similar to that of calicheamicins (21). The potency of esp A_1 against P388 leukemia in mice, on a molar basis, is approximately 3–6 fold more potent than CC-1065 (22, 23). Therefore esp A_1 is one of the most potent antitumor agents yet discovered. In addition to its potency, esp A_1 showed a broad spectrum of activity against tumors of murine and human origin and activity against tumors located distal to the site of drug administration (25). Esp A_1 admin-

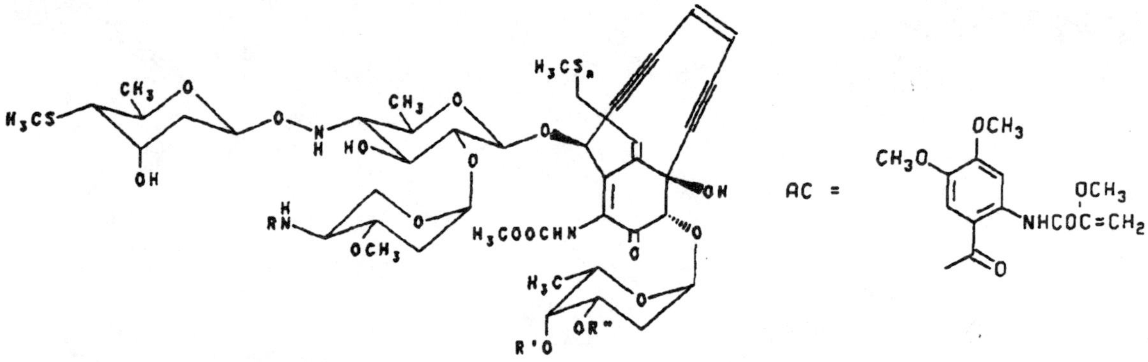

ESPERAMICIN	n	R	R'	R"
A_1	3	$CH(CH_3)_2$	H	AC
A_1b	3	CH_2CH_3	H	AC
A_1c	3	CH_2CH_3	H	AC
P	4	$CH(CH_3)_2$	H	AC
A_2	3	$CH(CH_3)_2$	AC	H
A_2b	3	CH_2CH_3	AC	H
A_2c	3	CH_3	AC	H

Figure 24.5 / Structures of naturally occurring esperamicins.

istered ip was active against ip implanted P388 leukemia, L1210 leukemia, B16 melanoma, M109 lung carcinoma, C26 colon carcinoma, M5076 sarcoma and Lewis lung carcinoma. Esp A_1 administered iv was active against iv implanted P388 and L1210 leukemia. In addition, esp A_1 was active against sc implanted B16 melanoma and M109. Esp A_1 was also active against the MX-1 human mammary xenograft implanted in the subrenal capsule of nude mice.

Table 24.5 shows the antimicrobial spectrum of esp A_1. Esp A_1 exhibited very potent activity against Gram-positive bacteria with most MIC values in the pg/ml range. Although the MIC values for Gram-negative bacteria and yeast were significantly higher than the ones observed for Gram-positives, the activity against Gram-negatives and yeast was still very significant (<1µg/ml).

Table 24.3 / Effect of supernatants from strains DG-108-9-3 and DG-111-10-6 on P388 leukemia

Culture	Dilution	Median Survival Time (Day)	%T/C
DG-108-9-3	1–20	19.5	195
	1–60	19.5	195
	1–180	17.0	170
DG-111-10-6	1–20	16.5	165
	1–60	21.0	210
	1–180	19.0	190
Control	-	10	100

Tumor inoculum: 10^6 ascites cells, ip
Host: CDF_1 mice
Evaluation: MST = median survival time
Effect: % T/C = (MST treated/MST control) × 100
Criteria: % T/C ≥125 considered significant antitumor activity

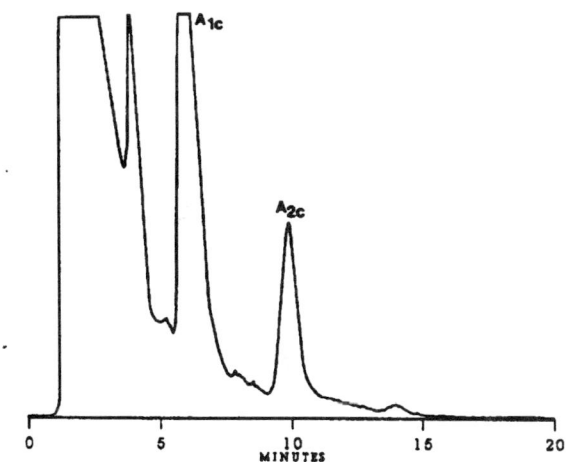

Figure 24.6 / HPLC of whole broth extract from cultures of *Actinomadura verrucosospora* DG-108-9-3 grown in medium H946.

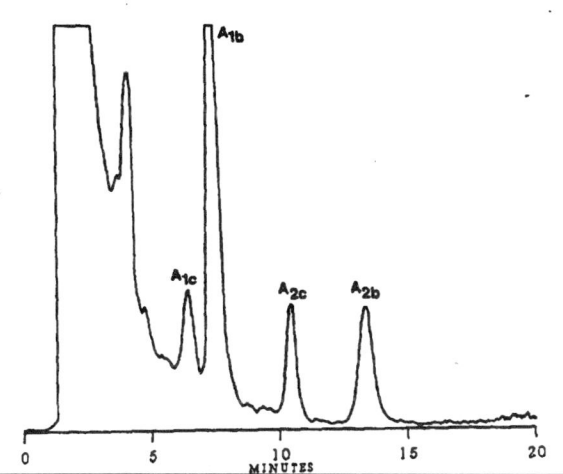

Figure 24.7 / HPLC of whole broth extract from cultures of *Actinomadura verrucosospora* DG-11-10-6 grown in medium H946.

Progress in Esperamicin Research 269

Table 24.4 / Antitumor activity of esperamicin A_1 against P388 leukemia in mice

Dosage (μg/kg/inj.)	Median Survival Time (Day)	% T/C
1.6	17.0	170
0.8	16.0	160
0.4	14.0	140
0.2	14.0	140
0.1	14.0	140
0.05	13.0	130
Control	10.0	100

Tumor inoculum: 10^6 ascites cells, ip
Host: CDF_1 mice
Evaluation: MST = Median Survival Time
Effect: % T/C = (MST treated/MST control) × 100
Criteria: % T/C ≥125 considered significant antitumor activity

Because esp A_1 fully met our criteria of novelty, broad spectrum of activity, and novel mechanism of action, a decision was made to proceed with clinical development. Phase I clinical trials were initiated in mid-1988 and early 1989 (2, 26). Seventy patients with a variety of advanced malignancies have been treated thus far. Phase I clinical trials of esp A_1 have just been completed and the start of phase II clinical trials is imminent.

BIOSYNTHESIS

We have initiated studies to establish the biosynthetic origin of the diyne-ene portion of esp A_1, the moiety responsible for the extremely potent activity of esp A_1. Recently, we have shown that the diyne-ene portion of esp A_1 is derived from acetate. Feeding the *Actinomadura* culture with sodium [^{13}C-1] acetate, we have demonstrated that carbons 3, 7, 9, 11 and 14 of the diyne-ene portion were clearly enriched by ^{13}C-nmr analysis (Fig. 24.8A). The intensity of the signals of these carbons was at least twice the intensity of the signals of the control (culture without addition of sodium [^{13}C-1]acetate). Using sodium [^{13}C-2] acetate, we have demonstrated that carbons 4, 6, 8, 10, 13

Table 24.5 / Antibacterial activity of esperamicins A_1

Strain	MIC (µg/ml)
Staphylcoccus aureus 209P	<0.0008
S. aureus Smith	<0.0008
Bacillus subtilis PCI 219	<0.0008
Micrococcus luteus 1001	0.0016
M. flavus	<0.0008
Mycobacterium 607	0.05
Escherichia coli NIHJ	0.1
Klebsiella pneumoniae D11	0.4
Pseudomonas aeruginosa D15	0.8
Bacteriodes fragilis A20928	0.2
Clostridium difficile A21675	0.4
C. perfringens A9635	0.05
Candida albicans IAM 4888	0.4
Cryptococcus neoformans	1.6

and 15 were clearly enriched (Fig. 24.8B). The signal of carbon 5 was buried underneath the signal of the solvent (chloroform). Other solvents (i.e., benzene) will be used in order to obtain the accurate integration of the signal for carbon 5. The signals of carbons 1, 2 and 12 were difficult to quantitate because they are very broad signals due to the presence of the carbamate moiety attached to carbon 2. Larger sample size (10–15 mg) of ^{13}C-enriched esp A_1 is required for ^{13}C-nmr analysis to obtain accurate integration of the signals from these carbons. However, we have clearly shown that at least 11 of the 15 carbons of the diyne-ene moiety were derived from acetate. The related C^{14} diyne-ene moiety of the chromophore from neocarzinostatin was also reported to be derived from acetate units (11).

DISCUSSION

Several years of fermentation development in our laboratories have reliably increased the production of esp A_1 from approximately 0.05 µg/ml to 25–30 µg/ml. This titer was judged adequate for commercialization given the potency of esp A_1 (mouse $LD_{10} = 9$ µg/m^2). Large scale fermentation and isolation of esp A_1 was carried out at National Cancer Institute's Frederick Cancer Research Facility (1). Over 40 grams of esp A_1 were isolated. The collaborative effort of National Cancer Institute's Frederick Cancer Research Facility and Bristol-Myers Squibb Company has ensured adequate drug supply for clinical evaluations. Phase I clinical trials of esp A_1 have just been completed and are awaiting for the start of phase II clinical trials.

Future work on esperamicins research in our laboratory will be concentrated on the isolation of congeners of esp A_1 and elucidation of the biosynthesis of esperamicins. We have demonstrated that the isolation of blocked mutants has resulted in the identification of several new esperamicin analogs in the fermentation. Another avenue of detecting novel esperamicin analogs in the fermentation is to use the media manipulation approach. We have demonstrated that the use of selective media has also led us to discover several new esperamicin congeners in the fermentation (data not shown). The formulation of these selective media was based on the information gained from the extensive media development for improving the production of esp A_1.

Knowledge of antibiotic biosynthesis can be used to: 1) provide information leading to increase in production of the target antibiotic; 2) exploit fermentation possibilities leading to the production of novel analogs; and 3) provide optimal route for the preparation of isotopically labelled drug for pharmacokinetic studies which could yield important data for the compounds, like esp A_1, in clinical studies. We have demonstrated that the diyne-ene portion of esp A_1 is derived from acetate. We are in the process of deducing the folding pattern of this C_{15} unit by examining the [^{13}C-1,2] acetate enrichment pattern of esp A_1. Understanding the biosynthesis of the allylic trisulfide is also important. In fact, mechanism of action studies

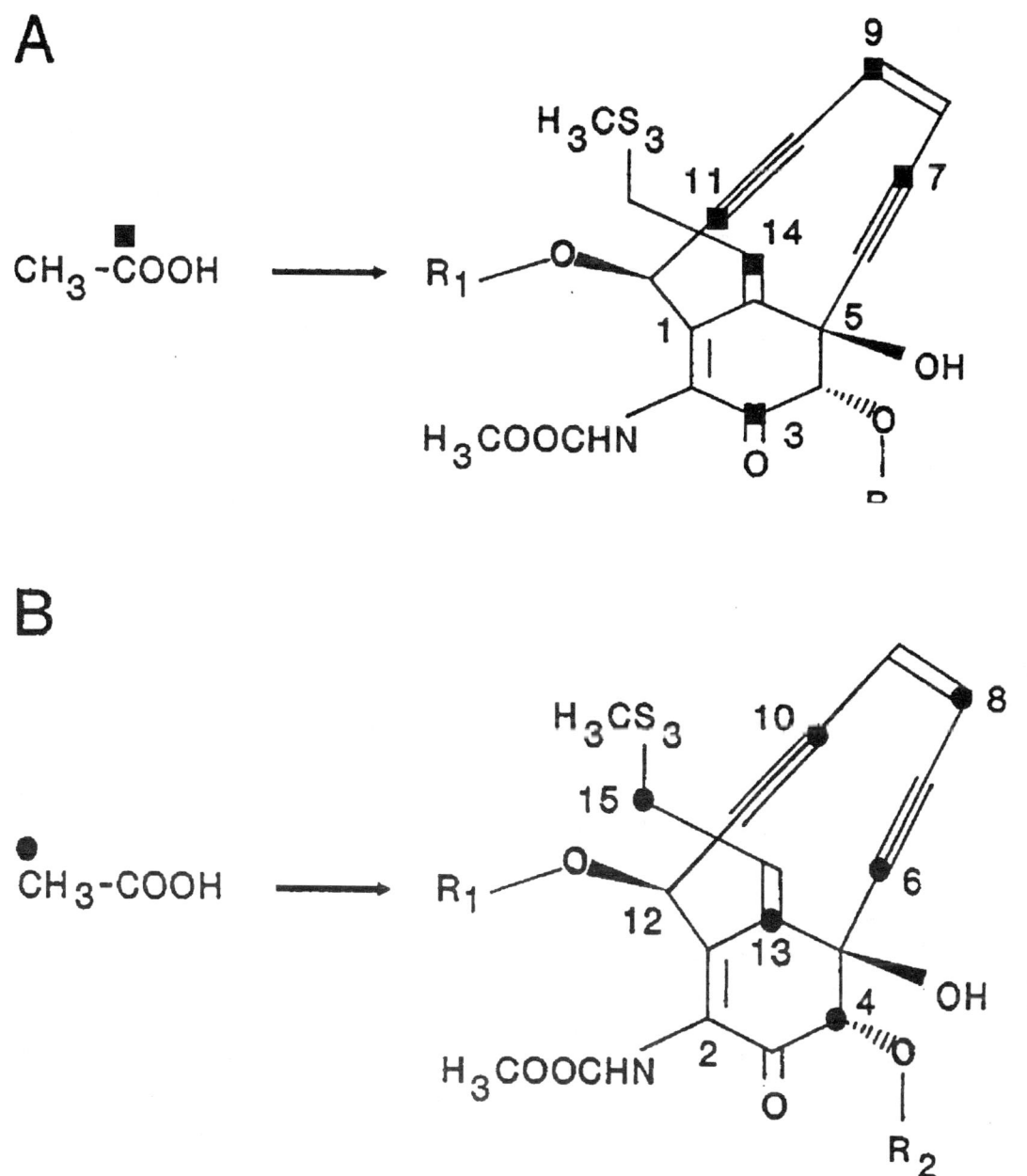

Figure 24.8 / 13C-Enrichment pattern of esperamicin A_1 from cultures of *Actinomadura verrucosospora* SA-25262 supplemented with (A) sodium [^{13}C-1] acetate and (B) sodium [^{13}C-2] acetate.

indicate that the reduction of the methyl trisulfide group to form a thiolate anion and its subsequent interaction with the bridgehead enone and the diyne-ene moiety results in a highly efficient DNA strand scission mechanism (20, 28). Further work will be carried out in our laboratories to identify the biosynthetic origin of the trisulfide using ^{35}S-precursor labelling technique.

The mechanism of action of esp A_1 and several related compounds, calicheamicins (31, 32) neocarzinostatin (13, 24) and dynemicins (27) has been reported. It appears that the above compounds are capable of cleaving DNA via direct carbon radical abstraction of deoxyribose hydrogen atoms. The presence of the diyne-ene function in these compounds is a prerequisite to their potent activities. Therefore we refer to the above compounds as members of the diyne-ene class of antitumor antibiotics.

The newest member of the diyne-ene class, dynemicin A, is produced by *Micromonospora chersina* ATCC 53710 (15, 17). The titer of dynemicin A in the initial fermentation was very low, about 0.1 µg/ml (29). Based on our experience in the medium development for esp A_1 fermentation, we quickly improved the production of dynemicin A to 3.5 µg/ml in the fermentation. We anticipate the isolation of more novel compounds in the diyne-ene class due to the experience we gained in working with esperamicins, calicheamicins, neocarzinostatin and dynemicins.

REFERENCES

1. Beutler, J. A.; Clark P.; Ross, J.; Roach, J.; Forenza, S.; Matson, J.; Lebherz, W.; and Muschik, G. 1989. Large scale isolation of esperamicins. Abs. 0–33. 30th Annual Meeting of American Society of Pharmacognosy, San Juan, Puerto Rico, 6–10 August.
2. Brown, T.; Havlin, K.; Weiss, K.; Rodriguez, G.; Cagnola, J.; Koeller, J.; Herndon, H.; Kelley, S.; Skrokov, M; and VonHoff, D. 1990. Proc. ASCO, Washington, D.C.
3. Bunge, R. H.; Hurley, T. R.; Smitka, T. A.; Willmer, N. E.; Brankiewicz, A. J.; Steinman, C. E.; and French, J. C. 1984. PD111, 759 and PD115, 028, novel antitumor antibiotics with phenomenal potency. I. Isolation and characterization. *J. Antibiot.* 37: 1566–1571.
4. Forenza, S.; Claridge, C. A.; Titus, J. A.; Veitch, J. A.; Tomita, K.; Hatori, M.; Miyaki, T.; and Kawaguchi, H. 1987. Esperamicins, a novel class of potent antitumor antibiotics: Taxonomy and fermentation. Abs. 0–23, 87th Annual Meeting of the American Society for Microbiology, Atlanta, Georgia, 1–6 March.
5. Golik, J.; Beutler, J. A.; Clark, P.; Ross, J.; Roach, J.; Lebherz, W. B. III; and Muschik, G. 1989. Esperamicin P, a novel antitumor antibiotic (BMY-41339). U.S. Patent applied, USSN323648.
6. Golik, J.; Clardy, J.; Dubay, G.; Groenewold, G.; Kawaguchi, H.; Konishi, M.; Krishnan, B.; Ohkuma, H.; Saitoh, K-I.; and Doyle, T. W. 1987. Esperamicins, a novel class of potent antitumor antibiotics. 2. Structure of esperamicin X. *J. Am. Chem. Soc.* 109: 3461–3462.
7. Golik, J.; Dubay, G.; Groenewold, G.; Kawaguchi, H.; Konishi, M.; Krishnan, B.; Ohkuma, H.; Saitoh, K-I.; and Doyle, T. W. 1987. Esperamicins, a novel class of potent antitumor antibiotics. 3. Structures of esperamicin A_1, A_2, and A_{1b}. *J. Am. Chem. Soc.* 109: 3462–3464.
8. Golik, J.; Doyle, T. W.; VanDuyne, G.; and Clardy, J. 1990. Stereochemical studies on esperamicin A_1: A single crystal x-ray structure of thiosugar moiety. *Tetrahedron Letters* 31: 6149–6150.
9. Golik. J.; Wong, H.; Krishnan, B.; Vyas, D. M.; and Doyle, T. W. 1991. Stereochemical studies on esperamicins: Determination of the absolute configuration of hydroxyamino sugar fragment. *Tetrahedron Letters* 32: 1851–1854.
10. Golik, J.; Wong, H; Vyas, D. M.; and Doyle, T. W. 1989. Stereochemical studies on esperamicins: Determination of the absolute configuration of isopropylamino sugar moiety. *Tetrahedron Letters* 30: 2497–2500.
11. Hensens, O. D.; Giner, J-L.; and Goldberg, I. H. 1989. Biosynthesis of NCS chrom A, the chromophore of the antitumor antibiotic neocarzinostatin. *J. Am. Chem. Soc.* 111: 3295–3299.

12. Iwami, M.; Kiyoto, S.; Nishikawa, M.; Terano, H.; Kohsaka, M.; Aoki, H.; and Imanaka, H. 1985. New Antitumor antibiotics, FR-900405 and FR-900406. I. Taxonomy of the producing strain. *J. Antibiot.* 38: 835–839.

13. Kappen, L. S. and Goldberg, I. A. 1983. Deoxyribonucleic acid damage by neocarzinostatin chromophore: Strand breaks generated by selective oxidation of C-5' of deoxyribose. *Biochemistry* 22: 4872–4878.

14. Kiyoto, S.; Nishikawa, M.; Terano, H.; Koshsaka, M.; Aoki, H.; Imanaka, H.; Kawai, Y.; Uchida, I.; and Hashimoto, M. 1985. New antitumor antibiotics, FR-900405 and FR-900406. II. Production, isolation, characterization and antitumor activity. *J. Antibiot.* 38: 840–848.

15. Konishi, M.; Ohkuma, H.; Matsumoto, K.; Tsuno, T.; Kamei, H.; Miyaki, T.; Oki, T.; Kawaguchi, H.; VanDuyne, G. D.; and Clardy, J. 1989. Dynemicin A, a novel antibiotic with the anthraquinone and 1,5-diyn-3-ene subunit. *J. Antibiot.* 42: 1449–1452.

16. Konishi, M.; Ohkuma, H.; Saitoh, K-I.; Kawaguchi, H.; Golik, J.; Dubay, G.; Groenewold, G.; Krishnan, B.; and Doyle, T. W. 1985. Esperamicins, a novel class of potent antitumor antibiotics. 1. Physico-chemical data and partial structure. *J. Antibiot.* 38: 1605–1609.

17. Konishi, M.; Ohkuma, H.; Tsuno, T.; and Oki, T. 1990. Crystal and molecular structure of dynemicin A: A novel 1,5-diyne-3-ene antitumor antibiotic. *J. Am. Chem. Soc.* 112: 3715–3716.

18. Lam, K. S.; Gustavson, D. R.; and Forenza, S. 1989. Isolation of blocked mutants of esperamicin A_1 from *Actinomadura verrucosospora*. Abs. P-84, 46th Annual Meeting of Society for Industrial Microbiology, Seattle, Washington, 13–18 August.

19. Lam, K. S.; Titus, J. A.; and Kimball, D. L. 1987. Isolation of hyperproducing strains of *Actinomadura verrucosospora*. Abs. 0-40, 87th Annual Meeting of the American Society for Microbiology, Atlanta, Georgia, 1–6 March.

20. Long, B. H.; Golik, J.; Forenza, S.; Ward, B.; Rehfuss, R.; Dabrowiak, J. C.; Catino, J. J.; Musial, S. T.; Brookshire, K. W.; and Doyle, T. W. 1989. Esperamicins, a class of potent antitumor antibiotics: Mechanism of action. *Proc. Nat. Acad. Sci. USA.* 86: 2–6.

21. Maiese, W. M.; Lechevalier, M. P.; Lechevalier, H. A.; Korskalla, J.; Kuck, N.; Fantini, A.; Wildey, M. J.; Thomas, J.; and Greenstein, M. 1989. Calicheamicins, a novel family of antitumor antibiotics: Taxonomy, fermentation and biological properties. *J. Antibiot.* 42: 558–563.

22. Martin, D. G.; Biles, C.; Gerpheide, S. A.; Hanka, L. J.; Krueger, W. C.; McGovern, J. P.; Mizsak, S. A.; Neil, G. L.; Stewart, J. C.; and Visser, J. 1981. CC-1065 (NSC298223), a potent new antitumor agent. Improved production and isolation, characterization and antitumor activity. *J. Antibiot.* 34: 1119–1125.

23. Martin, D. G.; Chidester, C. G.; Duchamp, D. J.; and Mizsak, S. A. 1980. Structure of CC-1065 (NSC298223), a new antitumor antibiotic. *J. Antibiot.* 33: 902–903.

24. Myers, A. G. 1987. Proposed structure of the neocarzinostatin chromophore-methyl thioglycolate adduct; a mechanism for the nucleophilic activation of neocarzinostatin. *Tetrahedron Letters* 28: 4493–4496.

25. Schurig, J. E.; Rose, W. C.; Kamei, H.; Nishiyama, Y.; Bradner, W. T.; and Stringfellow, D. A. 1990. Experimental antitumor activity of BMY-28175 a new fermentation derived antitumor agent. *Invest. New Drugs* 8: 7–15.

26. Sessa, C.; Drozd, E.; Gumbrell, L.; Durr, R.; Bruntsch, U.; Kelley, S.; and Cavalli, F. 1990. Proc. ASCO, Washington, D.C.

27. Sugiura, Y.; Shiraki, T.; Konishi, M.; and Oki, T. 1990. DNA intercalation and cleavage of an antitumor antibiotic Dynemicin that contains anthracycline and enediyne cores. *Proc. Nat. Acad. Sci. USA.* 87: 3831–3835.

28. Sugiura, Y.; Uesawa, Y.; Takahashi, Y.; Kuwahara, J.; Golik, J.; and Doyle, T. W. 1989. Nucleotide-specific cleavage and minor-groove interaction of DNA with esperamicin antitumor antibiotics. *Proc. Nat. Acad. Sci. USA.* 86: 7672–7676.

29. Titus, J. A.; Lam, K. S.; Dabrah, T. T.; Kimball, D. L.; Mattei, J. M.; Compton, B. J.; Matson, J. A.; and Forenza, S. 1990. Improved production of dynemicin A, a novel antitumor antibiotic, by media manipulation. Abs. P-46, 47th Annual Meeting of Society for Industrial Microbiology, Orlando, Florida, July 29-August 3.

30. Wittman, M. D.; Halcomb, R. L.; Danishefsky, S. J.; Golik, J.; and Vyas, D. 1989. A route to glycals in the Allal and Gulal series: Synthesis of the thiosugar of esperamicin A_1. *J. Org. Chem.* 55:1979–1981.
31. Zein, N.; Poncin, M.; Nilakantin, R.; and Ellestad, G. A. 1989. Calicheamicin γ_1^I and DNA: Molecular recognition process responsible for site-specificity. *Science* 244:697–699.
32. Zein, N.; Sinha, A. M.; MacGahren, W. J.; and Ellestad, G. A. 1988. Calicheamicin γ_1^I: An antitumor antibiotic that cleaves double-stranded DNA site specifically. *Science* 240:1198–1201.

25

Oxetanocins, Antiviral Nucleosides

Tomohisa Takita
Nippon Kayaku Co.,Ltd., Tokyo 102, Japan

Key Words: antiviral, cyclobutane, nucleoside, oxanosine, oxetanocin.

INTRODUCTION

In 1981, oxanosine was discovered from a screening for antibacterial antibiotics at the Institute of Microbial Chemistry [10]. It was an interesting compound from chemical and biological viewpoints rather than for potential usefulness. Our current studies of antiviral nucleosides were a result of the discovery of oxanosine. This review describes the discovery of oxetanocin A [12], which is a new lead compound for an antiviral agent, and the following developments. It will be clear that without the studies of oxanosine, which did not exhibit any antiviral activity, we would not have discovered oxetanocin A.

DISCOVERY AND STUDIES OF OXANOSINE

Oxanosine was discovered by conventional screening of antibacterial antibiotics. The most intriguing property of oxanosine was the strong IR absorption at around 1800 cm^{-1}, which may be assigned to β-lactam. The structure was finally determined by X-ray crystallography as shown in Fig. 25.1 [4]. Oxanosine is a kind of ribonucleoside and has a new chromophore with interesting chemical properties, such as reversible ring opening depending on the pH and ready transformation to xanthosine by treatment with a trace amount of NaOMe in MeOH.

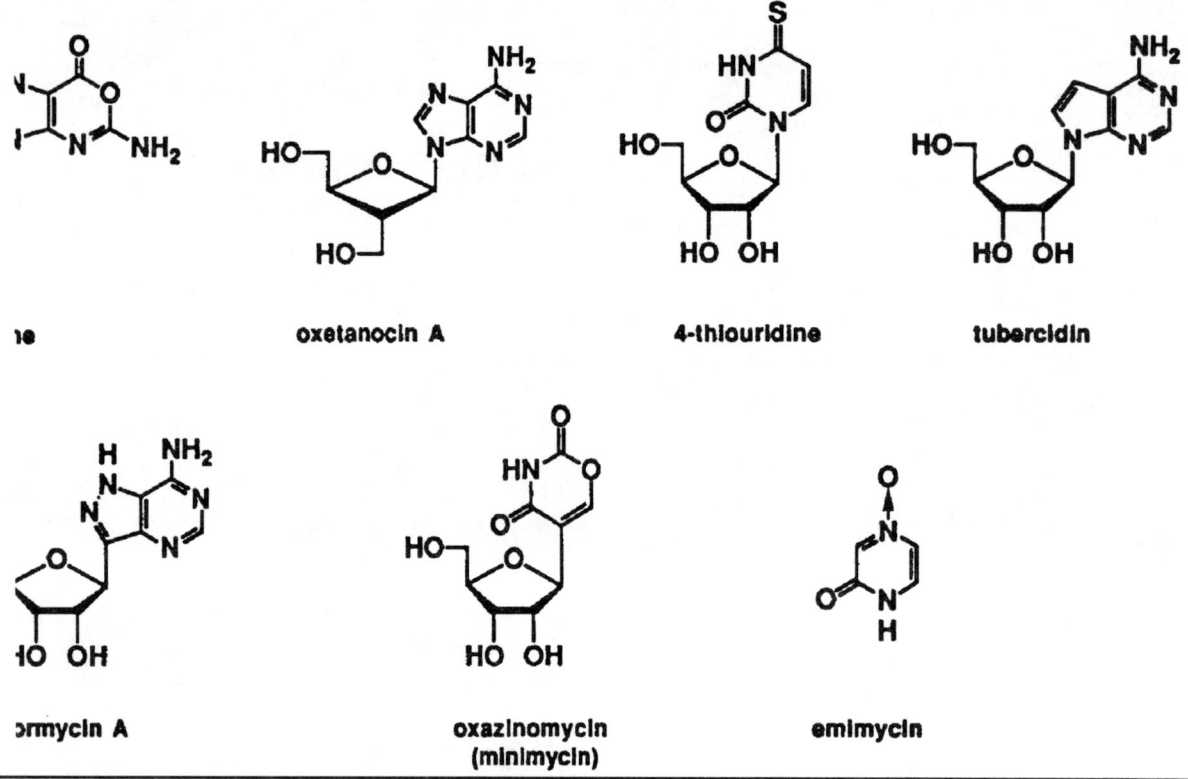

Figure 25.1 / Structures of oxanosine and nucleoside antibiotics isolated by a specific screening.

Oxanosine showed weak antibacterial activity against some strains of *E. coli*, *Shigella* and *Proteus* under special conditions such as on plain peptone agar, but it did not exhibit any antibacterial activity on nutrient agar containing meat extract. We examined the antagonistic effect of various kinds of nucleosides on the antibacterial activity of oxanosine. Addition of thymidine, cytidine, uridine and xanthosine at 50 mcg/ml to the peptone agar did not show any antagonistic effect. The addition of inosine and adenosine showed a weak antagonistic effect, and guanosine resulted in complete suppression of the antibacterial activity of oxanosine. Guanine and 5'-GMP also exhibited the same effect as guanosine. Oxanosine showed weak antitumor activity *in vitro* and *in vivo* and inhibited oncogene functions [14]. However, no antiviral activity was observed.

A SPECIFIC SCREENING FOR NOVEL NUCLEOSIDES AND DISCOVERY OF OXETANOCIN A

During the studies of oxanosine, aciclovir was developed as an anti-herpesvirus agent. Its mode of action led to the study of novel nucleosides as antiviral agents. To date, more than 150 kinds of nucleosides have been isolated from microbial metabolites. Our studies on oxanosine suggested the possibility of a specific screening for novel nucleosides from microbial metabolites.

Dr. Shimada designed a screening method in the laboratory at Nippon Kayaku Co. The principle of the screening is very simple; cultured filtrates of microorganisms were tested for antibacterial activity on plain peptone agar and peptone agar added with five different nucleosides: A, G, T, C and U at 50 mcg/ml. When a culture filtrate showed antibacterial activity on the peptone agar but not on the nucleoside-added agar, the culture filtrate was considered a potential positive and examined further.

This provided an effective way to isolate novel nucleosides. In Fig. 25.1, 7 nucleosides and related compounds thus discovered are shown. Tubercidin, formycin A, oxazinomycin and emimycin were discovered earlier, while oxetanocin A (OXT-A) and 4-thiouridine were newly discovered.

The structure of OXT-A was determined by X-ray crystallography [5]. The nucleobase is adenine and the sugar moiety has an oxetane-ring with two hydroxymethyl groups. The name oxetanocin A was derived from the structure having an oxetane and adenine. OXT-A is the first glycoside in nature having an oxetane-ring in the sugar moiety.

The antibacterial activity of OXT-A was antagonized by adenosine. OXT-A showed the expected antiviral activity; that will be described later.

DERIVATIVES OF OXETANOCIN A AND THEIR ANTIVIRAL ACTIVITIES

Recently, aciclovir and ganciclovir have been clinically used for the treatment of herpes and cytomegalovirus infections. Guanine is present in their base moiety. OXT-A is now sufficiently produced by fermentation. Thus, transformation of OXT-A to OXT-G, that has guanine in the base moiety, was studied. OXT-A was readily transformed quantitatively to OXT-H (H = hypoxanthine) by adenosine deaminase. OXT-A was also transformed to OXT-X (X = xanthine) via OXT-H by *Nocardia interforma* in more than 95% yield. Biotransformation of OXT-A to OXT-G or OXT-X to OXT-G was not successful. Transformation of OXT-X to OXT-G was achieved by combination of chemical and biological transformations (Fig. 25.2) [11]. OXT-X was first acetylated and then sulfonylated followed by ammonolysis to give 2-amino-OXT-A in 45% overall yield. Deamination

Table 25.1 / Anti-herpesvirus activity of OXTs

OXTs	IC$_{50}$ (µg/ml)	
	HSV-2*	HCMV**
OXT-A	10	13
OXT-H	>50	18
OXT-X	>50	>50
2-amino-OXT-A	4.2	2.1
OXT-G	3.5	1.0

*Herpes simplex virus type 2
**Human cytomegalovirus

of 2-amino-OXT-A by adenosine deaminase gave OXT-G in quantitative yield.

The five OXTs: OXT-A, OXT-H, OXT-X, 2-amino-OXT-A and OXT-G, thus obtained, were tested for their antiviral activities. In Table 25.1, the antiviral activities against herpesviruses are listed. These studies were carried out by Dr. Nishiyama and his collaborators in the Medical School of Nagoya University [6]. Against both herpes simplex virus type 2 (HSV-2) and human cytomegalovirus (HCMV), 2-amino-OXT-A and OXT-G showed stronger activity than the original compound, OXT-A. 2-Amino-OXT-A is a prodrug of OXT-G and is readily transformed to OXT-G by adenosine deaminase distributed ubiquitously.

The anti-herpesvirus activities of OXT-G were further examined in more detail and compared with aciclovir and ganciclovir (Table 25.2). Against the wild strain of HSV-2, that is the thymidine kinase plus (TK$^+$) strain, OXT-G showed about one tenth of the activity of aciclovir and ganciclovir, but against the TK$^-$ strain, OXT-G showed 10 times or greater activity. OXT-G exhibited almost the same activity against the TK$^+$ and TK$^-$ strains. This implys that the mode of action of OXT-G is different from that of aciclovir and ganciclovir. Against HCMV, OXT-G showed about equal activity as ganciclovir, but aciclovir exhibited only about one tenth activity of OXT-G.

In Table 25.3, the activities against hepatitis B virus (HBV) of OXTs are listed. This *in vitro* test

a: adenosine deaminase; b: *Nocardia interforma*; c: 1) Ac₂O, 2)TPSCl, 3)NH₃; d: adenosine deaminase

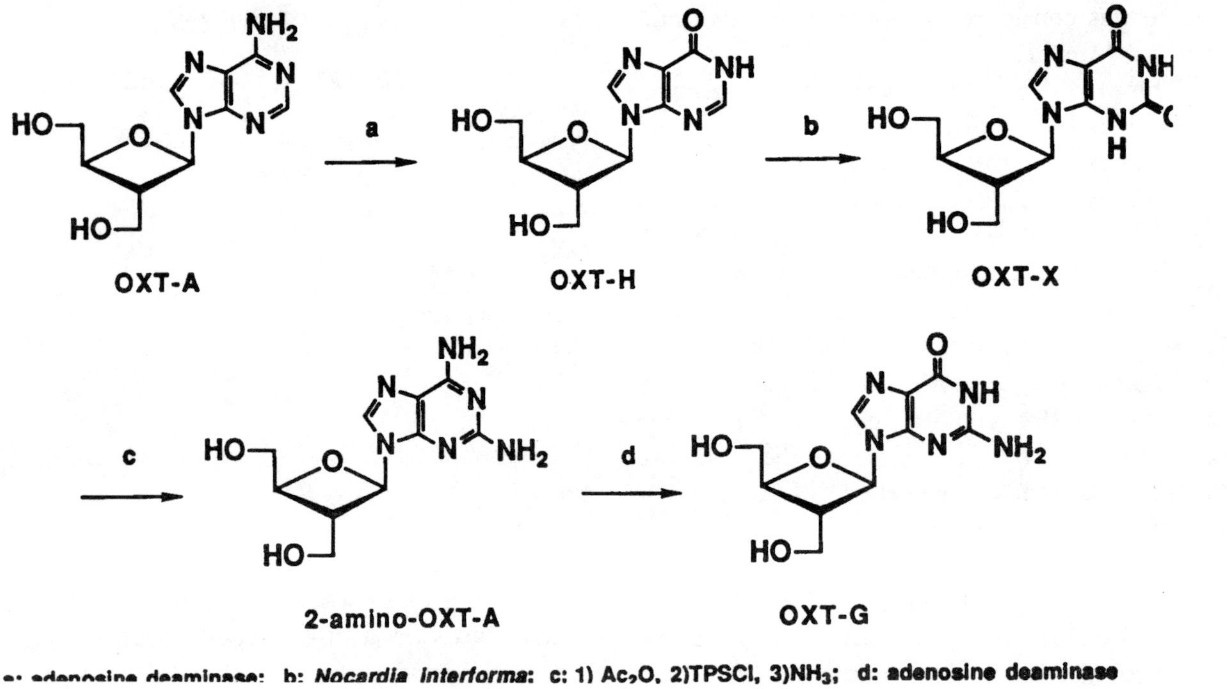

Figure 25.2 / Chemical and biological transformation of oxetanocin A.

Table 25.2 / Inhibitory Effects of OXT-G against HSV-2s and HCMV

Compound	Virus	Host Cell	EC_{50} (µg/ml)
OXT-G	HSV-2(TK⁺)	Vero	2.2
	HSV-2(TK⁻)	Vero	2.0
	HCMV	HEL	0.75
aciclovir	HSV-2(TK⁺)	Vero	0.16
	HSV-2(TK⁻)	Vero	18
	HCMV	HEL	6.8
ganciclovir	HSV-2(TK⁺)	Vero	0.30
	HSV-2(TK⁻)	Vero	69
	HCMV	HEL	0.60

Table 25.3 / Inhibitory Effects of OXTs against HBV

Compound	Cytotoxicity IC_{50} (µg/ml)	Anti-HBV activity ID_{50} (µg/ml)
OXT-A	79	9.1
OXT-H	>200	26.5
OXT-X	>200	–
2-amino-OXT-A	>200	0.32
OXT-G	>200	0.40
Ara-A	51	8.5
Aciclovir	>200	45

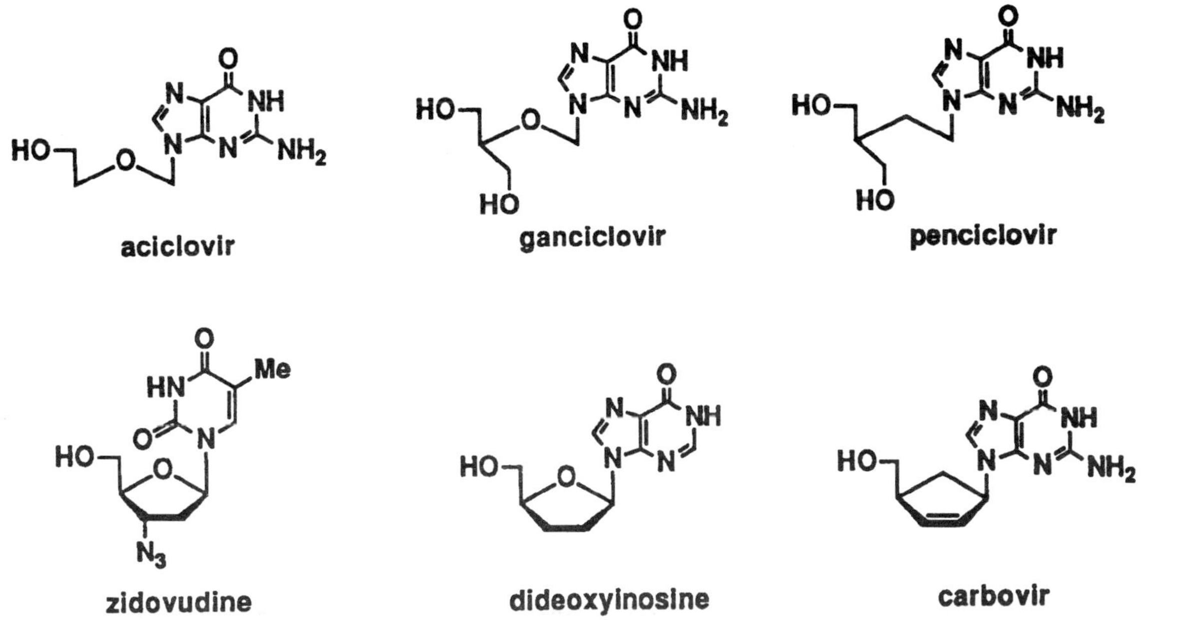

Figure 25.3 / Antiviral nucleoside analogues clinically used or under investigation.

was established by Prof. Matsubara and his collaborators of Osaka University. The experimental details were recently described elsewhere [13]. The antiviral activity is represented by ID_{50} on HBV DNA synthesis. 2-Amino-OXT-A and OXT-G exhibited the strongest activity and little cytotoxicity. Adenine arabinoside is now being clinically tested for HBV. However, in this test it showed weaker activity and stronger toxicity than OXT-G. Incidentally, aciclovir exhibited much weaker activity [3].

The activities against human immunodeficiency virus (HIV) were tested by Prof. Hoshino and his collaborators in the Medical School of Gunma University [9]. All of OXTs exhibited weaker activity than the control, dideoxyadenosine (DDA).

OXT-G exhibited *in vivo* activity against HSV-2 and CMV in mice [7]. It was decided to develop OXT-G further. One of the favorable characteristics of OXT-G is the excellent oral absorption. Using dogs, the oral absorption was compared with the intravenous administration.

The area under the concentration-time curve of the oral absorption was almost the same as that of the intravenous administration. More than 90% of the administered OXT-G was recovered from urine in the intact form (Yamashita, K., *et al.* Unpublished data). Detailed toxicological studies are now ongoing using mice, rats, dogs and monkeys.

CARBOCYCLIC ANALOGUES OF OXETANOCIN

In Fig. 25.3, six antiviral nucleoside analogues clinically used or under investigation are listed. Among them, penciclovir and carbovir have an unsubstituted methylene group in the place of oxygen atom of the furan ring. From these 6 nucleoside analogues, a common structure required for manifestation of antiviral activity can be deduced. Namely; there exist a nucleobase and an OH group corresponding to the 5'-OH of normal nucleoside, and these two groups are

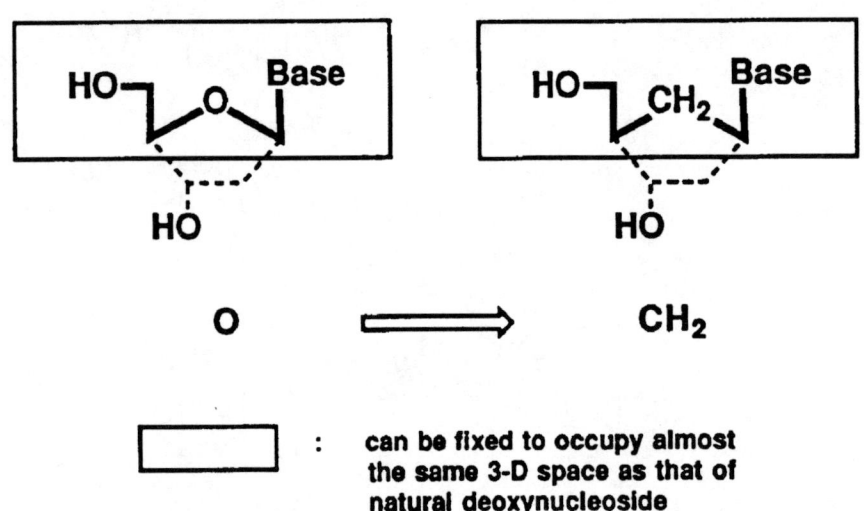

Figure 25.4 / Structure of nucleoside analogue required for manifestation of antiviral activity.

connected by four atoms: -C-O-C-C- or -C-isosteric CH$_2$-C-C-. The partial structure within the box as shown in Fig. 25.4 can be fixed to occupy almost the same three-dimensional space as that of natural deoxynucleosides.

It was very interesting to prepare the cyclobutane analogues of oxetanocin (see Fig. 25.5) and to test their antiviral activity. In this case, the most important thing was to prepare the stereoisomer that had the same absolute configuration as that of oxetanocin ("D"-isomer in Fig. 25.5). The "L"-isomer, the mirror image of "D"-isomer, can be drawn as illustrated in the far right figure in Fig. 25.5. This structure does not have the isosteric methylene required for antiviral activity. Therefore, the synthesis of the racemate means preparation of a mixture containing 50% impurity. Actually, both stereoisomers were synthesized independently by stereospecific synthesis as described later. As we expected, the L-isomer did not exhibit any antiviral activity.

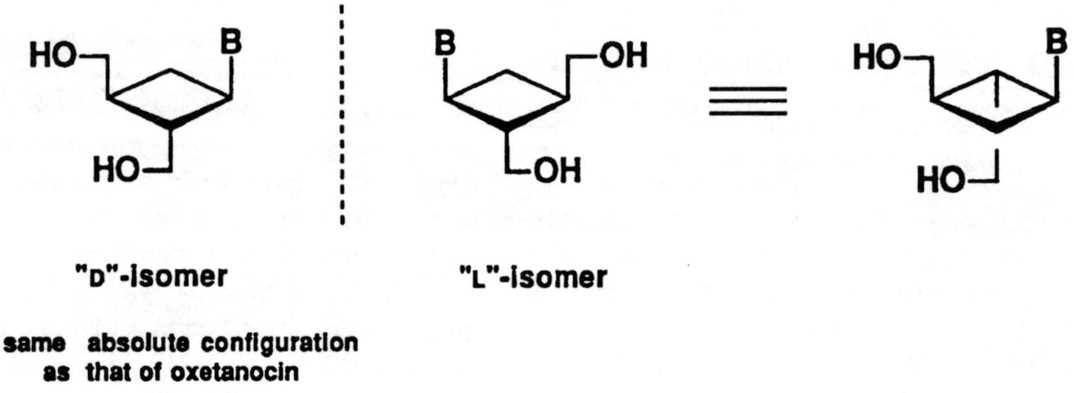

Figure 25.5 / Stereochemical relation between D- and L-isomers of carbocyclic oxetanocin.

The key feature of the synthetic route is stereoselective [2+2] formation of the cyclobutane skeleton using a catalytic amount of chiral titanium reagent developed by Prof. Narasaka and his collaborators of Tokyo University [1]. The synthetic scheme is shown in Fig. 25.6. The detail of the synthesis is described elsewhere [2]. Carbocyclic analogues of OXT-A and OXT-G were prepared from a common intermediate (mesylate of 1,2-cis-cyclobutanol, see Fig. 25.6).

The *in vitro* antiviral activities of carbocyclic analogues of OXT-A and OXT-G are described in Table 25.4. Carbocyclic OXT-A showed strong activity against HIV and HBV, and carbocyclic OXT-G exhibited strong activity against HSV-2 and HCMV[8]. Both compounds will be examined for further development.

Table 25.4 / Antiviral Activities of Carbocyclic Oxetancin Analogues

Compound	IC_{50} (μg/ml)			
	HSV-2	HCMV	HIV	HBV
carbocyclic OXT-A	7.5	12	0.03	0.024
carbocyclic OXT-G	0.04	0.4	0.3	0.86
aciclovir	0.16	6.8	—	47.0
ganciclovir	0.30	0.6	—	2.2

CONCLUSION

We found oxanosine by use of conventional screening methods for antibacterial antibiotics. Based on the studies of oxanosine, we designed a specific screening method for novel nucleosides from microbial metabolites. OXT-A was discovered by this screening method. As expected, OXT-A has exhibited antiviral activity. Starting from OXT-A, OXT-G was derived by combination of chemical and biological transformations. OXT-G showed strong antiviral activities against HCMV and HBV, and is now being studied in more detail. OXT-A is a new lead compound that has a four membered oxetane-ring in the sugar moiety of nucleoside. We synthesized the cyclobutane analogues of OXT with the same absolute configuration as the natural one stereospecifically. The synthesized carbocyclic OXT-A and OXT-G also exhibited strong antiviral activities. Some of the new OXT derivatives and analogues will hopefully contribute to future chemotherapy against viral infections.

SUMMARY

Oxanosine, a novel ribonucleoside having an imidazooxazinone chromophore, was found in 1981 by conventional and random screening of antibacterial antibiotics. A screen for novel nucleosides produced by microorganisms, was developed based on studies of oxanosine. Oxetanocin A (OXT-A) was isolated from a culture filtrate of *Bacillus megaterium* by using this screening in 1986. It is a novel nucleoside having expected antiviral activity and the first glycoside in nature having an oxetanoside structure in the sugar moiety. OXT-H, OXT-X, 2-amino-OXT-A, and OXT-G were derived from OXT-A, and the carbocyclic analogues of OXT-A were stereospecifically synthesized. In this paper, the specific screening method for novel nucleosides from microbial metabolites as well as the antiviral activities of OXT-A, its derivatives and the synthetic analogues are presented.

ACKNOWLEDGEMENTS

The studies on oxetanocins presented in this review article were performed by many researchers at Research Laboratories of Nippon Kayaku Co. in collaboration with Dr. Nishiyama's group at Nagoya University (HSV-2, HCMV), Prof. Matsubara's group at Osaka University (HBV), Prof. Hoshino's group at Gunma University (HIV)

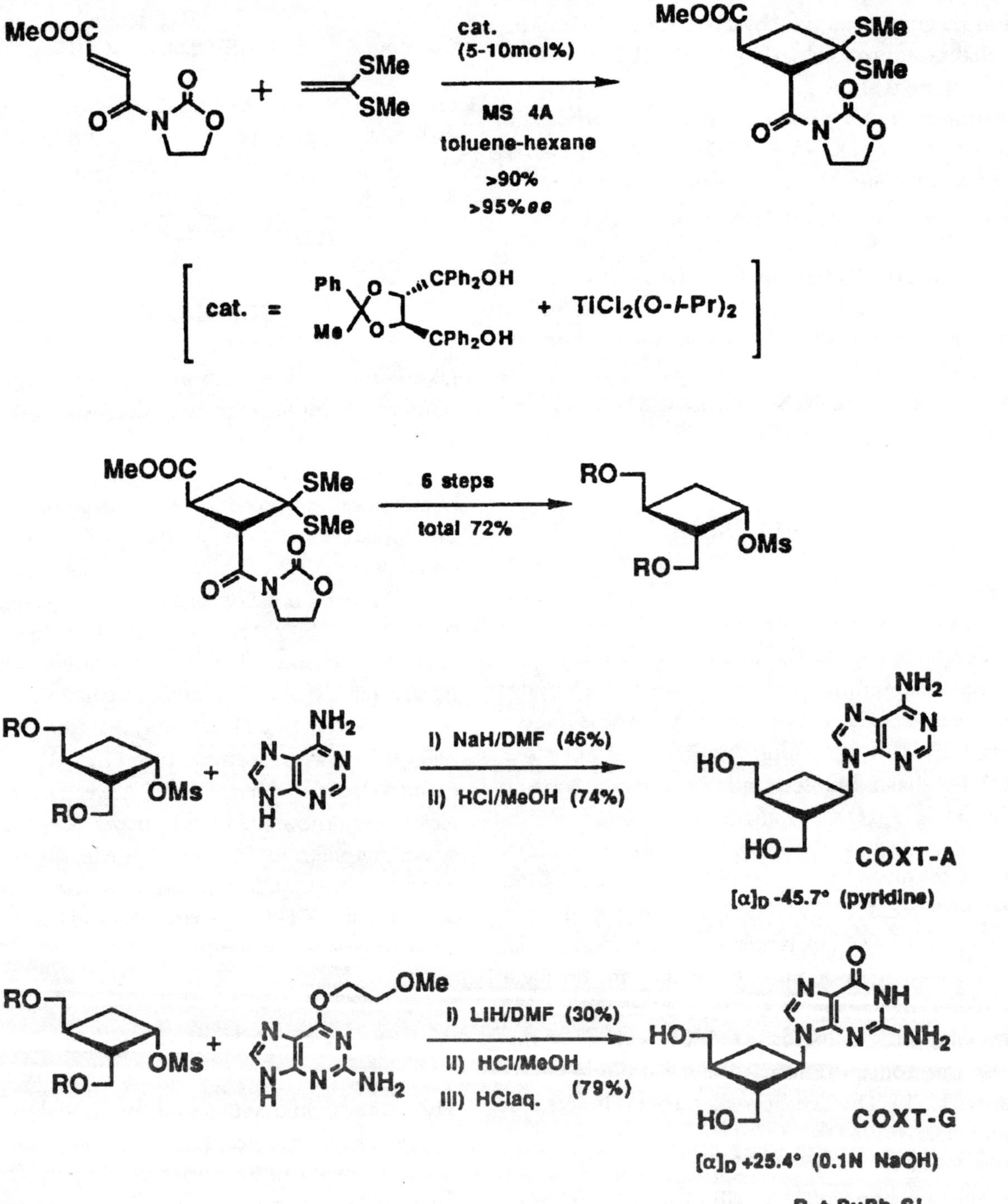

Figure 25.6 / Scheme of stereospecific syntheses of carbocyclic oxetanocin A and G, (COXT-A) and (COXT-G).

and Prof. Narasaka's group at Tokyo University (Synthesis). The studies on oxanosine were carried out at Institute of Microbial Chemistry. I would like to express sincere thanks to all of the contributors.

REFERENCES

1. Hayashi, Y. and K. Narasaka. 1989. Asymmetric [2 + 2] cycloaddition reaction catalyzed by a chiral titanium reagent. *Chem. Lett.:* 793–796
2. Ichikawa, Y., A. Narita, A. Shiozawa, Y. Hayashi and K. Narasaka. 1989. Enantio- and diastereo-selective synthesis of carbocyclic oxetanocin analogues. *J. Chem. Soc, Chem. Commun.:* 1919–1921.
3. Nagahata, T., K. Ueda, T. Tsurimoto, O. Chisaka and K. Matsubara. 1989. Anti-hepatitis B virus activities of purine derivatives of oxetanocin A. *J. Antibiot.* 42: 644–646.
4. Nakamura, H., N. Yagisawa, N. Shimada, T. Takita, H. Umezawa and Y. Iitaka. 1981. The X-ray structure determination of oxanosine. *J. Antibiot.* 34: 1219–1221.
5. Nakamura, H., S. Hasegawa, N. Shimada, A. Fujii, T. Takita and Y. Iitaka. 1986. The X-ray structure determination of oxetanocin. *J. Antibiot.* 39: 1626–1629.
6. Nishiyama, Y., N. Yamamoto, K. Takahashi and N. Shimada. 1988. Selective inhibition of human cytomegalovirus replication by a novel nucleoside, oxetanocin G. *Antimicrob. Agents Chemother.* 32: 1053–1056.
7. Nishiyama, Y., N. Yamamoto, Y. Yamada, H. Fujioka, N. Shimada and K. Takahashi. 1989. Efficacy of oxetanocin G against herpes simplex virus type 2 and murine cytomegalovirus infections in mice. *J. Antibiot.* 42: 1308–1311.
8. Nishiyama, Y., N. Yamamoto, Y. Yamada, T. Daikoku, Y. Ichikawa and K. Takahashi. 1989. Anti-herpesvirus activity of carbocyclic oxetanocin G *in vitro. J. Antibiot.* 42: 1854–1859.
9. Seki, J., N. Shimada, K. Takahashi, T. Takita, T. Takeuchi and H. Hoshino. 1989. Inhibition of infectivity of human immunodeficiency virus by a novel nucleoside, oxetanocin, and related compounds. *Antimicrob. Agents Chemother.* 33: 773–775.
10. Shimada, N., N. Yagisawa, H. Naganawa, T. Takita, M. Hamada, T. Takeuchi and H. Umezawa. 1981. Oxanosine, a novel nucleoside from actinomycetes. *J. Antibiot.* 34: 1216–1218.
11. Shimada, N., S. Hasegawa, S. Saito, T. Nishikiori, A. Fujii and T. Takita. 1987. Derivatives of oxetanocin: oxetanocins H, X and G, and 2-Aminooxetanocin A. *J. Antibiot.* 40: 1788–1790.
12. Shimada, N., S. Hasegawa, T. Harada, T. Tomisawa, A. Fujii and T. Takita. 1986. Oxetanocin, a novel nucleoside from bacteria. *J. Antibiot.* 39: 1623–1625.
13. Ueda, K., T. Tsurimoto, T. Nagahata, O. Chisaka and K. Matsubara. 1989. An *in vitro* system for screening anti-hepatitis B virus drugs. *Virology* 169: 213–216.
14. Uehara, Y., M. Hasegawa, M. Hori and H. Umezawa. 1985. Increased sensitivity to oxanosine, a novel nucleoside antibiotic, of rat kidney cells upon expression of the integrated viral *src* gene. *Cancer Res.* 45: 5230–5234.

26

High-Volume Screening of Natural Products for IL-1 Receptor Level Antagonists

A.L. Laborde*, J.A. Shelly, S.E. Truesdell, V.P. Marshall, J.I. Cialdella, W.F. Liggett, D.A. Yurek,
D.G. Chirby, J.W. Paslay, C.K. Marschke, M.S. Kuo
Chemical and Biological Screening, The Upjohn Company, Kalamazoo, MI 49001

One of the objectives of The Upjohn Company during the past year has been to screen fermentation broths for IL-1 receptor antagonists. In order to increase the chances of success, a high-volume screen was required. This report will describe development, automation, and evaluation of such a screen. The IL-1 antagonist screen is able to assay 2000–3000 samples per day for inhibition of IL-1 binding to YT-NCI cells. The procedures for this screen were modified to accommodate laboratory robotics and multi-sample handling systems. The LKB Betaplate harvester/counter system was used to allow for rapid sample throughput and automatic data collection for transfer to data analysis programs in The Upjohn Company computer system. The statistical variations inherent within the system were determined as well as the minimal level of detectable antagonist. Protocols were implemented to ensure that the isolated activities were of low molecular weight, could be reproduced in subsequent refermentations, and were amenable to realistic chemical isolation procedures. This report will also address the problems presented by the production of proteases and polyene antibiotics in fermentation broths and how these nuisance activities were circumvented. Assays to assess toxicity and antagonist specificity for the IL-1 receptor were also developed. In the course of screening, several small molecular weight antagonists were isolated. The evolved system to screen for IL-1 antagonists accommodated large numbers of samples early in detection, thereby creating a broad-based screening effort.

Keywords: microbial products, screening, IL-1 antagonists

INTRODUCTION

IL-1 is a polypeptide that has been implicated as playing a major role in the pathogenesis of rheumatoid arthritis (5,7). It has been reported that IL-1 levels in arthritic patients were significantly higher in this group as compared to nonarthritic individuals. Moreover, many of the consequences of IL-1 cellular interactions are associated with the symptomology exhibited in this disease. Synovial cells release vasoactive agents that mediate the inflammatory response and promote bone resorption. Because of the pathology associated with increased levels of IL-1 in arthritic joints (7), The Upjohn Company along with other pharmaceutical companies undertook a project to discover an IL-1 antagonist. The program at The Upjohn Company involved dual strategies. In the first, an attempt was made to isolate and purify an IL-1 receptor level antagonist that was produced by a human monocytic cell line. The second strategy involved using fermentation broths as a potential source of a novel IL-1 receptor level antagonist.

This report will describe the development and evolution of a high-volume screen using fermentation broths as the potential source of therapeutic agent. The problems that were encountered and the subsequent solutions are also discussed with respect to the special problems that arise due to the fermentation broths themselves.

MATERIAL AND METHODS

CELL CULTURE

YT-NCI, a human lymphoma cell line, was used in experiments described here. Cultures were maintained at 37°C in a humidified atmosphere of air/CO_2 (95:5) in RPMI-1640 containing 10% heat-inactivated fetal calf serum (Gibco), 2 mM glutamine, 20 mM Hepes buffer (pH 7.3), and 10 µg/ml gentamicin.

For use in the receptor binding assay, cells were harvested by centrifugation and washed once with receptor binding (RB) buffer. The RB buffer (pH 7.3) was RPMI-1640 supplemented with BSA (1 mg/ml), sodium azide (0.1%), and 20 mM Hepes buffer. The cell pellet was resuspended in RB buffer to a cell density of 1×10^7 cells/ml and used in the binding assays.

^{125}I-IL-1 RECEPTOR BINDING ASSAY

Assays were conducted in 96-well microtiter plates and contained 1×10^6 YT-NCI cells, 50 pM ^{125}I-IL-1, and 50 µl centrifuged fermentation broth in a final volume of 200 µl. Total Bound (TB) counts were determined by replacing the 50 µl of fermentation broth with 50 µl RB buffer. Nonspecific binding (NSB) was determined in the presence of 1000-fold excess unlabelled IL-1 and never exceeded 15% of total binding. Control curves for inhibition of IL-1 binding were also generated. A set of samples containing various concentrations of cold IL-1 (37.5 pM-1200 pM) and 50 pM ^{125}I-IL-1 were assayed to determine the % inhibition detected for each concentration of cold IL-1.

Assays were incubated at ambient temperature for one hour and the cells harvested with a Skatron cell harvester. Filtermats were soaked in cold PBS containing nonfat dry milk (5%) and sodium azide (0.1%) prior to use. After harvesting, the filtermats were washed five times with PBS containing 0.05% Tween-20 and dried overnight. The number of bound counts was determined by counting the filtermats in a LKB-Betaplate liquid scintillation counter.

1A5/HT2

This assay is an IL-1 driven IL-2 proliferative assay conducted as described by Tracey et al (12).

IL-1

Recombinant human IL-1β was obtained from The Upjohn Company. Preparation of labelled ^{125}I-IL-1β was by the procedure of Bolton and Hunter (1,8). Average specific activities of the labelled material were 3000 Ci/mmol.

TOXICITY ASSAY

The tetrazolium salt MTT [3-(4,5-dimethylthiazol-2-yl)-2,5-diphenyl tetrazolium bromide] was used in conjunction with the target cell line, YT-NCI, in a colorimetric assay for cytotoxicity (9). The MTT assays were conducted in 96-well microtiter plates or deep-well blocks (Beckman) and contained 5×10^5 cells, 100 µg MTT, and the appropriate drug or RB buffer in a final volume of 220 µl. Wells containing only cells and MTT were used to establish baseline viable cell values. After briefly mixing, the assays were incubated without agitation for one hour at 37°C. Following the removal of 135 µl of supernatant, 135 µl of 0.04 N HCl/Isopropanol was added to dissolve the formazan crystals, and 135 µl of this solution was transferred to a 96-well plate for reading. Absorbance values were recorded at 570 nm using a V_{max} microplate reader. Wells containing only MTT were used as a control blank.

PROTEASE ASSAY

The presence of proteases was determined with azocoll (2,6). Assays routinely contained azocoll (0.5 mg), RB buffer, and known concentration of protease or 50 µl fermentation broth in a final volume of 200 µl. Incubation was overnight at ambient temperature. Dissolution of azocoll particles was scored as positive for protease.

BIOMEK ROBOTIC WORKSTATION

The Biomek system is a robotic laboratory workstation whose capabilities include micro-scale liquid handling (pipetting, diluting/dispensing), plate washing, and visible-light photometry. Software for the system's computer control allowed flexible programming for a wide variety of operations and labware configurations. A particular advantage of this system was its capability of eight-channel pipetting and dispensing, which allowed rapid processing of samples in standard 96-well microtiter plates and thus greatly facilitated high sample throughput. The instrument's flexibility in configuration and programming allowed adaptation to any protocol which was developed for standard micro-scale labware.

DATA COLLECTION AND ANALYSIS

Count data from the Betaplate counter were stored as IBM DOS text files on disk. Data was transferred to The Upjohn Company mainframe on a PC terminal. Data analysis was performed using the 1PTINHIB and RODBRD programs in the mainframe's BINDING software package.

CHEMICAL CHARACTERIZATION METHODS

Automation was necessary due to the large number of fermentations to be characterized and the time-consuming process of scouting techniques by hand. The scouting process described in the text was thought to be amenable to automation due to the repetitive techniques involved. The Waters Millilab Workstation was selected to perform the automation since it can utilize resin cartridges as well as pipetting and mixing functions.

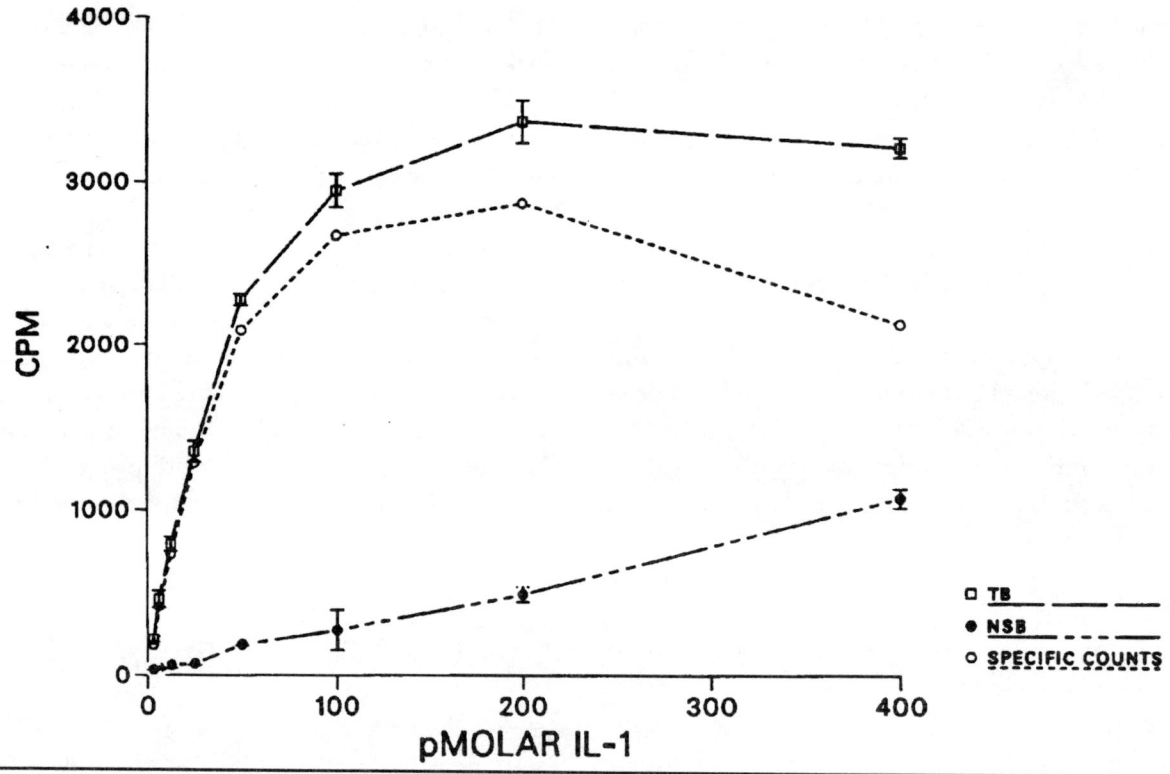

Figure 26.1 / Saturation curve and Scatchard analysis for binding of ^{125}I-IL-1β to the YT-NCI receptor.

RESULTS

ASSAY DEVELOPMENT

The IL-1 receptor ligand assay developed by using YT-NCI cells was modified for use in a high volume screening situation. Various experimental parameters were determined under screening conditions. The concentration-dependent equilibrium binding of ^{125}I-IL-1β to YT-NCI cells is shown in Figure 26.1. IL-1 binding was saturable at 200 pM IL-1; Scatchard analysis of the saturation data indicated an equilibrium dissociation constant (Kd) of 46pM with 452 receptors per cell. When IL-1 binding was examined as a function of time, the number of counts bound appeared to plateau after a three-hour incubation period. The results of the kinetic studies suggested that for optimum assay sensitivity, the assay should contain ^{125}I-IL-1 at a concentration equal to the Kd. Although the number of counts plateaued after three hours, it was decided to incubate the assay for one hour at room temperature. At one hour, sufficient counts were bound so that detection of inhibition was not compromised. Furthermore, a three-hour incubation period was thought not to be feasible for a high-volume screen.

In order to determine the lowest concentration of IL-1 antagonist that can be detected under screening conditions, the following study was performed. Fermentation broths, shown to be inactive in the IL-1 and protease assays, were randomly spiked with a known amount of cold IL-1 to result in a final concentration between 25 and 300 pM. Our results (Table 26.1) indicate that as little as 25 pM IL-1 may be detected in a fermentation broth. This is equivalent to 85 pgrams of

Table 26.1 / Detectable Levels of Cold IL-1 in Fermentation Broths

pM Cold IL-1	*Bound ^{125}I-IL-1 (cpm)	% Inhibition of Binding
300	280 ± 55 (N = 6)	88 ± 4 (N = 6)
200	391 ± 80 (N = 6)	79 ± 6 (N = 6)
100	614 ± 104 (N = 6)	59 ± 6 (N = 6)
50	870 ± 62 (N = 6)	35 ± 5 (N = 6)
25	1096 ± 121 (N = 6)	15 ± 10 (N = 6)
Total Bound*	1201 ± 156 (N = 6)	

*Corrected for Nonspecific Counts

IL-1. However, at this level of inhibition, i.e. 15%, there is considerable standard deviation. At 50 pM IL-1, the standard deviation is much less. These results suggest that as little as 170 pgrams of IL-1 can be detected with confidence. Moreover, defining a positive "hit" as exhibiting >20% inhibition appears to exclude those positives which originate due to assay variation.

PROBLEMS WITH FERMENTATION BROTHS

Table 26.2 shows the various problems encountered in using fermentation broths to screen for IL-1 antagonists and the subsequent solution to the problem.

1. *Proteases.* It became apparent very early in the development of the IL-1 antagonist screen that the presence of proteases in fermentation broths would present obstacles to the discovery of such an antagonist. Their presence would lead to false IL-1 antagonist activities being detected, presumably via the proteolytic degradation of the ligand. It was found that heat treatment of the fermentation broth or solvent extraction effectively removed these nuisance activities.
2. *Polyene Antibiotics.* After processing several fermentations, it became apparent that polyene macrolide compounds were creating a false-positive response in the assay. These

Table 26.2 / Problems encountered in the IL-1 antagonist screen as a result of using fermentation broths as the source of novel compounds and the solutions to these problems.

Problem	Proteases
Solution	Heat inactivation *or* solvent extraction
Problem	Polyenes
Solution	Early detection
Problem	Toxicity
Solution	Early use of MTT cytotoxicity assay
Problem	Specificity for the IL-1 receptor
Solution	Assay in the whole cell IL-1 driven Il-2 proliferative assay (1A5/HT2 Assay)

molecules exert their toxic effects by binding very strongly to cholesterol found in the membrane of the YT-NCI cells used in the IL-1 assay. Therefore, a method of identifying polyenes in the fermentation was developed. Since polyenes are normally found in the mycelia of the fermentation, an extraction of the centrifuged cake is performed using a solution of 1:1 methanol:acetone. A UV spectrum of this extract is then taken. The presence of polyenes is ascertained by their characteristic UV spectrum. Currently, the entire cake from a 100 ml fermentation is used to identify polyenes before any other chemical characterization is performed. However, since the chromophore of the polyene molecules is very strong, a smaller amount of cake may be used.
3. *Toxicity.* The MTT assay for cytotoxic materials (9) was used to detect fermentations containing cytotoxic compounds. This colorimetric assay was automated to allow for high sample throughput.
4. *Specificity.* In order to determine whether or not the interaction was specific for the IL-1

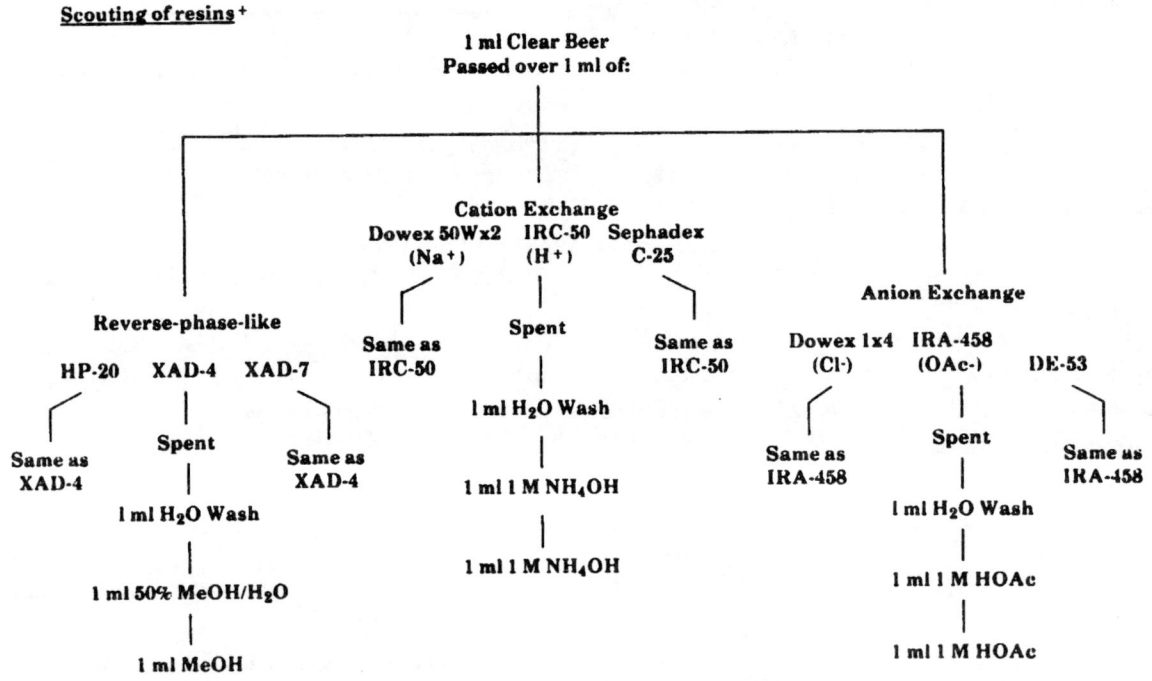

Figure 26.2 / IL-1 antagonist screen: chemical characterization methods.

receptor, isolated compounds were submitted for testing in the IL-1 1A5/HT2 assay. Stimulation of 1A5 cells with IL-1 results in *de novo* synthesis and release of IL-1 which in turn stimulates proliferation of HT2 cells. This assay can assess whether or not inhibition of HT2 proliferation is a consequence of a receptor mediated event or the result of metabolic inhibition (12).

CHEMICAL CHARACTERIZATION METHODS

The development of chemical characterization methods for leads generated from the IL-1 antagonist screen was based on general procedures developed for the purification of microbial metabolites. The purpose of such a system is to provide a fast, easy way of identifying possible techniques for purification and chemical characterization of activities found in fermentation broths. These procedures also yield a means of dereplication of possible leads. Currently, these methods provide data on the stability of the activity of interest, its polarity, and the suitability of a broad range of resins commonly used to purify natural products. In the case of the IL-1 antagonist screen, the chemical methods used must take into account the requirements of the assay and its limitations. These include the amount and concentration of submit

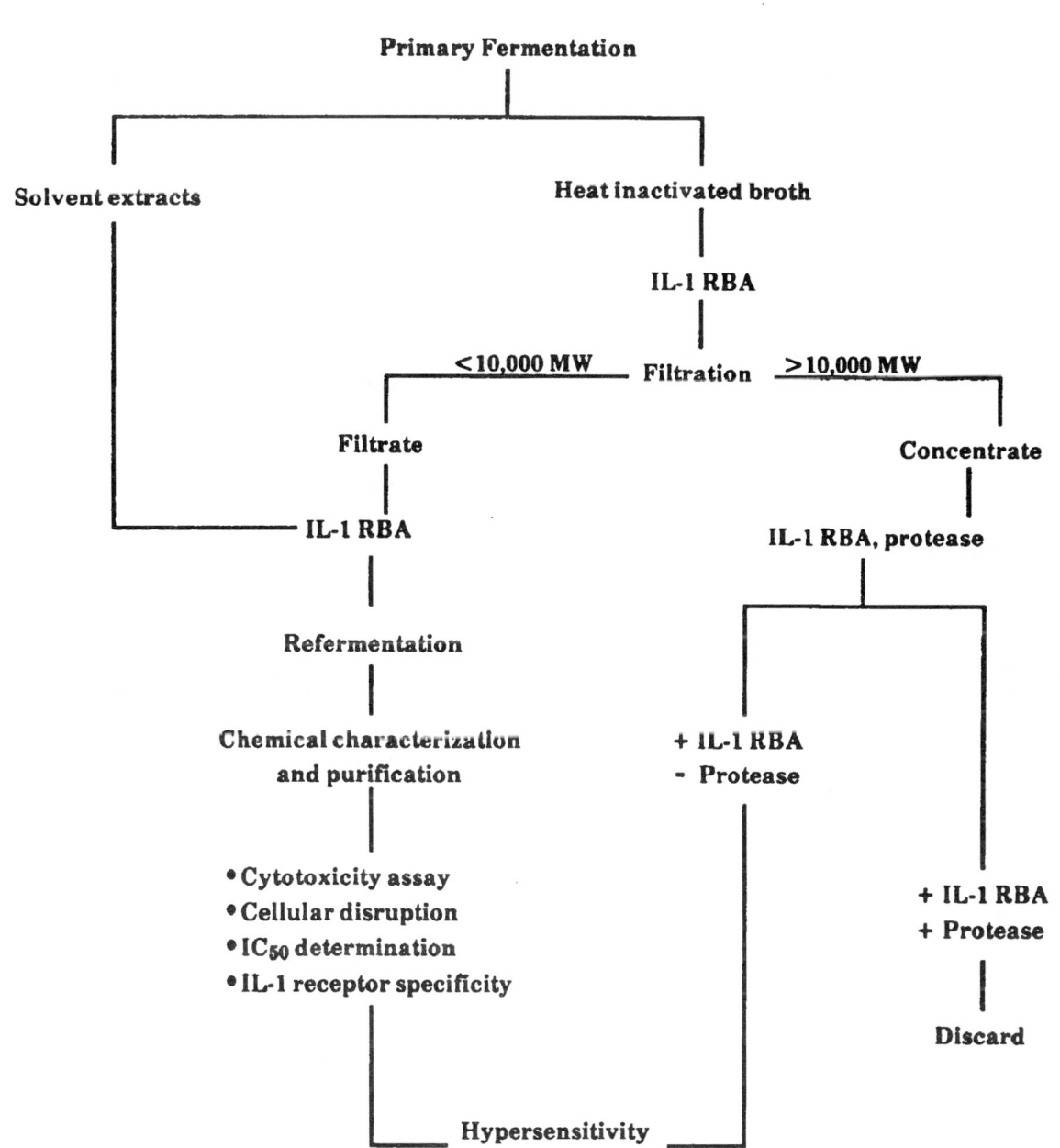

Figure 26.3. / Extraction scouting: heat and pH stability testing.

ted samples, and what may not be included in the samples (such as organic solvents and salts). A detailed description of the "scouting" procedures is given and these are summarized in the flow charts in Figures 26.2 and 26.3.

Three 5 ml samples of clear beer (the supernatant produced by centrifuging the fermentation) are taken and the pH is adjusted to 3, 7, and 10, respectively. Two 1 ml samples of pH-adjusted clear beer are removed for a control and for heat stability testing. The control sample is refrigerated, while the other is heated in a water bath at 50°C for 1–3 hours. These samples, after assay, will give information on the stability of the activity under acidic, neutral, and basic conditions. The remaining 3 ml of beer at each pH are divided into three 1 ml portions. These nine samples are then extracted with 1 ml of methylene chloride (CH_2Cl_2), ethyl acetate (EtOAc), and n-butanol (BuOH), respectively. These solvents cover a broad range of polarities, are easily removed by evaporation, and may be employed in a large-scale purification. Each extraction is separated into the organic and aqueous phases. Since the IL-1 assay is affected by these solvents, all organic solvent is removed from both phases by drying under a stream of nitrogen. After assay, the data will give a measure of the polarity of the activity, as well as indicating the best extraction conditions.

The utility of resins for purification of the activity of interest is ascertained by scouting a broad range of resins. These include reverse-phase-like resins (HP 20, XAD-4, XAD-7), anion exchange resins [Dowex 1x4 (Cl^-), IRA 458 (OAc^-), and DE-53] and cation exchange resins [Dowex 50Wx2 (Na^+), IRC 50 (H^+) and Sephadex C-25]. The ion exchange resins include both strong and weak forms to distinguish between different ionic types. (Also included are agarose and cellulose based resins that will minimize loss of peptide activities.) The resins are placed in individual reservoirs (Analytichem International) at a bed volume of about 1 ml. Clear beer (1 ml) is drawn over the mini-column using Analytichem's vacuum system and collected in a test tube. The resin is then washed with water followed by the eluting solutions. For the reverse-phase-like resins, 1 ml each of a 1:1 methanol:water solution followed by pure methanol is used. For the anion exchange resins, two 1 ml portions of 1 M acetic acid are used to elute the activity. For cation exchange resins, two 1 ml portions of 1 M ammonium hydroxide are used. Each mini-column therefore results in four fractions for assay; spent, wash, eluate #1, and eluate #2. Samples containing methanol, acid, or base are dried under a stream of nitrogen to remove the organic solvent and the volatile acid or base prior to assay since the assay is sensitive to pH and the presence of organic solvents. Note that salt solutions were not employed to elute activities from ion exchange resins due to their false positive responses in the assay. The scouting results provided by the mini-column experiments indicate resins suitable for purification of the activity.

Therefore, the combination of solvent extraction and mini-column experimental data provides us with an array of methodologies that could be deployed to purify the activity(ies) of interest. Furthermore, valuable information regarding the stability, polarity, ionizable functional groups, and the plurality of the activity(ies) is obtained.

The chemical characterization methods utilized to determine techniques for the purification of leads from the IL-1 receptor binding antagonist screen have been automated using a Waters Millilab Workstation. This automation handles the large number of fermentations to be characterized in an efficient manner, freeing the operator for other tasks. The Workstation also improves the accuracy of the procedures, since each operation is performed identically.

NAME	STRUCTURE	SOURCE
Epiepoformin	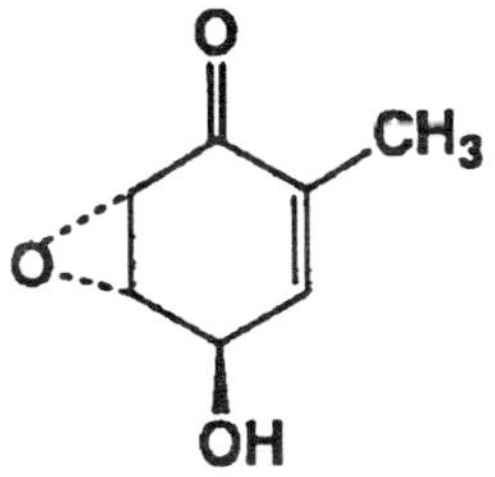	*Penicillium patulum*
Terrein	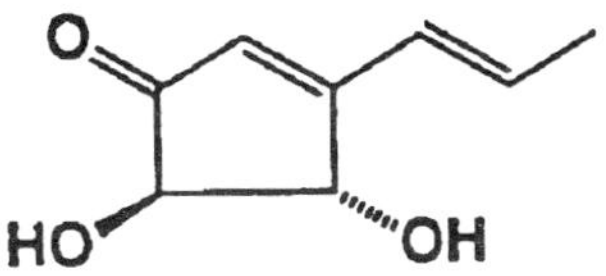	*Aspergillus terreus*
Geodin	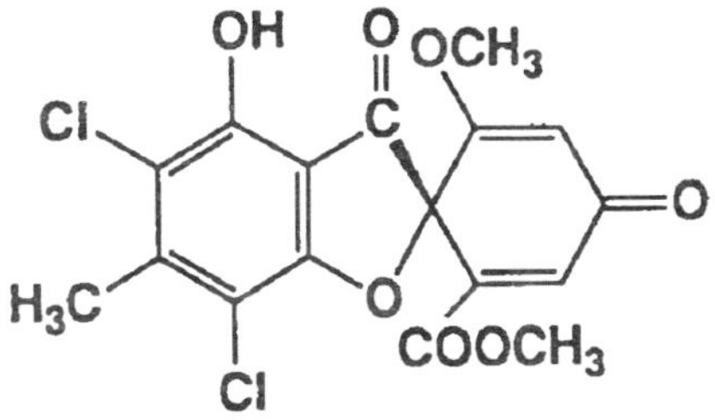	Unidentified Fungus

Figure 26.4

AUTOMATION PROCEDURES

Automation of the biological and chemical procedures used in the screening of fermentation broths for IL-1 antagonists allowed the assay of large numbers of samples, thereby, increasing the potential for success. The receptor binding assay used in the IL-1 antagonist screen was automated by interfacing the Beckman Biomek robotic workstation and the LKB Betaplate system.

The procedures of the IL-1 receptor antagonist screen have been modified to accommodate laboratory robotics and multi-sample handling systems. Programs for the Beckman Biomek robotic workstation were written to allow the instrument to handle a large proportion of the liquid transfer steps in the protocol. Procedures were developed for the LKB Betaplate harvester/counter system to allow rapid sample throughput and automatic data collection for transfer to data analysis programs on The Upjohn Company mainframe computer. The BINDING receptor binding analysis software is utilized for result calculation and output, and an active/inactive analysis feature was developed to facilitate handling of screen results. These improvements have streamlined the assay to the point where processing of 2000–3 000 samples per day can be routinely performed.

Automation of the chemical characterization procedures used to determine methods of purification of leads from fermentation beers by the Chemical and Biological Screening IL-1 receptor binding antagonist screen was accomplished utilizing a Waters Millilab Workstation.

IL-1 ANTAGONIST SCREENING PROTOCOL

Presented in Figure 26.5 is the flow chart of screening procedures used in the IL-1 antagonist screen. Primary fermentations include bacterial, actinomycete, and fungal cultures. Fermentation broths are extracted with an organic solvent or undergo heat treatment prior to assay in the IL-1 RBA to remove protease enzymes. All heat-treated broths that test positive in the IL-1 RBA are then filtered and both filtrate and concentrate solutions are assayed for the presence of IL-1 antagonists. This filtration step was introduced to distinguish between high and low molecular weight activities. The filter used has an exclusion limit of 10,000 daltons. Only those cultures producing a low molecular weight activity (< 10,000 daltons) are initially submitted for refermentation; the presumptive high molecular weight activities are examined for protease activity in addition to IL-1 RBA activity. Those found to be negative for protease enzymes were subsequently submitted to The Upjohn Company department of Hypersensitivity Diseases Research (HDR) for assay as potential inflammation modulators.

Following demonstration of activity upon refermentation, the broths are submitted for chemical characterization and purification of the desired low molecular weight activity. Once the compound has been isolated, the potential IL-1 antagonist is further evaluated with respect to toxicity and receptor specificity. In addition, an IC50 is determined. Once these assays are complete, the compound is given to Hypersensitivity Diseases Research for assay in their rheumatoid arthritis models.

The time table of events for the IL-1 antagonist screen at The Upjohn Company is given in Table 26.3. For a given two-year time frame for IL-1 antagonist screening, three months were spent developing and modifying assays as well as "problem shooting." In the one and a half years in which the screen was conducted, 62,000 fermentation broths or extracts were assayed in the IL-1 RBA. This screen had a hit rate of ca. 1%.

ISOLATION OF POTENTIAL IL-1 ANTAGONISTS

Among the compounds isolated in the IL-1 antagonist screen were epiepoformin, terrein, and geodin (Figure 26.4). All are produced by fungi and all were previously isolated because they demonstrated weak antibacterial activity (6,10,11). None were reported as having IL-1 antagonist activity. For each compound 100% inhibition of IL-1 binding to the IL-1 receptor was never achieved irregardless of antagonist concentration. As a consequence,

Table 26.3 / Timetable of Events for the IL-1 Antagonist

Assay development, modification, and automation	3 months
Problem shooting	3 months
Screening IL-1 antagonists	1.5 years
Total fermentation broths/extracts	62,000
Total IL-1 RBA positives	558

Extraction Scouting; Heat and pH Stability Testing

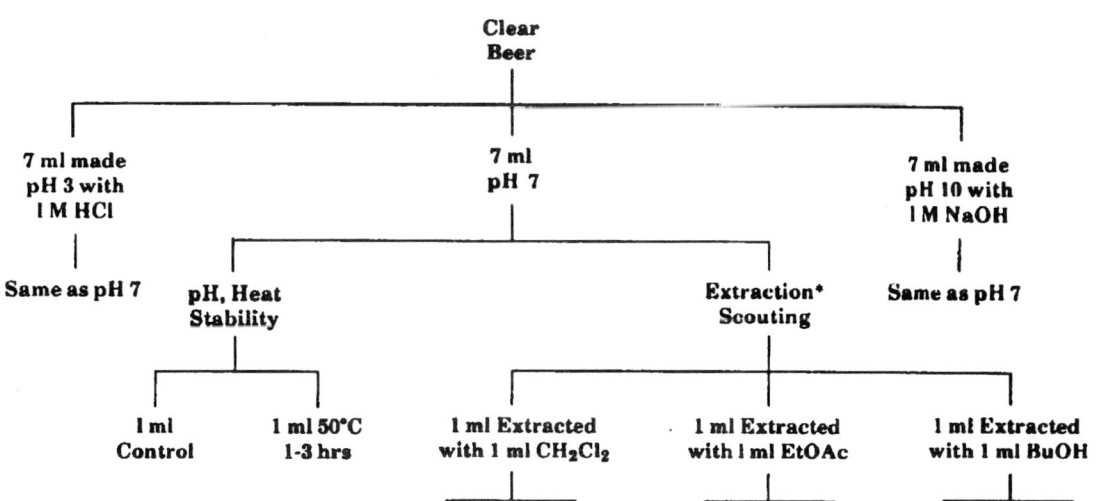

*All fractions dried to remove organic solvents prior to IL-1 assay.

Figure 26.5

inhibitory concentrations (IC) were calculated in terms of 30% inhibition since the maximum amount of inhibition detected for all higher concentrations was 60% (Table 26.4). These compounds were subsequently assayed in the 1A5/HT-2 assay which is an IL-1 driven IL-2 proliferative assay which measures both toxicity and specificity of IL-1 antagonism with regard to the receptor. All of the isolated compounds were found to be toxic and the inhibition of IL-1 binding shown not to be receptor mediated.

DISCUSSION

This report has described the development and evaluation of a receptor-ligand screen for detection of IL-1 antagonists in microbial fermentations at The Upjohn Company, Kalamazoo, Michigan. The establishment of this screen reflects the successful integration of chemical, microbiological, and biological disciplines. It also reflects how automation with robotics enhances a screen's potential for high-volume throughput. The screen that was developed was rapid and time efficient. Procedures were implemented to ensure reliability and repeatability in all aspects of the screening system. At the end, this system was capable of handling 4,000 assays per day and can serve as a model for other enzyme receptor-based screens.

The overall success of this particular screen is dubious at best since known structures were isolated as IL-1 antagonists and later shown not to act at the receptor level. However, this screen was able to facilitate purification of an IL-1 antagonist protein from a human monocytic cell line (3) because of the ability to handle large numbers of samples and because of the relatively quick turnover of results.

Table 26.4 / IL-1 Activities of Isolated Compounds

Compound	IL-1 RBA		1A5/HT-2 Assay		
	IC_{30} (µg/ml)	Toxicity	Activity	Specificity	Toxicity
Epiepoformin	70	-	+	-	+
Geodin	30	-	+	-	+
Terrein	500	-	+	-	+

ACKNOWLEDGEMENTS

We wish to thank John Chosay for running the 1A5/HT2.

REFERENCES

1. Bolton, A. E. and W. M. Hunter. 1973. The labelling of proteins to high specific radioactivities by conjugation to a ^{125}I-containing acylating agent. Application to the radioimmunoassay. *Biochem. J. 133:* 529–539.
2. Calbiochem Biochemical/Immunochemical Catalog. 1989. pp. 227–228.
3. Carter, D. B., M. R. Deibel, Jr., C. J. Dunn, et al. 1990. Purification, cloning, expression and biological characterization of an interleukin-1 receptor antagonist protein. *Nature. 346*(6267): 633–638.
4. Chavira, R. Jr. 1984. Assaying proteinases with azocoll. *Anal. Biochem. 136:* 446.
5. Eastgate, J. A., N. C. Wood, F. S. DiGiovine, J. A. Symons, F. M. Grimlinton, G. W. Duff. 1988. Correlation of plasma interleukin-1 levels with disease activity in rheumatoid arthritis. *Lancet*, Sept. 24, 706–709.
6. Fujimoto, H., H. Flasch, and B. Franck. 1975. Biosynthese der seco-antarachivone geodin und dihydrogeodin aus emodin. *Chem. Bev. 108:* 1244–1288.
7. Hamerman, D. 1989. The Biology of Osteoarthritis. *N. Eng. Med. 320*(20): 1322–1330.
8. Instructions for Bolton-Hunter Reagent ^{125}I Kit ICN Radiochemicals.
9. Mosmann, T. 1983. Rapid colorimetric assay for cellular growth and survival: Application to proliferation and cytotoxicity assays. *J. Immunol. Methods 65:* 55–63.
10. Nagasawa, H., A. Suzuki and S. Tamura. 1978. Isolation and structure of (+)-desoxyepiepoxydon and (+)-epiepoxydon, phytotoxic fungal metabolites. *Agric. Biol. Chem. 42*(6): 1303–1304.
11. Raistrick, H. and G. Smith. 1935. The metabolic products of *Aspergillus terreus* THOM. A new mould metabolic product-terrein. *Biochem. J. 29:* 606–611.
12. Tracey, D. E., M. M. Hardee, K. A. Richard and J. W. Paslay. 1988. Pharmacological inhibition of interleukin-1 activity in T-cells by hydrocortisone, cyclosporine, prostaglandins, and cyclic nucleoxides. *Immunopharmacology 15:* 47–62.